# Welding Engineering

# Welding Engineering

Edited by **Howard Currant**

NY RESEARCH
P R E S S

New York

Published by NY Research Press,
23 West, 55th Street, Suite 816,
New York, NY 10019, USA
www.nyresearchpress.com

**Welding Engineering**
Edited by Howard Currant

International Standard Book Number: 978-1-63238-509-3 (Hardback)

The publisher's policy is to use permanent paper from mills that operate a sustainable forestry policy. Furthermore, the publisher ensures that the text paper and cover boards used have met acceptable environmental accreditation standards.

**Trademark Notice:** Registered trademark of products or corporate names are used only for explanation and identification without intent to infringe.

Printed in the United States of America.

# Contents

# Preface

It is often said that books are a boon to mankind. They document every progress and pass on the knowledge from one generation to the other. They play a crucial role in our lives. Thus I was both excited and nervous while editing this book. I was pleased by the thought of being able to make a mark but I was also nervous to do it right because the future of students depends upon it. Hence, I took a few months to research further into the discipline, revise my knowledge and also explore some more aspects. Post this process, I began with the editing of this book.

This book provides comprehensive insights into the field of welding engineering. As a branch of engineering, welding engineering deals with the study and practice of welding, cutting and brazing. The applications of this field are spread across various areas like, ship building, aerospace, nuclear power and mining, etc. It encompasses the elements of mathematics, chemistry, physics, physical metallurgy, thermodynamics, engineer mechanics, etc. This book elucidates the concepts and innovative models around prospective developments with respect to this discipline. It provides readers with diverse topics, which address the varied branches and applications of welding engineering. It aims to serve as a resource guide for readers and contribute to the growth of this area. Through this book, we attempt to further enlighten engineers, welders, scientists, builders, practitioners and students about the new techniques in this field.

I thank my publisher with all my heart for considering me worthy of this unparalleled opportunity and for showing unwavering faith in my skills. I would also like to thank the editorial team who worked closely with me at every step and contributed immensely towards the successful completion of this book. Last but not the least, I wish to thank my friends and colleagues for their support.

**Editor**

# High Effective FE Simulation Methods for Deformation and Residual Stress by Butt Welding of Thin Steel Plates

**Mikihito Hirohata, Yoshito Itoh**

Graduate School of Engineering, Nagoya University, Nagoya, Japan
Email: hirohata@civil.nagoya-u.ac.jp

## Abstract

In order to propose high effective simulation using finite element method (FEM) for predicting deformation and residual stress generated by one pass butt welding, a series of numerical analyses were carried out. By idealizing the movement of heat source (the instantaneous heat input method), the tendency of welding out-of-plane deformation and the residual stress distribution could be predicted. The computing time was around 9% of that by the precise model with considering the movement of heat source. On the other hand, applicability of two dimensional shell elements instead of generally used three dimensional solid elements was examined. The heat input model with considering the temperature distribution in the thickness direction was proposed for the simulation by using the shell elements. It was confirmed that the welding out-of-plane deformation and residual stress could be predicted with high accuracy by the model with shell elements and the distributed heat input methods. The computing time was around 8% of that by the precise model with solid elements.

## Keywords

Welding Deformation, Residual Stress, Butt Welding, FEM, Shell Element

## 1. Introduction

In constructing steel structures, welding is generally used for joining and assembling members. Then, welding deformation and residual stress are inevitably generated due to expansion and shrinkage of welded parts caused by local heating/cooling. Welding deformation and residual stress influence the accuracy of manufacturing, load carrying capacity and fatigue strength of members [1]. In the case that the welding deformation becomes larger than the acceptable limit, it is required to be straightened. On the other hand, high residual stress should be released by annealing in the welded joints of important parts of structures. A high cost and a long time will be necessary due to these processes. Therefore, it is important that welding deformation and residual stress are predicted and controlled.

For predicting welding deformation and residual stress generated in structural steel members, numerical simulation by thermal elasto-plastic analysis based on FEM is an effective method [2]. Welding is a complicated phenomenon involving non-steady heat transfer by movement of heat source, temperature dependency of physical constants and mechanical properties of material and three dimensional elasto-plastic problems [3]. Therefore, it takes huge computing times for welding simulation of large steel structures even though the performance of computer currently becomes faster and faster.

In this study, two calculation methods are examined by which the simulation of welding deformation and residual stress becomes more effective. One is to idealize the movement of heat source which requires many calculation steps and the other is to apply the two dimensional elements instead of three dimensional elements which enlarge the numbers of the nodes and elements of models.

A butt welding of thin steel plates [4] is simulated by these methods. The analytical results by these methods are compared with analytical results modelled precisely. The effectiveness of these methods is examined from the viewpoints of an accuracy of analytical results and computing time. Furthermore, the possibility of application of these methods on simulation for welding deformation and residual stress are investigated.

## 2. Deformation and Residual Stress Generated by One Pass Butt Welding

### 2.1. Precise Analysis Model

An analysis model in this study is one pass butt welding of thin steel plates. **Figure 1** shows the shapes and dimensions of the model. Kim *et al.* performed welding experiment for this model [4]. The base metal and the weld metal are SM400A and YGW11 specified by JIS. They are a general structural carbon steel and a corresponding welding wire. The mechanical properties and physical constants with temperature dependencies are shown in the reference [5]. This welding experiment is simulated by thermal elasto-plastic analysis by FEM. **Figure 2** shows the precise FE analysis model. A commercial FE program, ABAQUS Ver. 6.10 is used. A half model is adopted by considering symmetric conditions along the weld line. The temperature dependencies of the mechanical properties and physical constants are considered in the analysis. In order to simulate the experiment with high accuracy, three dimensional 8-nodes solid elements are used for modeling the shape of the weld groove. And then, for modeling the movement of weld heat source, heat input elements are generated step by step in the calculation considering the welding speed. In this model, there are 30 elements in the welding direction. The length of each heat input element: $L$ is 10 mm. Therefore, the number of the calculation steps for heat input is 30. And also, additional calculation step for cooling is required after the heating steps. Of course, each calculation step is divided into many fine time increments.

The heat input of welding, $Q$ (J/mm) is calculated by Equation (1) [6]. The heat energy, $q_m$ (J/mm$^3$) by

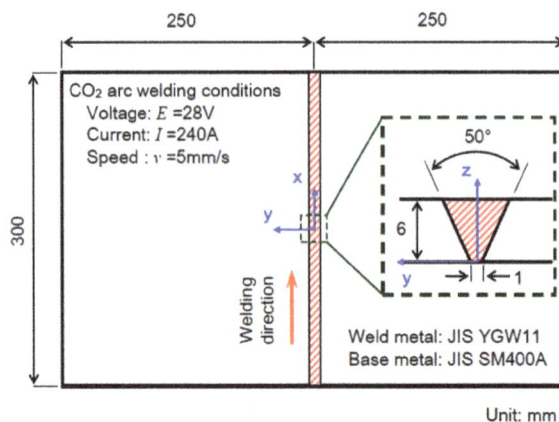

**Figure 1.** One pass butt welding model.

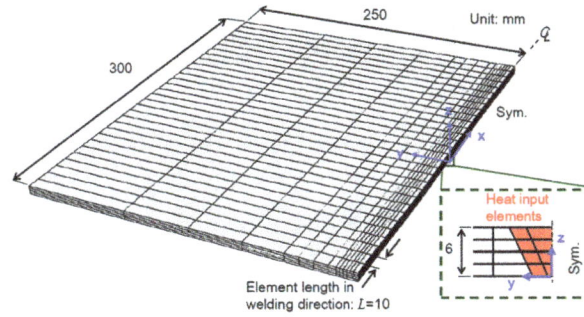

**Figure 2.** FE analysis model by 8-nodes solid elements.

Equation (2) is given into the heat input elements as a body heat flux. The heating time per each heat input element is decided by dividing the length of each heat input element, $L$ (mm) by the welding speed, $v$ (mm/s).

A heat transfer from the surface of model is considered as thermal boundary conditions. A rigid body displacement is fixed as mechanical boundary conditions.

$$Q = \eta E I / v \qquad (1)$$

$$q_m = Q v / A L \qquad (2)$$

Here, $Q$ : Heat input (J/mm),
$\eta$ : Heat efficiency (65% to 80% in arc welding [6]),
$E$ : Welding voltage (V),
$I$ : Welding current (A),
$v$ : Welding speed (mm/s),
$A$ : The sectional area of the heat input elements,
$q_m$ : Heat energy (J/mm$^3$), and
$L$ : The length of each heat input element (mm).

## 2.2. Analysis Results

### 2.2.1. Temperature Histories
Non-steady thermal conduction analysis was carried out for the butt welding model. **Figure 3** shows the temperature histories in the welding. The symbols in the figure represent the experimental results [4]. Measured positions of the temperature were at the bottom of the plates of the center in the welding direction. The distances from the weld line were 15 mm, 30 mm, 50 mm and 80 mm ( $y = 15$ , 30, 50 and 80). In order to simulate the temperature histories obtained by the experiment as accurate as possible, the calculations were tried in some times with varying the heat efficiency, $\eta$ . The analytical results almost agreed with the experimental results when the heat efficiency, $\eta$ was 0.65.

In the thermal elasto-plastic analysis, the temperature data obtained by the non-steady thermal conduction analysis is used as input data for the thermal elasto-plastic stress analysis. Therefore, it is indispensable that the temperature data are simulated with high accuracy.

### 2.2.2. Welding Deformation and Residual Stress
By using the obtained thermal conduction analysis results, thermal elasto-plastic stress analysis was carried out. **Figure 4** shows the welding out-of-plane deformation and the distributions of welding residual stress at the cross section of the center in the welding direction $(x)$ . The symbols in the figures represent the experimental results [4]. The residual stress components were obtained by a stress relaxation method. The values of the stress are the average in the thickness direction. The analytical results simulated the experimental results well.

### 2.2.3. Computing Time
The results indicated that the welding deformation and residual stress could be simulated with high accuracy by the precisely modeled analysis in which the movement of heat source was considered and three dimensional solid elements were used. When using a general personal computer (CPU 2.93 GHz), the computing time of this

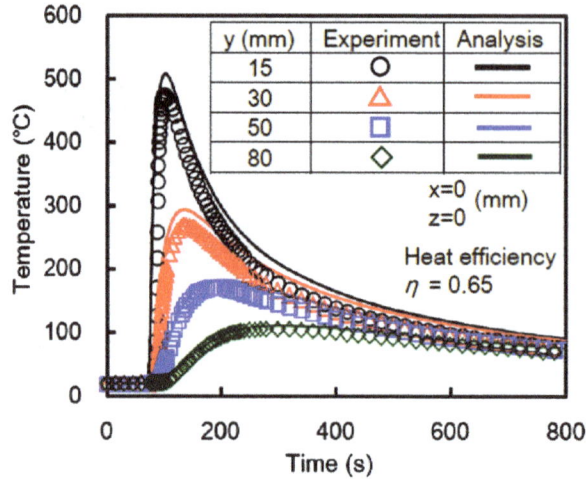

**Figure 3.** Temperature histories by precise analysis model.

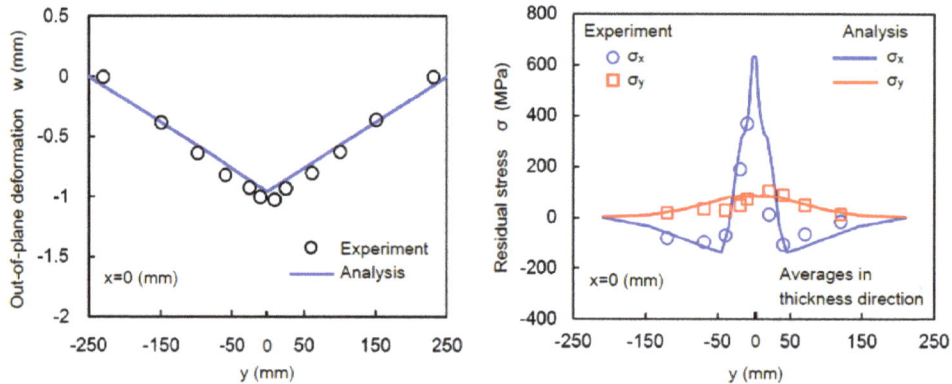

**Figure 4.** Results of precise analysis model. (a) Out-of-plane deformation; (b) Residual stress.

model was 4416 seconds (around 74 minutes).

## 3. Idealization for Movement of Heat Source

### 3.1. Instantaneous Heat Source Model

As shown in Chapter 2, many calculation steps are required for considering the movement of heat source in welding. The longer the weld line becomes, the more calculation steps are necessary. Therefore, it is examined to idealize the movement of heart source in welding. That is, the total heat energy of welding is given into the weld line at one time. This method is well known as instantaneous heat source model [7]. Here, the instantaneous heat source method is applied on the butt welding model (namely, the instantaneous heat source model). The accuracy of analysis model and the computing time are compared with the results by the precise analysis model, that is, the moving heat source model.

In the case of instantaneous heat source model, the heat energy, $q_i$ (J/mm$^3$) by Equation (3) is given into the all heat input elements in one second. The calculation step for heat input is only one time. And also, making the complicated analysis modeling becomes relatively simple and easy.

$$q_i = Q/A \tag{3}$$

### 3.2. Analysis Results

#### 3.2.1. Temperature Histories

**Figure 5(a)** shows the temperature histories. In the case of the instantaneous heat source model, the temperature

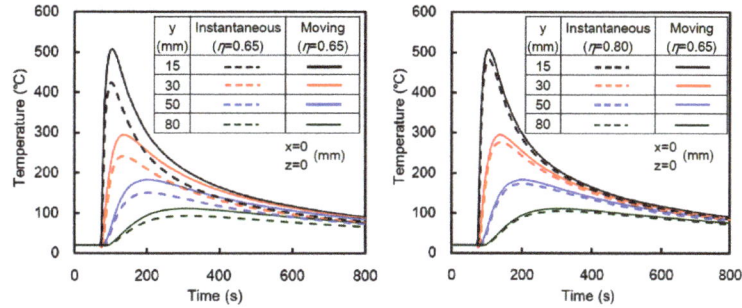

**Figure 5.** Temperature histories obtained by instantaneous heat source model.
(a) Heat efficiency $\eta = 0.65$; (b) Heat efficiency $\eta = 0.80$.

rose rapidly just after the heat input. On the other hand, the temperature at the center of the weld line rose just after the heat source passes that section in the case of the moving heat source model. For comparing with the moving heat source model, the welding start time of the instantaneous heat source model was shifted.

Even though the total heat energy was the same, the maximum temperature of the instantaneous heat source model was lower than that of the moving heat source model. This was because the temperature of the instantaneous heat source model rose in the short time and decreased after that. On the other hand, the temperature of the moving heat source model gradually rose with the passage of the heart source during the relatively long time.

In order to simulate the temperature histories by the instantaneous heat source model as close to those by the moving heat source model as possible, the heat efficiency, $\eta$ was changed from 65% to 80%. **Figure 5(b)** shows the temperature histories in the case that the heat efficiency, $\eta$ was 80%. The maximum temperature of the instantaneous heat source model was almost agreed with that of the moving heat source model.

### 3.2.2. Welding Deformation and Residual Stress

By using the temperature data in the case that the heat efficiency, $\eta$ was 80%, the thermal elasto-plastic analysis was carried out. **Figure 6** shows the welding out-of-plane deformation and the distributions of welding residual stress at the cross section of the center in the welding direction $(x = 0)$. Even though the magnitude of the welding out-of-plane deformation of the instantaneous heat source model was smaller than that of the moving heat source model, their tendencies were the same. The magnitude of the out-of-plane deformation of the instantaneous heat source model was around 72% of that of the moving heat source model. On the other hand, the residual stress distributions of both models were agreed with each other.

In order to investigate the reason why the magnitude of welding out-of-plane deformation of the instantaneous heat source model was smaller than that of the moving heat source model, the generation histories of the welding out-of-plane deformation of both models are shown in **Figure 7**.

In the case of the moving heat source model, the out-of-plane deformation at the center of the weld line occurred when the heat source passed that section (the time was 30 s). Because of the $V$-shaped groove, the heat input at the upper side was larger than that at the lower side in the thickness direction. Therefore, the shrinkage in the direction perpendicular to the weld line was larger at the upper side rather than at the lower side. As a result, $V$-shaped out-of-plane deformation occurred [6]. On the other hand, in the case of the instantaneous heat source model, the large expansion in the direction perpendicular to the weld line occurred when the heat input started. Because the heat input at the upper side was larger than that at the lower side in the thickness direction due to the $V$-shaped groove, the expansion at the upper side was larger than that at the lower side. Therefore, the inverted $V$-shaped out-of-plane deformation occurred firstly. After that, the expanded material shrunk and the $V$-shaped out-of-plane deformation was generated in the cooling process. As a result, the finally generated out-of-plane deformation of the instantaneous heat source model became smaller than that of the moving heat source model.

### 3.2.3. Applicability of Instantaneous Heat Source Model

The computing time of the instantaneous heat source model was 387 seconds (around 6.5 minutes). It was around 9% of that of the moving heat source model.

The generation mechanism of the out-of-plane-deformation of the instantaneous heat source model differed

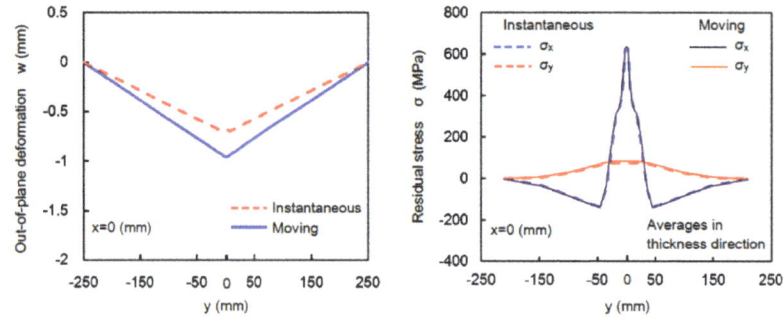

**Figure 6.** Welding out-of-plane deformation and residual stress by instantaneous model. (a) Welding out-of-plane deformation; (b) Welding residual stress.

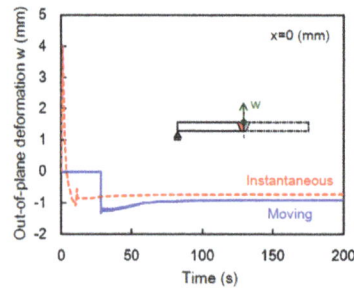

**Figure 7.** Generation histories of welding out-of-plane deformation.

from that of the moving heat source model. However, the tendency of the out-of-plane deformation could be predicted by the instantaneous heat source model even if the prediction accuracy of magnitude of out-of-plane deformation was not required so high. Furthermore, the welding residual stress distribution could be accurately predicted by the instantaneous heat source model. When the required accuracy of welding out-of-plane deformation is not so high or when only the prediction of residual stress distribution is required, the instantaneous heat source model will be sufficiently available.

## 4. Application of Two Dimensional Shell Elements

### 4.1. Shell Model

One of the reasons why the numerical simulation of welding by FEM takes a huge computing time is the use of three dimensional solid elements. For modeling the groove shape, solid elements should be used and some layers should be made in the thickness direction even though the thickness of the plate is thin. Therefore, the numbers of nodes and elements becomes large.

Here, application of two dimensional shell elements on the welding simulation is examined. By using the shell elements, the number of nodes and elements in the thickness direction can be decreased. **Figure 8(b)** shows the image of the analysis model using 4-nodes shell elements (namely, the shell model). In the case of the precise analysis model using 8-nodes solid elements (namely, the solid model), the shape of groove can be simulated. Even though the shape of groove cannot be considered in the shell model, the sectional area of heat input element is made to be the same as that of the solid model. And then, heat energy calculated by Equation (4) is given into the heat input elements as concentrated heat flux. For calculating the heat energy, the heat efficiency, $\eta$ of 0.65 is used as well as the case of the solid model. The movement of heat source is considered in both the solid model and the shell model.

$$q_s = q_m AL/n \qquad (4)$$

Here, $n$, the number of nodes assembling the heat input elements.

By the way, the out-of-plane deformation occurs due to a difference of temperature between upper and lower surfaces of plates in the case of one pass butt welding of thin steel plates with $V$-groove [6]. In other words, heat energy at upper surface is larger than that at lower surface. Therefore, larger shrinkage in cooling process of welding occurs at upper surface rather than at lower surface. By the above proposed shell model, the difference of the heat energy between the upper and the lower surfaces is not considered because the heat input into the elements is uniform in the thickness direction. Therefore, the out-of-plane deformation may not occur.

To solve this problem, a heat input method for the shell model shown in **Figure 8(c)** is also proposed in this study. The linear distribution of the heat energy in the thickness direction calculated by Equation (5) is considered. The difference of the heat energy between the upper and the lower surfaces is determined by the geometry of $V$-groove shape. The distributed heat energy is given into each integration point in the heat input elements.

$$q_d = \mathrm{d}(z)\big/\mathrm{ds} \cdot q_m AL\big/n_i \tag{5}$$

Here, $q_d$ : The distributed heat energy (J/mm$^3$),
$\mathrm{d}(z)$ : The $z$-coordinate of each integration point,
$\mathrm{ds}$ : The width of heat input elements, and
$n_i$ : The number of integration points in the thickness direction.

## 4.2. Analysis Results

### 4.2.1. Temperature Histories

**Figure 9** shows the temperature histories in the welding. The maximum temperatures at the position of $y = 15$ of the both shell models with uniform and distributed heat inputs were higher than that of the solid model by around 40 degrees Celsius. The reason of it was possibly the difference of the shapes of the heat input elements. That is to say, the width of the heat input elements at the bottom of the shell model was wider than those of the solid model. Therefore, the larger heat energy reached to the temperature measured points in the shell model

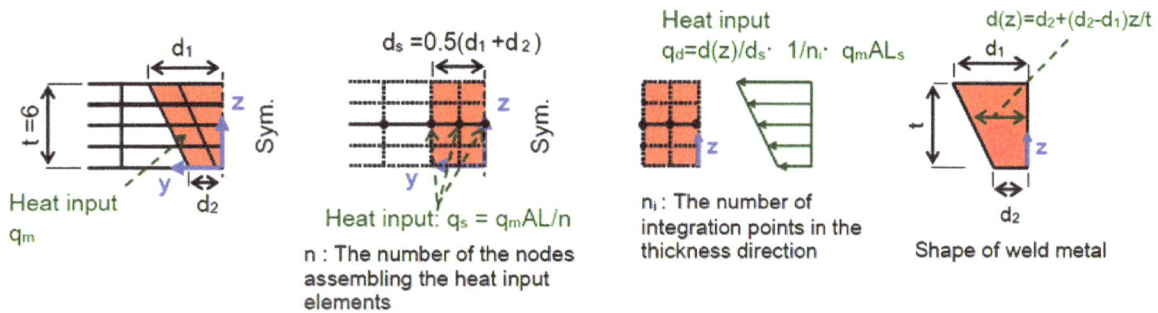

**Figure 8.** Image of shell model. (a) Solid model; (b) Shell model with uniform heat input; (c) Shell model with distributed heat input.

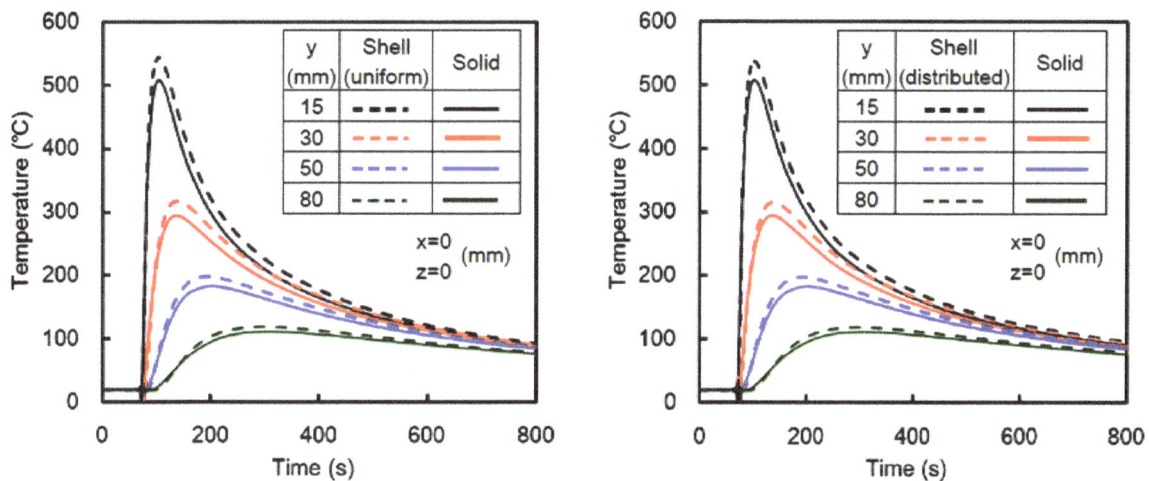

**Figure 9.** Temperature histories obtained by shell models. (a) Results by uniform heat input; (b) Results by distributed heat input.

compared with the solid model.

Figure 10 shows the temperature differences between the upper and the lower surfaces of heat input elements of each model. Even though the temperature differences did not occur in the shell model with uniform heat input, it occurred in the shell model with distributed heat input as well as the solid model.

### 4.2.2. Welding Deformation and Residual Stress

Figure 11(a) shows the welding out-of-plane deformations at the center of the welding direction. In the case of the shell model with uniform heat input, the out-of-plane deformation scarcely occurred as expected. On the other hand, the out-of-plane deformation by the shell model with distributed heat input almost agreed with that by the solid model. The difference of magnitude of the out-of-plane deformation between the shell and the solid models was around 7%.

Figure 11(b) shows the distributions of welding residual stress at the cross section of the center in the welding direction. Both of the analytical results by the solid model and by the shell models with uniform and distributed heat inputs were almost the same with each other.

By considering the distribution of the heat energy in the thickness direction in the shell model, the welding out-of-plane deformation could be simulated with high accuracy. It could be said that the proposed heat input

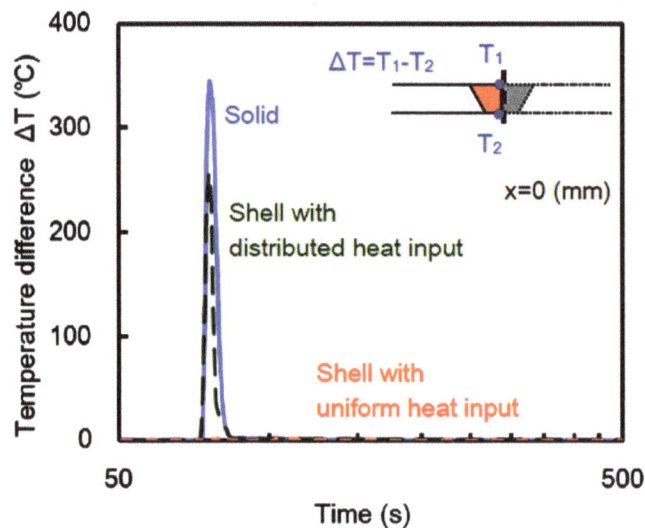

Figure 10. Temperature differences between upper and lower surfaces.

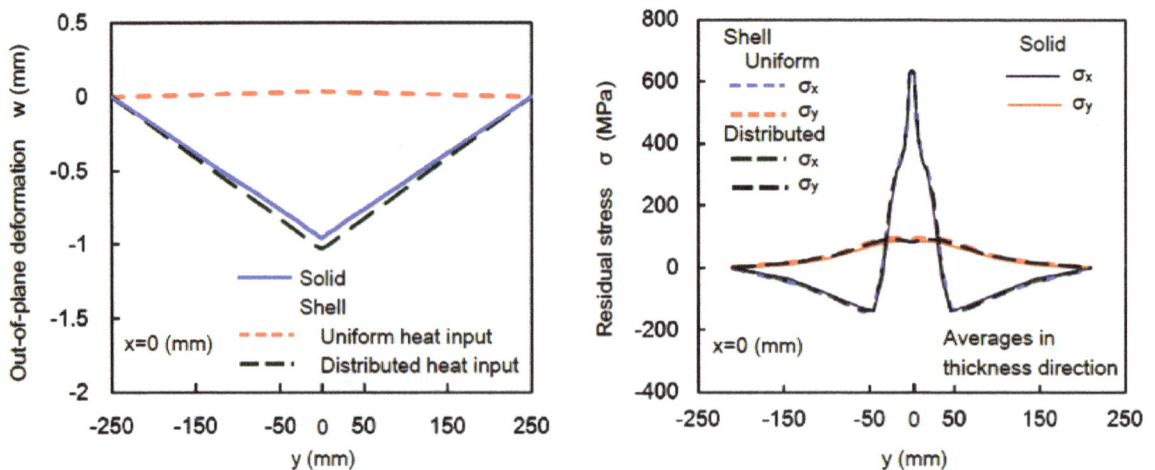

Figure 11. Welding out-of-plane deformation and residual stress by shell models. (a) Welding out-of-plane deformation; (b) Welding residual stress.

method was valid. By the way, the computing time of the shell model with uniform heat input was 342 seconds (5.7 minutes). That with distributed heat input was 368 seconds (around 6.1 minutes). The computing times of the shell models were around 8% of that of the solid model.

## 5. Conclusions

In order to propose the effective calculation methods by FEM for predicting deformation and residual stress generated by one pass butt welding of thin steel plates, a series of numerical analyses were carried out.
The obtained main results are as follows.

The effect of idealizing the movement of heat source in welding (the instantaneous heat input model) was examined. It was confirmed that the generation mechanism of welding out-of-plane deformation by the instantaneous heat input model differed from that by the precise model considering the movement of heat source.

Even though the tendency of out-of-plane deformation could be predicted, the magnitude of out-of-plane deformation of the instantaneous heat source model was around 72% of that of the moving heat source model. On the other hand, the residual stress distribution of the instantaneous heat source model was almost the same as that of the moving heat source model.

The computing time by the instantaneous heat source model was around 9% of that of the moving heat source model. When the required accuracy of welding out-of-plane deformation is not so high or when only the prediction of residual stress distribution is required, the instantaneous heat source model will be sufficiently available.

Applicability of two dimensional shell elements instead of generally used three dimensional solid elements was examined. When the temperature distribution in the thickness direction was not considered, the welding out-of-plane deformation could not be simulated well. Therefore, the heat input method with considering the temperature distribution in the thickness direction was proposed for the simulation by using the shell elements.

It was confirmed that the welding out-of-plane deformation and residual stress could be predicted with high accuracy by the model with shell elements and the distributed heat input method. The difference of magnitude of the out-of-plane deformation between the shell model and the solid model was around 7%.

The computing time by the shell model was around 8% of that by the solid model.

The obvious effectiveness could be confirmed even in the simple analysis model in this study. The results indicated that the larger analysis models such as actual steel structures are simulated, the higher effectiveness by using the proposed simulation methods will be expected.

## Acknowledgements

This research was partly supported by the Sasakawa Scientific Research Grant from the Japan Science Society.

## References

[1]  Japan Society of Civil Engineers (2005) Guidelines for Stability Design of Steel Structures. MARUZEN Co., Ltd., Tokyo.

[2]  Lindgren, L.-E. (2006) Numerical Modelling of Welding. *Computer Methods in Applied Mechanics and Engineering*, **195**, 6710-6736. http://dx.doi.org/10.1016/j.cma.2005.08.018

[3]  Zhu, X.K. and Chao, Y.J. (2002) Effects of Temperature Dependent Material Properties on Welding Simulation. *Computers and Structures*, **80**, 967-976. http://dx.doi.org/10.1016/S0045-7949(02)00040-8

[4]  Kim, Y.-C., Lee, J.-Y., Sawada, M. and Inose, K. (2007) Verification of Validity and Generality of Dominant Factors in High Accurate Prediction of Welding Deformation. *Quarterly Journal of Japan Welding Society*, **25**, 450-454.

[5]  Kim, Y.-C., Lee, J.-Y. and Inose, K. (2007) Dominant Factors for High Accurate Prediction of Distortion and Residual Stress Generated by Fillet Welding. *International Journal of Steel Structures*, **7**, 93-100.

[6]  Japan Welding Society (2003) A Handbook of Welding and Joining. 2nd Edition, MARUZEN Co., Ltd., Tokyo.

[7]  Satoh, K. (1967) On Heat Conduction by Moving Heat Source. *Journal of the Japan Welding Society*, **36**, 154-159.

# An ICME Approach for Optimizing Thin-Welded Structure Design

**Guoqing Gou[1], Yuping Yang[2*], Hui Chen[1]**

[1]School of Materials Science and Engineering, Southwest Jiaotong University, Chengdu, China
[2]EWI, Columbus, USA
Email: [*]yyang@ewi.org

## Abstract

Integrated computational materials engineering (ICME) is an emerging discipline that can speed up product development by unifying material, design, fabrication, and computational power in a virtual environment. Developing and adapting ICME in industries is a grand challenge technically and culturally. To help develop a strategy for development of this new technology area, an ICME approach was proposed and implemented in optimizing thin welded structure design. The key component in this approach is a database which includes material properties, static strength, impact strength, and failure parameters for a weld. The heat source models, microstructure model, and thermo-mechanical model involved in ICME for welding simulation were discussed. The shell elements representing method for a fusion weld were introduced in details for a butt joint, lap joint, and a Tee joint. Using one or multiple solid elements representing a spot weld in a shell model was also discussed. Database building methods for resistance spot welding and fusion welding have been developed.

## Keywords

Welding, Finite Element Analysis, Modeling and Simulation, ICME

## 1. Introduction

Welding process has been widely used in structure fabrication to join materials. However, welding inevitably induces material properties change, residual stress and deformation due to intense localized heating during welding and fast cooling after welding, especially for heat treatment strengthened high strength steels and aluminum alloys. Although post-weld heat treatment (PWHT) can be used to recover the material strength, it is not

---

[*]Corresponding author.

practical to conduct PWHT for all welds. It is common that welding effects on materials are ignored during designing welded structures. The same material properties as base materials were used in the welding area, which could result in structure failures during service because a softening zone is often formed near a weld of high strength steels and joint strength is typical lower than base materials for heat-treatment strengthened aluminum alloys.

ICME has been gradually used in industries to develop products and produce new alloys [1]-[3] because it creates significant cost benefit and time reduction. Recently, an ICME study led by The Minerals, Metals, and Materials Society (TMS) [4] was conducted to identify, priority, and make detailed recommendations for the frameworks and key steps needed to implement ICME. The frameworks included a processing tool, a microstructure tool, and a property tool. Developing these tools for manufacturing process simulation remains significant challenges [5].

This paper introduces ICME implementation framework for developing a product involving welding process for thin welded structures such as car bodies of automotive vehicles and high speed trains. An ICME approach to optimize welded structure design was outlined. The detailed heat source models, microstructure model, thermomechanical model, and property model used in welding simulation were discussed. The weld representations in a shell model for fusion welds and spot welds are introduced. A database creation method which is a key component in the ICME approach was discussed using modeling examples. This ICME approach allows considering the weld effects including material properties, joint strength, and residual stress during optimizing thin welded structures, which could greatly improve the safety of automotive vehicles and high-speed trains.

## 2. Optimizing Welded Structure Design

### 2.1. ICME Framework

An ICME implementation framework (**Figure 1**) for developing a product involving manufacturing processes such as welding was developed by generalizing the framework for automotive, aerospace, and maritime in [4]. This framework has been successfully used in designing automotive spot-weld structures [6] [7], repairing aeroengine blades [8], and designing composite-to-steel adhesive joined ship structures [9].

As shown in **Figure 1**, the framework includes three major blocks: product requirements, manufacturing processes modeling tools, and final product. The product requirements include property requirement, design, and materials. The manufacturing process modeling tools include modeling inputs (process parameters), process tool (a thermal model, a microstructure model, and a property model), and performing prediction tools. The final product includes product validation and verification and final product configuring (material and design). There are 12 steps in the framework in which steps 8 and 11 are decision-making step.

### 2.2. Optimization Process Flow

The framework shows in **Figure 1** is an ideal case. It is not practical and time consuming for each weld to go

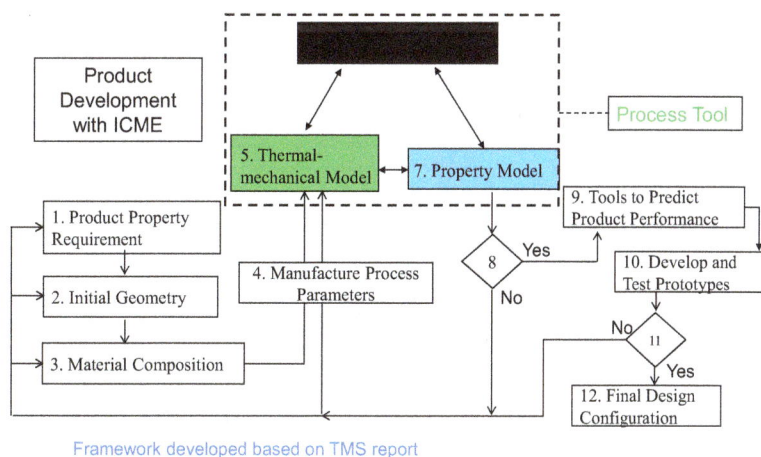

Figure 1. ICME implementation framework.

through the calculation since a welded structure could include hundreds and thousands welds. Thus, a database approach was proposed to model each weld joint before a structure design as shown in **Figure 2**.

**Figure 2** shows the optimization process flow to include weld effect in load, crash, and fatigue modeling. For a given prototype design such as a high speed train car body, weld joints are identified first and then the database will be checked for each weld to find the joint parameters (material properties, static strength, impact strength, and failure parameters). The joint parameters will be input to the model for the prototype design to conduct load, crash, and fatigue analyses. If analysis results show the design achieves the requirements, the optimization process is complete. A prototype based on the optimized design will be fabricated and tested, as shown in the step 10 of **Figure 1**. Otherwise, a new analysis will be conducted by changing welding parameters, plate thickness, and materials.

The database is the key component in the optimization process. Currently, welding processes in the database include fusion welding such as gas metal arc welding (GMAW) and resistance spot welding (RSW). Other welding processes such as laser welding and hybrid laser arc welding (HLAW) can be added to the database. The RSW database for mild steels and high strength steels with 0.5 mm to 2 mm thickness has been built for automotive applications. A database for aluminum alloys welded with laser welding and HLAW can be built for high-speed train design application. The database is strain-rate dependent and should cover at least three strain rates: low, middle and high since crash analyses for an automotive vehicle and high-speed train are conducted in high speed. Each weld will have a different speed during crash. A corresponding data for each weld can be obtained by interpolating the data in the database.

## 3. Models in the Process Tool

The process tool shown in **Figure 1** includes a thermal-mechanical model, a microstructure model, and a property model. This section discusses the mathematical equation and physics in each model.

### 3.1. Welding Heat Source Models

Welding heat source models are a part of the thermal-mechanical model. Welding heat source models have been developed to predict temperature history during welding and cooling by inputting welding parameters. Based on the type of welding process, a corresponding model was developed to simulate the physics involved in the process. As an example, an incremental coupled thermal-electrical-mechanical model was developed for resistance spot welding (RSW) to predict the nugget size and temperature history [6].

For fusion welding such as gas metal arc welding (GMAW), a moving-arc solution was developed using

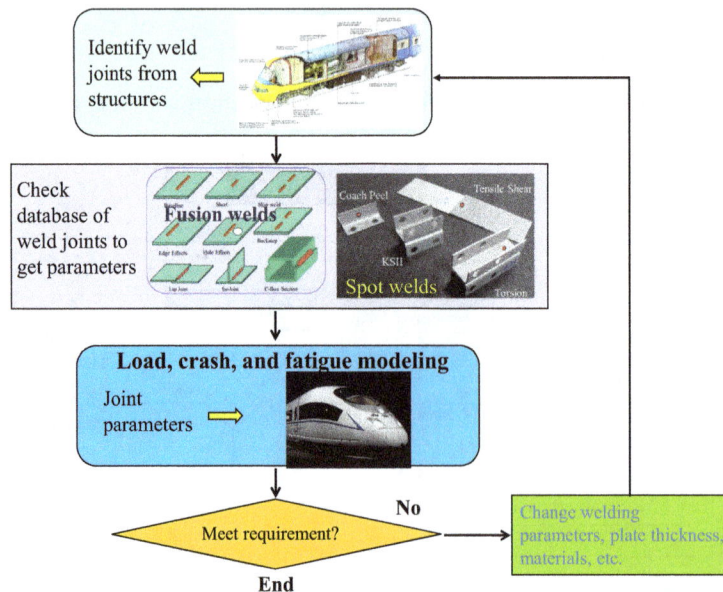

**Figure 2.** ICME application in optimizing welded structure design.

Goldak double ellipsoidal heat source model which could be expressed as follows [10]:

$$q(x,y,z,t) = f \frac{6\sqrt{3}Q\eta}{abc\pi\sqrt{\pi}} e^{\frac{-3x^2}{a^2}} e^{\frac{-3y^2}{b^2}} e^{\frac{-3[z+v(\tau-t)]^2}{c^2}}$$

(1)

$$Q = IU$$

where $f$ is a factor, $\eta$ is arc Efficiency (0.85), $I$ is welding current (255 A), $U$ is welding voltage (25 V), and $v$ is welding torch travel speed (4.969 mm/sec). $a$, $b$, $c$ are the semi-axes of the ellipsoid.

**Figure 3** shows a weld-cross macrograph, a Tee-joint mesh, and a predicted maximum temperature distribution using Goldak heat source model [10]. The material is DH36 and welding process is GMAW. The grey color shows the predicted fusion zone which is close to the weld fusion zone in the macrograph.

## 3.2. Microstructural Model

Many microstructure models were developed to predict microstructure and hardness near a weld such as Ashby microstructure model [11], Bhadeshia thermodynamic and kinetic model [12], and Gould analytical-based thermal and microstructure model [13].

Microstructural models are used to predict the volume fraction of each microstructure phases and then calculate the hardness based on the predicted microstructure. **Figure 4** shows an example of predicted hardness distribution in the weld and heat-affected zone (HAZ) for a four-pass pipe girth weld. The pipe material is a low alloy high strength steel and the filler material is ER120. The prediction was conducted using a standalone microstructure code [14] which was developed based on Ashby microstructure model. Because of fast cooling in the heat-affected zone (HAZ), martensite phase was formed in HAZ which results in higher hardness than inside the weld.

## 3.3. Thermal-Mechanical Model

Thermal-mechanical model was developed to predict thermal induced stress and deformation evolution during welding. **Figure 5** shows a frame work of the thermal-mechanical model which couples with the thermal model and the microstructure model. The thermal model predicted temperature history as a function of location and time is input to the thermal-mechanical model. The microstructure change induced plasticity including volume change and phase transformation is considered in the thermal-mechanical model. Temperature dependent mechanical properties are input to the model. Strain-hardening, large deformation and melting/re-melting mechanisms are modeled for welding simulation.

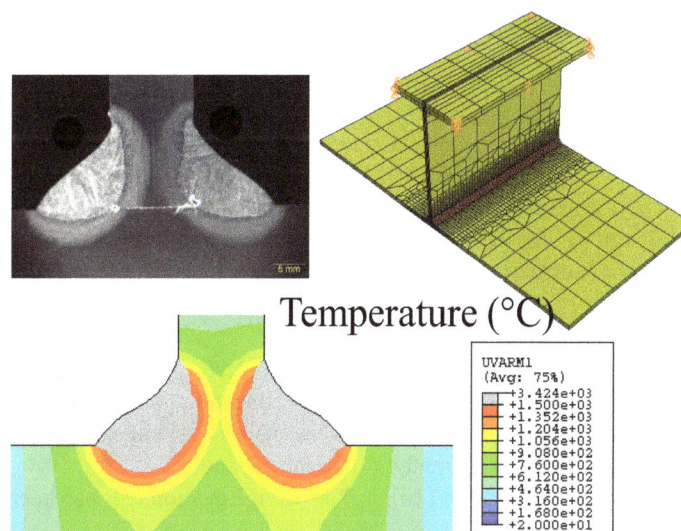

**Figure 3.** Predicted temperature distributions for welding a tee joint using thermal model.

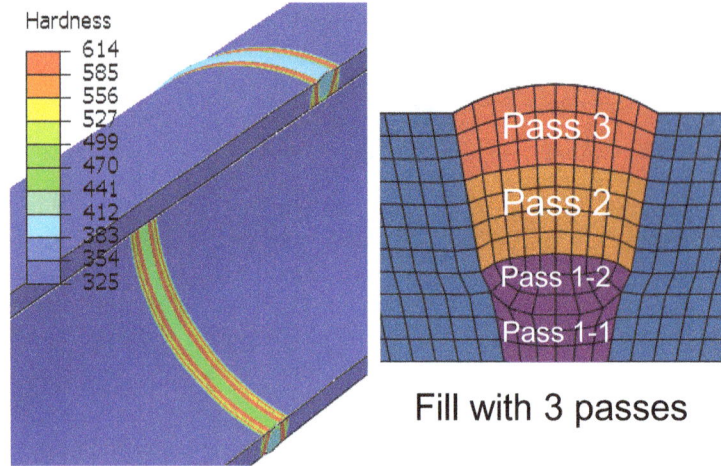

**Figure 4.** Predicted Vickers Hardness distributions for a multi-pass weld.

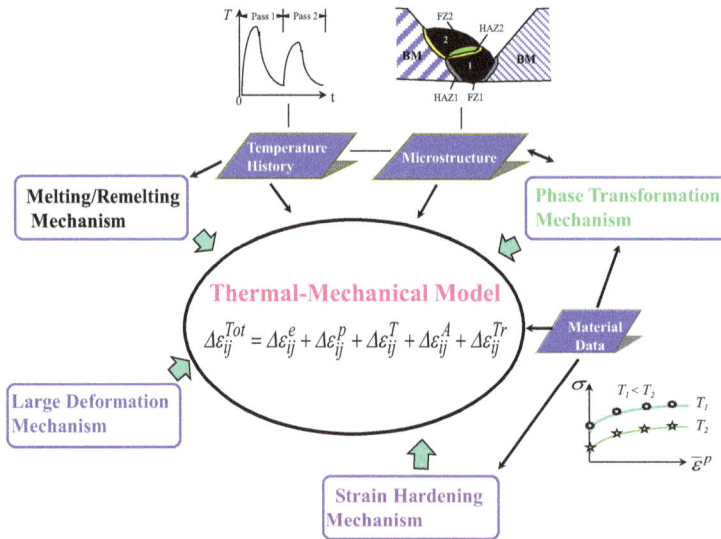

**Figure 5.** Frame work of thermal-mechanical model.

## 3.4. Property Model

A property model was developed to predict weld material properties, static strength, impact strength, and failure parameters of a weld joint. The joint property and failure parameters will be input to Step 9 as shown in **Figure 1** to predict the product performance by applying loads.

Weld static tensile strength can be estimated from the predicted hardness using the microstructure model. For example, static tensile strength of steels ($\sigma_{\text{static}}$, MPa) can be calculated from the following equation.

$$\sigma_{\text{static}} = 6.895\left(0.44\text{HV} + 11.15\right) \tag{2}$$

The equation was developed by fitting the curve in [15] in which HV is the Vickers hardness with 10 kgf.

The impact strength depends on the impact speed. Experimental testing at a series of strain rate can be conducted for a material to establish the relationship between material tensile strength and strain rate. Based on the relationship, the impact strength can be estimated for an impact speed.

Wilkins cumulative strain damage model [16], as shown in Equation (3), has been used to predict the failure in the weld joint for fusion welding [17].

$$D_p = \int w_1 w_2 \mathrm{d}\varepsilon_p \tag{3}$$

where $\varepsilon_p$ is the equivalent plastic strain, $w_1$ is the hydrostatic-pressure weighting term, and $w_2$ is the asymmetric-strain weighting term. When the cumulative damage strain $D_p$ exceeds a critical value $D_c$, discontinuous macro-crack will occur and grove step-wisely. The critical value $D_c$ can be determined by experimental testing and modeling [17].

Daimler Chrysler failure model, as shown in Equation (4), has been used to predict resistance spot weld failure [7]. It is a stress-based failure model considering tension stress, bending stress, and shearing stress.

$$ f = \left( \frac{\sigma_N}{S_N\left(\dot{\varepsilon}_{eff}\right)} \right)^{n_N} + \left( \frac{\sigma_B}{S_B\left(\dot{\varepsilon}_{eff}\right)} \right)^{n_B} + \left( \frac{\tau}{S_S\left(\dot{\varepsilon}_{eff}\right)} \right)^{n_S} \geq 1 \tag{4} $$

where $\sigma_N$, $\sigma_B$, and $\tau$ are axial (normal) stress, bending stress, and shear stress, respectively. $\dot{\varepsilon}_{eff}$ is effective strain rate and $S_N\left(\dot{\varepsilon}_{eff}\right)$, $S_B\left(\dot{\varepsilon}_{eff}\right)$, and $S_S\left(\dot{\varepsilon}_{eff}\right)$ are strain-rate dependent axial strength, bending strength, and shear strength at failure. The model includes six failure parameters ($S_N$, $S_B$, $S_S$, $n_N$, $n_B$, and $n_S$) which can be determined by testing and modeling [7].

## 4. Representing Welds Using Shell Elements

### 4.1. Fusion Weld Representation

Shell models are typically used by thin welded structure because of thin material thickness. Weld detailed geometry cannot be included in a shell model. To allow industries to include a weld joint in their structure design, methods to represent a weld joint using shell elements were developed [18] [19]. Analysis results showed that the shell-element model can predict the similar mechanical behavior as a detailed solid-element model [18].

**Figure 6** shows an example how to model a butt joint using shell elements. The middle surface, a dot line in the weld cross section, was used as reference surface to create the shell element mesh. The weld bead area, showing with red color, is thicker than other area. Fine meshes are used near a weld and coarse meshes are used in the area far away from a weld. Similar welding distortion was predicted by a shell model and a solid model although the weld bead shape is greatly simplified in the shell model.

**Figure 7** shows how to represent a lap joint using shell elements. The yellow dot line on the weld cross section macrograph was used as a reference surface to create the shell elements. Five integration points were used in a weld simulation to model the temperature gradient in the thickness direction.

**Figure 8** shows two methods to represent a Tee joint using shell elements. Both methods were evaluated to predict weld induced distortion. For a thick structure, both methods works well if the thickness is defined correctly in the weld shell elements. For a thin automotive structure, Method 1 is working better than Me-

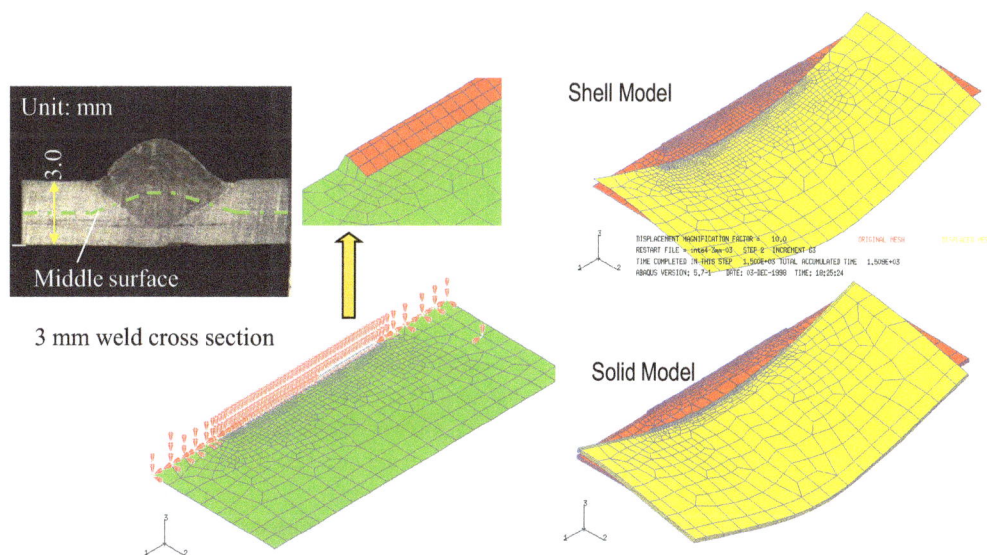

**Figure 6.** Butt joint shell element modeling method [18].

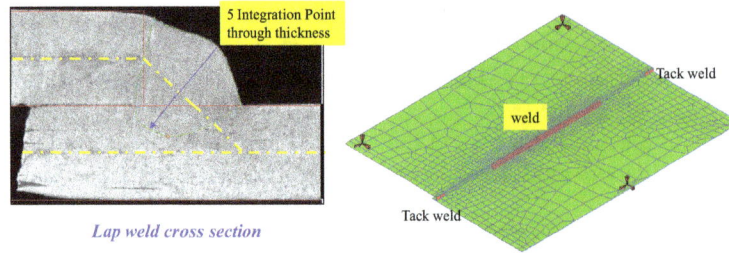

**Figure 7.** How to represent a lap joint with shell elements.

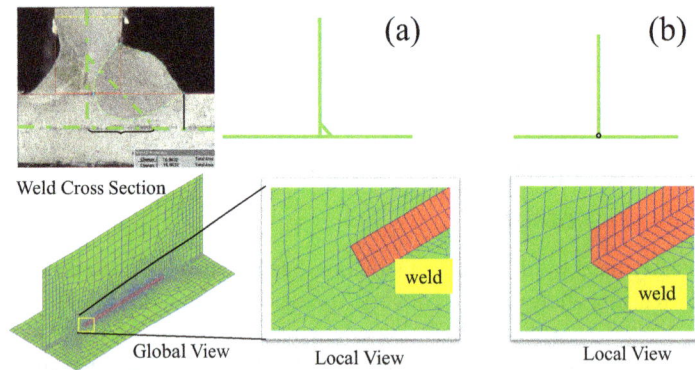

**Figure 8.** How to represent a Tee joint with shell elements (a) Method 1 (b) Method 2.

thod 2. However, to simplify the mesh generation in a large automotive structure, Method 2 could also be selected, as shown in **Figure 9**. **Figure 9** also includes examples on how to represent a butt joint, and a lap joint using shell elements in a thin-welded structure.

## 4.2. Spot Weld Representation

Spot welds were used to model by rigid links or beam elements in early days. Spot weld mechanical performance cannot be accurately predicted. With high strength steels increasingly used in automotive structures, the failure prediction of spot welds becomes critical to predict the mechanical performance correctly. Solid-element approach was developed and implemented in commercial software LS-Dyna [20]. A spot weld can be represented by one solid element, as shown in **Figure 10(a)**, for KSII, coach peel, and lap shear sample. A spot weld can also be represented by 4, 8, and 16 solid elements [21]. The component in **Figure 10(b)** includes about 40 spot welds in the structure. Each spot weld was represented by 8 solid elements.

## 5. Building Database

### 5.1. Spot-Weld Database

Spot-weld database includes weld properties and failure parameters for Equation (4). To create the database, testing and modeling were conducted in three strain rates: low, middle, and high. **Figure 11** shows a spot welding setup, process parameters (force and current), weld macrograph, and process modeling method and results. Spot weld process was modeled by a coupled thermal-electrical-metallurgical-mechanical model. The predicted nugget size is close to the weld macrograph. By inputting the predicted temperature history to the microstructure model, the spot hardness distribution was predicted which was very close to the measured one.

With the predicted hardness, the tensile strength of spot weld was estimated using Equation (2). Then the yield strength of spot weld can be estimated from the tensile strength. By assuming the spot weld has the same elastic and tangential modulus as the base material, a bi-linear stress strain curve can be established and save in the spot weld database.

Spot-weld failure parameters were developed by conducting six tests: KSII loading in 0, 30, 60, and 90 degree,

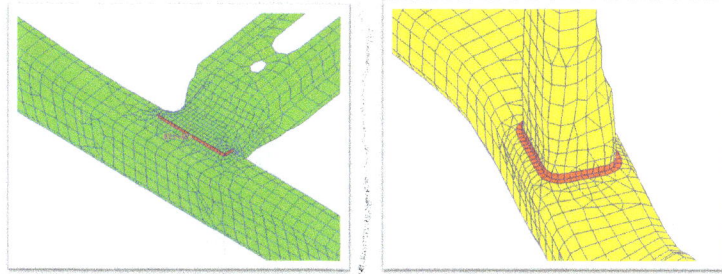

**Figure 9.** Applying shell modeling method in a thin welded structure.

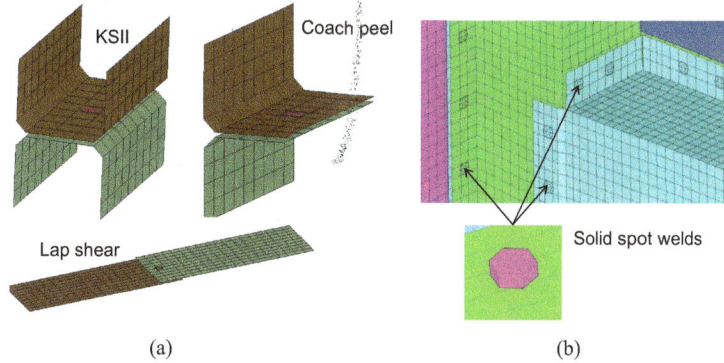

(a)        (b)

**Figure 10.** Spot-weld representation in a component (a) Sample modeling; (b) Component modeling.

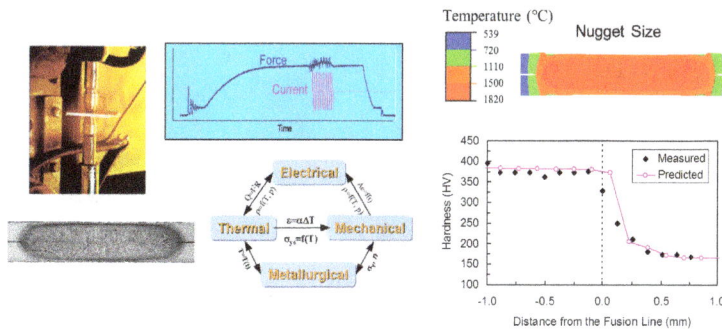

**Figure 11.** Spot-weld process modeling.

coach peel, and lap shear [7]. The six tests were modeled with the same mesh size (see **Figure 10**) as a global automotive structure since stress depends on the mesh size. The model predicts a peak stress (strength) in tension, bending, and shear. By solving set sets of equations, the six failure parameters in Equation (4) can be obtained.

**Figure 12** shows the testing and modeling of KSII sample with 90-degree loading. The model predicts the deformation evolution during load and the failure of spot welds. **Figure 13** shows the testing and modeling of lap shear sample. The testing fixtures were included in the model. The material is 50 ksi steel with 1 mm thickness for one sheet and 2 mm thickness for another sheet. Two analyses were conducted to check the effect of spot weld material properties. By inputting the base material properties for the spot weld, the analysis under predict the load. By inputting the predicted spot weld material properties from the process model, the predicted load can match with the experimental measurement.

The testing and modeling developed material properties and failure parameters in samples were verified in a component level before applying to a large thin welded structure. **Figure 14** shows a designed T-component made with about 40 spot welds. The finite element mesh of the component is shown in **Figure 12**. Analysis results show that the spot welds fail during loading. It should be pointed out that spot-weld failure modes cannot

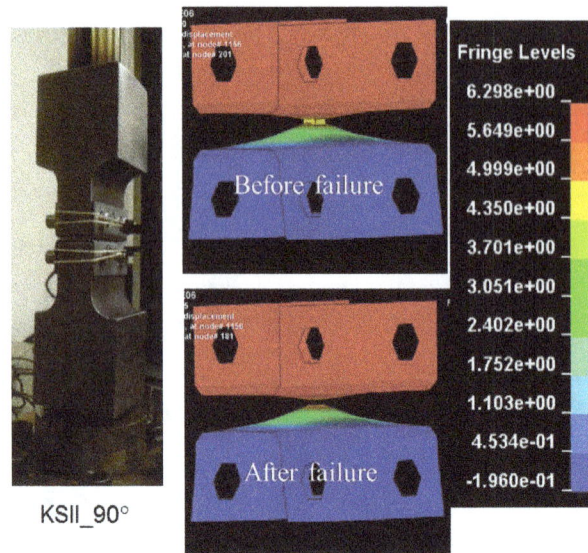

**Figure 12.** Failure prediction of KSII-90 sample.

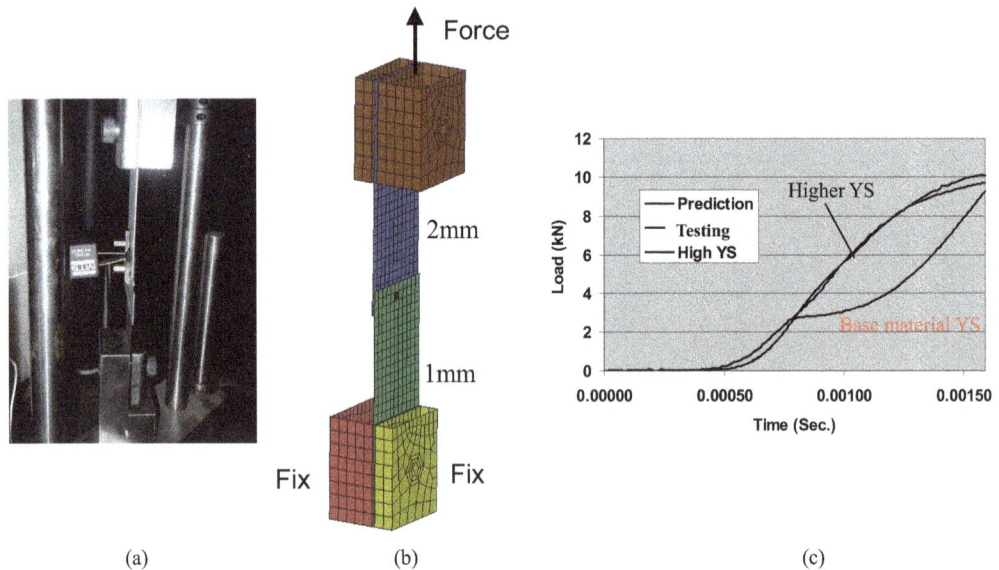

**Figure 13.** Lap-shear sample modeling (a) Testing set up; (b) Sample mesh; (c) Load-time curves.

predict because of coarse mesh which requires find mesh in the spot weld as discussed in [6]. Once cracks initiate, the model thinks the spot weld fail completely. Currently, the damage parameters are developed to refine the model.

## 5.2. Fusion-Weld Database

Fusion-weld database as a function of strain rate can be built. Welding process models for a butt joint, a lap joint, and a Tee joint have been developed to predict hardness and weld residual stress. Mechanical models have been developed to obtain the failure parameter for Equation (3).

As an example, **Figure 15** shows predicted temperature and stress distributions and comparisons of temperature and displacement between testing and modeling for a lap-joint. The lap joint joins two 125 mm wide and 300 mm long plates with a 150 mm long weld in the middle. High tension stresses were predicted on the area away from the weld toe because of weld shrinkage induced bending deformation. Six models were analyzed to identify the best one close to experiment. It was found that the model shown in **Figure 7** produces the best result

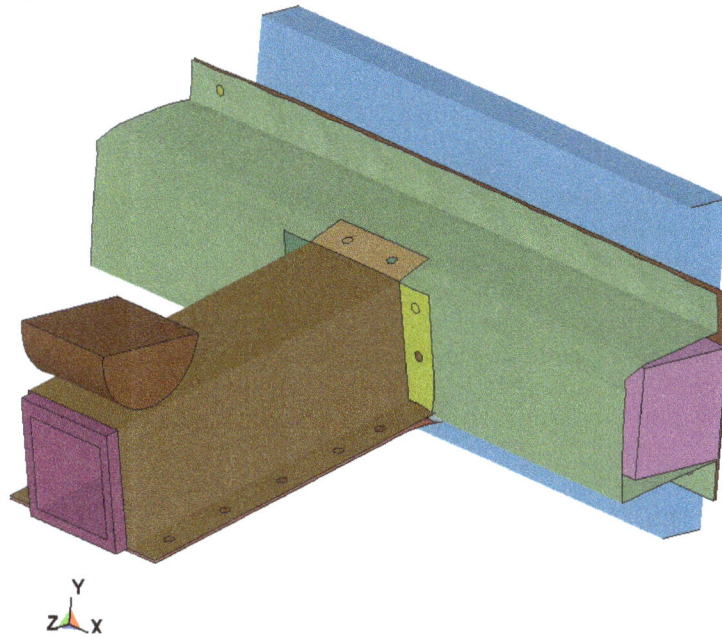

**Figure 14.** Database verification test and modeling on a component.

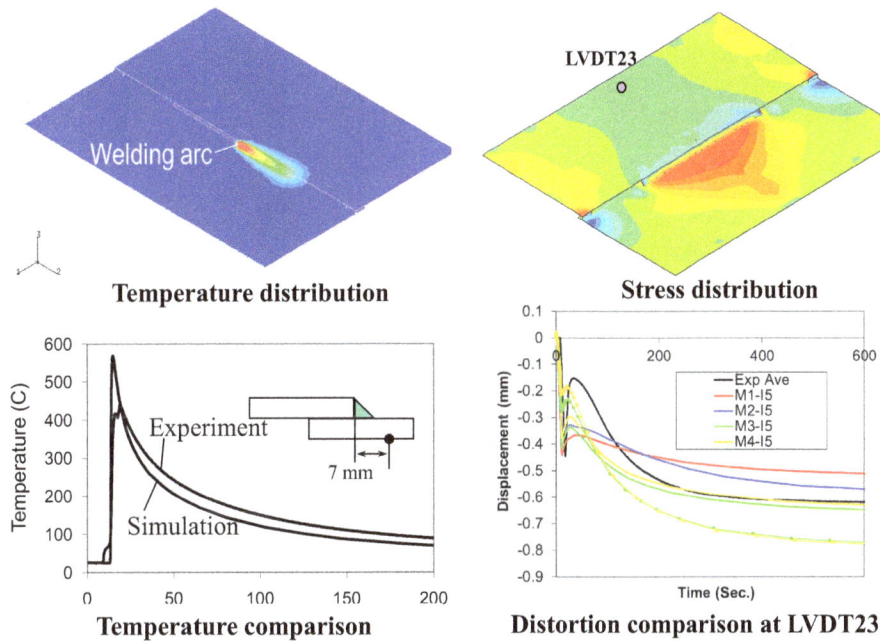

**Figure 15.** Lap-joint welding simulation.

which can be used to build the fusion-weld database. Similar excises were conducted for butt joint and lap joint to identify the best model.

## 6. Load, Crash, and Fatigue Analysis

Once the database is built for spot welds and fusion welds, weld effects such as material property change, joint strength, and residual stress can be modeled in the load, crash, and fatigue analysis of thin welded structures such as automotive vehicles and high-speed train car bodies. The long-term goal is to connect the database to the commonly used commercial software such as LS-Dyna to automate the simulation process. This ICME approach will improve the current structure design in industries to result in a better product and reduce the weld failures during service.

## 7. Summary

Weld effects on material property change, joint strength, and residual stress are ignored in industries during designing welded structures because of lack of methods and data. ICME has been gradually accepted by industries to speed up product development and produce new materials. To help apply ICME in designing weld structures, an ICME approach was proposed and implemented by building a weld database through modeling and testing. The ICME approach mainly focus on thin structures such as automotive vehicles and high-speed train car bodies. This approach can be used to optimize structure design for better and safe product.

The weld database as a function of strain rate can be built by testing and modeling. The weld modeling uses heat source models for heat transfer analysis, a microstructural model for microstructure and hardness prediction, a thermo-mechanical model to predict residual stress and deformation, and a property model to estimate the weld joint strength.

Since thin structures are often modeled using shell elements, detailed weld geometry cannot be represented using shell elements. Fusion weld and spot weld representation methods with shell elements were developed to allow designers to include weld effects during load, crash, and fatigue analysis.

Spot-weld database as a function of strain rate has been established by modeling and testing which include material properties and failure parameters for a spot weld as a function of materials, thickness, and weld size. Fusion weld database as a function of strain rate can also be built by testing and modeling.

## Acknowledgements

The authors want to thanks the Project, called "Influence on the corrosion behavior of residual stress and control technology of aluminum alloys trains carbodies (2682014CX003)" of the scientific and technological innovation projects of Chinese universities.

## References

[1]    Allison, J., Li, M., Wolverton, C. and Su, X.M., (2006) Virtual Aluminum Castings: An Industrial Application of ICME. *JOM*, **58**, 28-35. http://dx.doi.org/10.1007/s11837-006-0224-4

[2]    Schafrik, B. (2012) ICME-Promise and Future Directions. *TMS Annual Meeting*, Orlando, 14 March 2012. http://materialsinnovation.tms.org/web_events/TMS2012SpecialPlenary/Schafrik/Schafrik.html

[3]    Kuehmann, C.J. and Olson, G.B. (2009) Computational Materials Design and Engineering. *Materials Science and Technology*, **25**, 472-478. http://dx.doi.org/10.1179/174328408X371967

[4]    Spanos, G., Allison, J., Cowles, B., Deloach, J. and Pollock, T. (2013) Integrated Computational Materials Engineering (ICME): Implementing ICME in the Aerospace, Automotive, and Maritime Industries, the Minerals, Metals, and Materials Society (TMS). http://nexightgroup.com/wp-content/uploads/2013/09/icme-implementation-study.pdf

[5]    Cowles, B.A., Backman, D.G. and Dutton, R.E. (2011) The Development and Implementation of Integrated Computational Materials Engineering (ICME) for Aerospace Applications. *Proceedings of Materials Science & Technology 2010 Conference* (*MS & T* 2010), Houston, 17-21 October 2010, 44-60.

[6]    Yang, Y.P., Babu, S.S., Orth, F. and Peterson, W. (2008) Integrated Computational Model to Predict Mechanical Behavior of Spot Weld. *Science and Technology of Welding and Joining*, **13**, 232-239. http://dx.doi.org/10.1179/174329308X283901

[7]    Yang, Y.P., Gould, J., Peterson, W., Orth, F., Zelenak, P. and Al-Fakir, W. (2013) Development of Spot Weld Failure Parameters for Full Vehicle Crash Modeling. *Science and Technology of Welding and Joining*, **18**, 222-231. http://dx.doi.org/10.1179/1362171812Y.0000000082

[8]    Yang, Y.P. and Babu, S.S. (2010) An Integrated Model to Simulate Laser Cladding Manufacturing Process for Engine Repair Applications. *Welding in the World*, **54**, r298-r307. http://dx.doi.org/10.1007/BF03266743

[9]    Yang, Y.P., George, W.R. and David, R.S. (2011) Finite Element Analyses of Composite-to-Steel Adhesive Joints. *Advanced Materials and Processes*, **169**, 24-28.

[10]   Goldak, J., Charkravarti, A. and Bibby, M. (1984) New Finite Element Model for Welding Heat Sources. *Metallurgical Transactions B*, **15B**, 300-305.

[11]   Ion, J.C., Easterling, K.E. and Ashby, M.F. (1984) A Second Report on Diagrams of Microstructure and Hardness for

Heat-Affected Zones in Welds. *Acta Metallurgica*, **32**, 1949-1955.

[12]  Bhadeshia, H.K.D.H. and Svensson, L.E. (1993) Modelling the Evolution of Microstructure. In: Cerjak, H. and Easterling, K.E., Eds., *Steel Weld Metal, Mathematical Modelling of Weld Phenomena*, Institute of Materials, London, 109-182.

[13]  Gould, J.E., Khurana, S. and Li, T. (2006) Predictions of Microstructures When Welding Automotive Advanced High-Strength Steels. *Welding Journal*, **85**, 111-s-116-s.

[14]  Zhang, W. and Yang, Y.P. (2009) Development and Application of On-Line Weld Modeling Tool. *Welding in the World*, **53**, 67-75.

[15]  England, G. (2014) Vickers Hardness Scale and Tensile Strength Comparison. www.gordonengland.co.uk/hardness/brinell_conversion_chart.htm

[16]  Wilkins, M.L. (1999) Computer Simulation of Dynamic Phenomena. Springer Publication, Berlin.

[17]  Robin, V., Pyttel, T., Christlein, J. and Strating A. (2014) Fracture Analyses of Welded Components. http://www.esigmbh.de/downloads/ESI/Dokumente/PAM-CRASH_SAFE/Papers/final_Weldline_vincent.pdf

[18]  Yang, Y.P., Brust, F.W., Fzelio, A. and McPherson, N. (2004) Weld Modeling of Thin Structures with VFT Software. *Proceeding of ASME Pressure Vessels and Piping Conference*, San Diego, 25-29 July 2004, 99-107.

[19]  Yang, Y.P., Cao, Z., Zhang, J., Brust, F.W., Fisher, A., Broman, R. and Thakkar, R. (2000) Weld Simulation Technology Development on Automotive Thin Gauge Structure. *Presented at* 81*st American Welding Society Annual Meeting*, Chicago, 25-27 April 2000.

[20]  Seeger, F., Feucht, M., Frank, Th., Keding, B. and Haufe, A. (2005) An Investigation on Spot Weld Modeling for Crash Simulation with LS-Dyna. LS-Dyna Anwenderforum, Bamberg.

[21]  Malcom, S. and O'Hara, B. (2009) Application of Spot Weld and Sheet Metal Failure Prediction to Non-Linear Transient Finite Element Analysis of Automotive Structures. *SAE Int. J. Mater.Manf.*, **2**, 172-177.

# 3

# Studies on Effects of Welding Parameters on the Mechanical Properties of Welded Low-Carbon Steel

**M. A. Bodude[1]\*, I. Momohjimoh[2]**

[1]Department of Metallurgical and Materials Engineering, University of Lagos, Lagos, Nigeria
[2]Department of Metallurgical Engineering, Yaba College of Technology, Lagos, Nigeria
Email: \*mbodude@unilag.edu.ng, momohjimoh_ibrahim@yahoo.com

## Abstract

In this work, the effect of heat input on the mechanical properties of low-carbon steel was studied using two welding processes: Oxy-Acetylene Welding (OAW) and Shielded Metal Arc Welding (SMAW). Two different edge preparations on a specific size, 10-mm thick low-carbon steel, with the following welding parameters: dual welding voltage of 100 V and 220 V, various welding currents at 100, 120, and 150 Amperes and different mild steel electrode gauges of 10 and 12 were investigated. The tensile strength, hardness and impact strength of the welded joint were carried out and it was discovered that the tensile strength and hardness reduce with the increase in heat input into the weld. However, the impact strength of the weldment increases with the increase in heat input. Besides it was also discovered that V-grooved edge preparation has better mechanical properties as compared with straight edge preparation under the same conditions. Microstructural examinations conducted revealed that the cooling rate in different media has significant effect on the microstructure of the weldment. Pearlite and ferrite were observed in the microstructure, but the proportion of ferrite to pearlite varied under different conditions.

## Keywords

Low Carbon Steel, Shielded Metal Arc Welding (SMAW), Oxy-Acetylene Welding (OAW), Heat Affected Zone (HAZ)

## 1. Introduction

Fusion welding processes are widely used for fabrications in many engineering applications such as aerospace,

---

\*Corresponding author.

automobile and ship building industries. Oxy-Acetylene Welding (OAW) is the most commonly used gas welding process because of its high flame temperature. Gas welding is a welding process that melts and joins the metals by heating them with flame caused by the reaction between a fuel gas and oxygen [1].

The main advantage of the OAW process is that the equipment is simple, portable and inexpensive. The total heat input per unit length of the weld in OAW process is rather high, resulting in large Heat Affected Zone (HAZ) which causes severe distortion [2]. Shielded Metal Arc Welding (SMAW) is a process that melts and joins metals by heating them with an arc established between a stick-like covered electrode and the metals. The welding equipment is also simple, portable and inexpensive [3].

One of the limitations of the two welding processes mentioned above as reported by Sindo [1] is the low power density of the heat source. The power density increases from a gas flame to an electric arc and a high-energy beam, and as the power density of the heat source increases, the heat input to the work piece that is required for welding decreases. The portion of the work piece material exposed to a gas flame heats up so slowly that before any melting occurs, a large amount of heat is already conducted away into the bulk of the work piece. Excessive heating can cause damage to the work piece, including weakening and distortion [4]. However, control of welding parameters such as welding current, voltage and speed can reduce the amount of heat input into the weld. Besides, edge preparation prior to welding can significantly affect the quality of the weld.

In view of the fact that arc welding processes like SMAW offer a wide spectrum of thermal energy for joining different thicknesses of steel, it was considered important that undertaking the present study would be very beneficial in gaining an understanding of the mechanical properties of low-carbon steel that influence the service performance of the welded joints under different heat input combinations *i.e.* low heat input and high heat input [5] [6].

The chemical composition of steel has impact on the weldability and the mechanical properties of the material. According to Monika *et al.* [7], several elements are purposefully added in the production of structural steel, but other undesirable elements may equally be present arising from the scrap materials charged during the steel-making process. Carbon, manganese, tungsten and other elements increase strength and may increase the risk of cold cracking and therefore higher preheat and inter pass temperatures, better hydrogen control and sometimes post heat are necessary to avoid cracking [7].

Ueji [3] noted that weld fusion zone microstructure of low-carbon steel depends on the chemical composition of the material and the cooling rate. For steels, the required critical cooling rate for achieving martensite in the microstructure can be estimated using the following equation

$$\log v = 7.42 - 3.13C - 0.71Mn - 0.37Ni - 0.34Cr - 0.45Mo \qquad (1)$$

where $v$ is the critical cooling rate in kmol/hr.

Volume of melted metal is a function of heat input which in turn is governed by the welding parameters such as welding current and welding time. Increasing current decreases the fusion zone size while increasing time increases the fusion zone size [8] [9].

Ultrafine grained refinement from conventional grained size (10 μm) to less than 1 μm provides a preferable high strength to metallic materials used for structural applications as reported by Lowe and Zhu [10]. Ultrafine grained steel structural component in engineering application can be done by welding to produce complex structure which is difficult to manufacture directly. The major disadvantage of fusion welding of this ultrafine grained steel is the occurrence of grained growth during welding leading to reduction in strength [3]. Careful control of welding parameters can reduce the amount of heat input into the weld thereby avoiding the excessive grained growth and allotropic transformation in metals, thus improving the mechanical properties of the welded steel metal.

This work is therefore aimed at investigating the effects of welding current and voltage at two different edge preparations on the mechanical properties of low-carbon steel, using oxy-acetylene and shielded metal arc welding processes.

## 2. Experimental

### 2.1. Materials

Hot-rolled plate of low-carbon steel was obtained from a local metal market (Owode-Onirin) in Lagos, Nigeria. 10 mm thickness samples of the steel plate were cut and prepared for welding. E6013 electrode with gauge size

10 and 12 were selected for the shielded metal arc welding (SMAW) process.

## 2.2. Procedure

Two edge preparations were done to perform welding on Butt Straight and single V groove joints. The test specimen was prepared in two 200 × 50 × 10 mm thick steel plates pieces and then welded together to give a finished test plate 200 × 100 × 10 mm with a weld down the middle as shown in **Figure 1**. Three stages were followed for the experiment. In the first stage the metals plate with the prepared edges were taken for forming the welded joint. The material pairs selected for forming the welded joints are shown in **Figure 1**.

The second stage was the welding process and the joint formation using different process parameters.

The third stage was undertaken to study the effect of welding parameters on the mechanical properties and microstructure of welded joint.

The pieces of steel cut were paired and aligned on a table by the use of an angular Iron and then setting up of welding circuit before welding. Welding of samples in horizontal position was performed continuously by SMAW using the varied welding parameters: dual voltage device 100 V and 220 V, various welding current at 100, 120 and 150 Amperes.

## 2.3. Hardness Testing

Hardness testing of welding joints was performed in accordance to ASTM A370-14 standard using Rockwell Testing Machine. The sample specimen was placed with the surface on the anvil, and slowly turning the hand wheel until the specimen was raised to touch the indenter. The numbers were read directly from the dial indicator and converted to the Rockwell number.

Hardness Test was done in a row to ensure that the Base Material, Heat Affected Zone and Weld Metal at a distance ≤2 mm from surface and 2 mm from fusion line were captured. Schematic representation of Butt and Groove welded joint with the locations of hardness testing and examination of microstructure is shown in **Figure 2**.

Rockwell Micro Hardness traverses (1-mm increments) were produced across the weld regions using a 150-g load and a 10 s dwell time.

## 2.4. Tensile Testing

Tensile Test was conducted on the welded steel sample using a hydraulic extensometer with prepared machined specimen dimensioned 85 × 50 × 10 mm. The tensile testing was done in accordance to ASTM A370-14 standard. Tensile samples were prepared by milling of the top and bottom surfaces to remove flashing and other surface irregularities. The standard shape is obtained using lathe machine as shown in the **Figure 3**.

## 2.5. Impact Testing

Test specimens with a V notch in the fusion zone for impact test fracture were performed according to ASTM E23 standard.

The prepared specimen was placed on the Anvil with V-notch gauge. The Pendulum was set to predetermined level 220ft.Ib (298.642 J) and released to strike and fracture the positioned specimen. The value of Energy Impacted was taken and recorded.

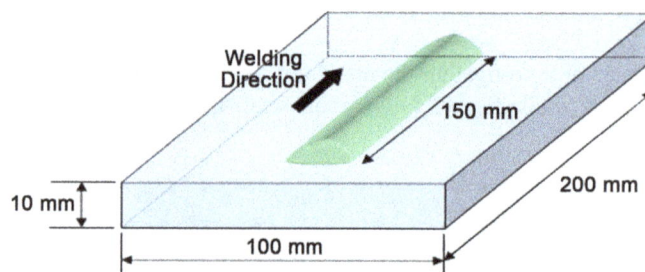

**Figure 1.** Schematic illustration of the bead on plate location and dimensions.

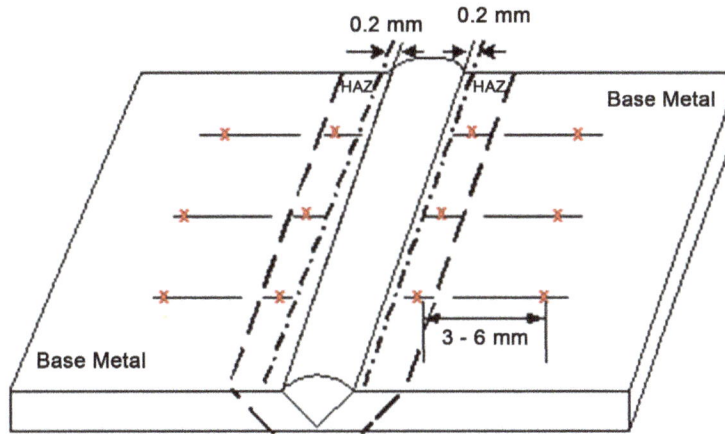

**Figure 2.** Butt welded joint with the sample locations of hardness testing.

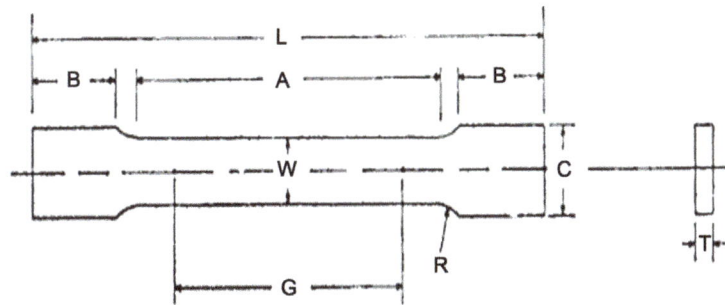

**Figure 3.** Showing the standard machined for tensile test.

## 2.6. Metallographic Test

Metallographic samples were produced from welds in accordance to ASTM E 23. Grinding, Polishing, etching with 2% Nital and Metallographic examination of butt and groove welds were performed using optical microscope Olympus PMG3 with magnification ×500.

## 3. Results and Discussion

**Figure 4** clearly shows the mechanical properties of the welded steel at fixed Voltage (220 V) and varying Current in Straight Edge preparation welded sample. It was discovered that current adjustment affects the weld bead size, appearance, penetration and strength of the weld. There was a decrease in Hardness and Tensile strength of the welded steel as the current was increased. With increased current, the heat generated increased causing the grain to recrystallized and grow in size. Increased in grain size increases the impact strength (toughness) of the weld but reduction in hardness and tensile strength of weldment and heat affected zone as depicted by the graph. This was in agreement with [4]. Similarly, the amount of heat input from the heat source is given by

$$H = \frac{(I \times V)60}{S}.$$

This shows that as the current increases more heat is produce into the weld causing more expansion and contraction between the weldment and the base metal. This increased the residual stresses resulting in lower mechanical properties.

**Figure 5** shows the mechanical properties of the welded steel at fixed voltage (220 V) and varying current in the V groove edge welded sample. From this plot, the hardness as well as the tensile strength of the weldment and heat affected zone decreased with an increase in the current. This again is due to increase in heat input into the weldment. The heat persisted for a period of time and is quickly conducted away into the base metal as a result of low power density of the welding process (SMAW) as reported by Sindo [1]. Consequently, the based metal was affected due to high heat input into the weld resulting in grain size increased. This case is also similar

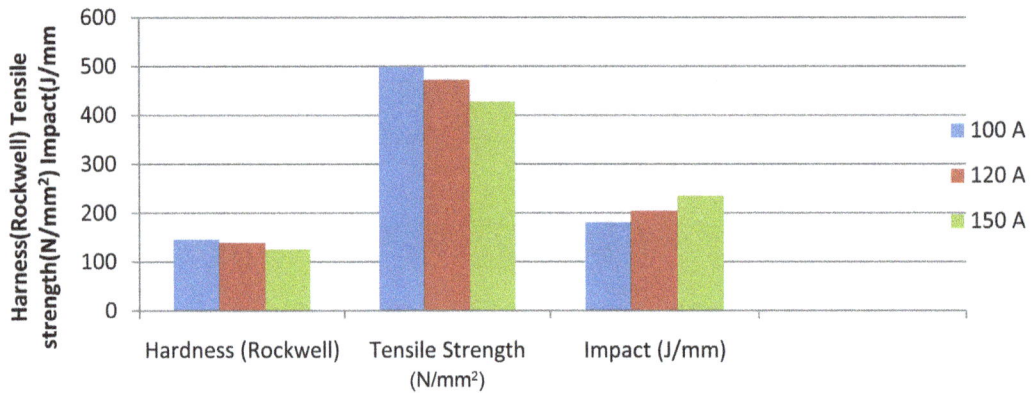

**Figure 4.** Fixed voltage (220 V) and varying current with straight edge.

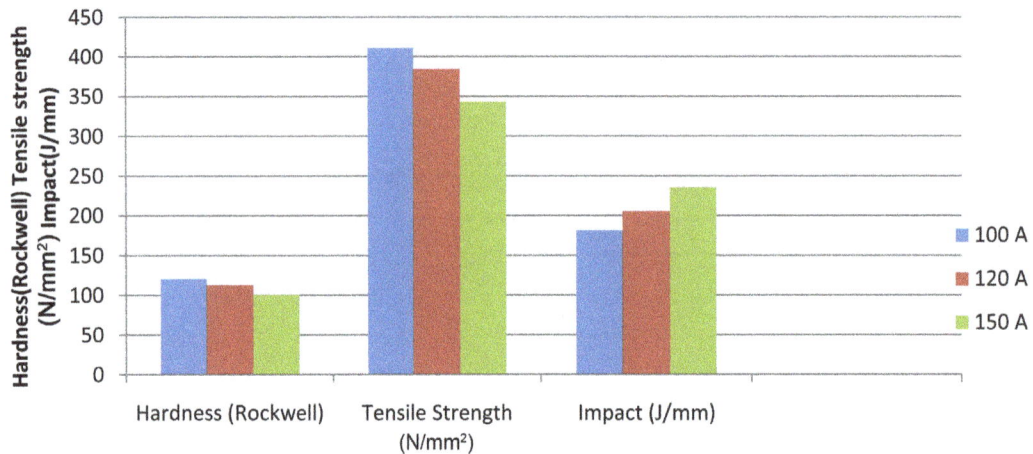

**Figure 5.** Fixed voltage (220 V) and varying current with V groove edge.

to what happened in **Figure 4**. The only exception is that **Figure 5** is with V-grooved edge with high amount of weldment deposit. This increased the residual stresses in the weldment and heat affected zone. Residual stresses can significantly affect the engineering properties of the material and this has contributed to reduction in strength and hardness of the weldment and heat affected zone as shown in the **Figure 5**.

In **Figure 6**, the current was fixed at 100 A while the voltage was varied from 100 V to 200 V. The hardness and the tensile strength of the weldment and the heat affected zone reduced as the voltage increases. This is similar to that of **Figure 4** and **Figure 5** as they all conformed to the mathematical model equation relating heat source and welding parameters (current and voltage) [5] [6]. Besides it is difficult to maintain constant gap between the arc and the weldment in manual welding, and as such the system is self-adjusting such that increase in voltage will not cause significant increase in welding current as reported by Tariq in his lecture 2014 at King Fard University of Petroleum Saudi Arabia.

**Figure 7** shows similar phenomenon as that in **Figure 6**, though the edge preparation differences caused significant changes in the mechanical properties. It was observed in **Figure 6** that the tensile strength at 100 A is about 500 N/mm$^2$ while the corresponding tensile strength in **Figure 7** is about 420 N/mm$^2$. This is because grove edge accommodates more weld deposit than straight edge and therefore the bonding strength should have been better than that of straight edge. As reported in several literature, the more the weld deposit, the higher the expansion and contractions consequently increase in residual stresses in the weldment and the heat affected zone. Gery *et al.* [4] noted that, residual stresses can significantly affect the mechanical properties of the weldment. It was also noted that the impact strength is slightly higher in **Figure 7** than in **Figure 6** with the similar reasons explained above.

**Figure 8** and **Figure 9** depict fixed current and voltage with different edge preparation. The phenomenon in both figures was almost the same as there were no changes in the heat input into the weldment. However differ-

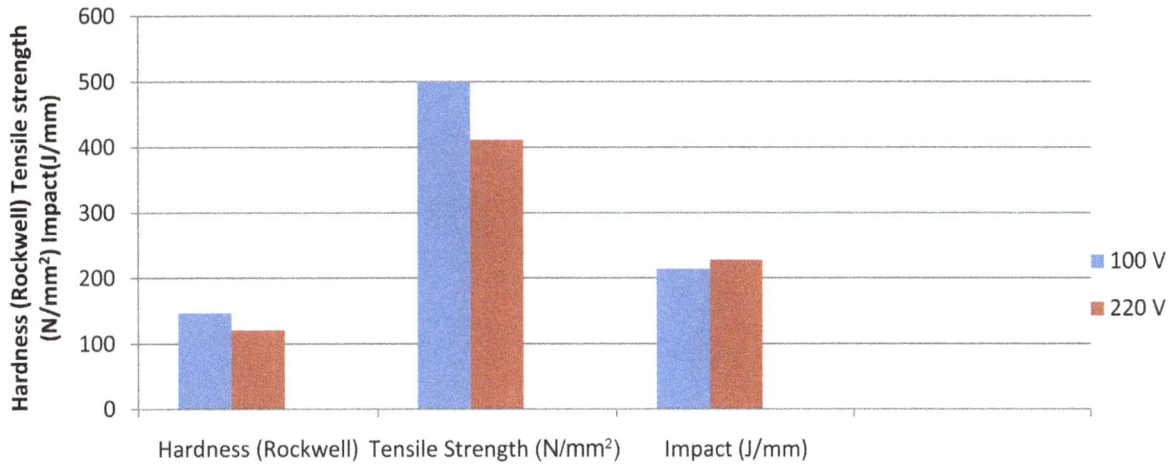

**Figure 6.** Fixed current (100 A) varying voltage with straight edge.

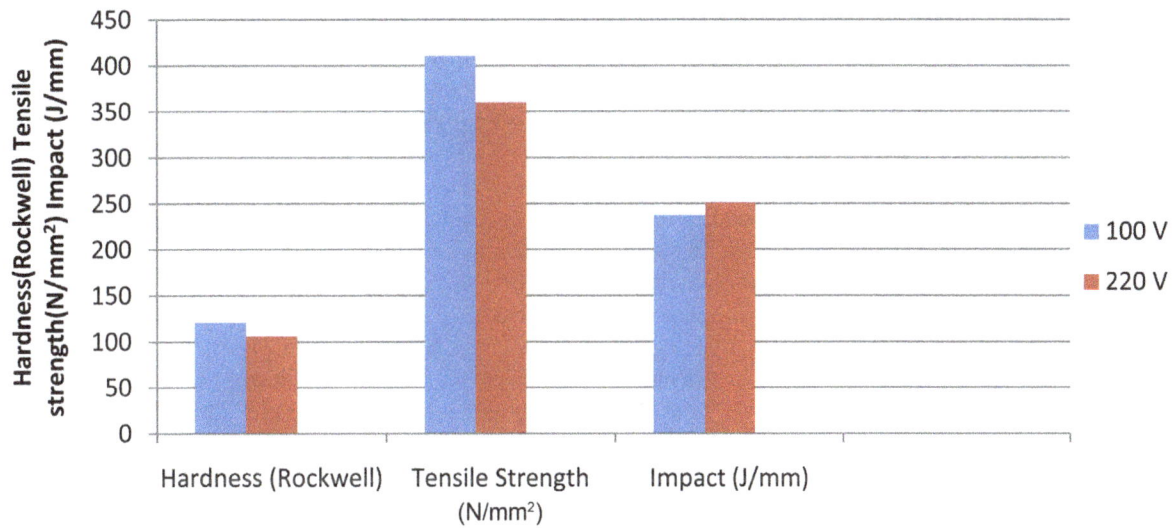

**Figure 7.** Fixed current (100 A) varying voltage with V groove edge.

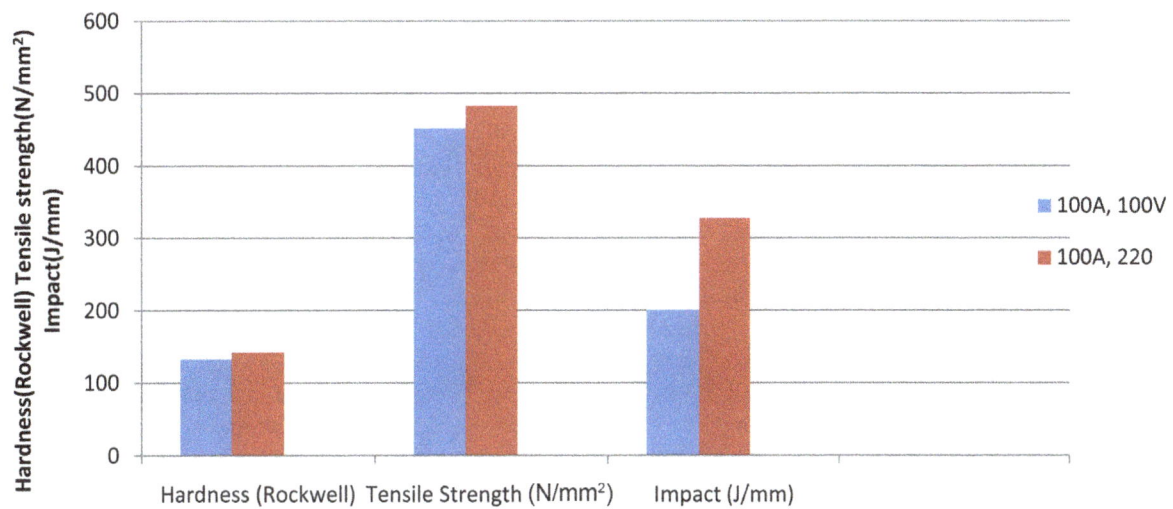

**Figure 8.** Fixed voltage and current with straight edge preparation.

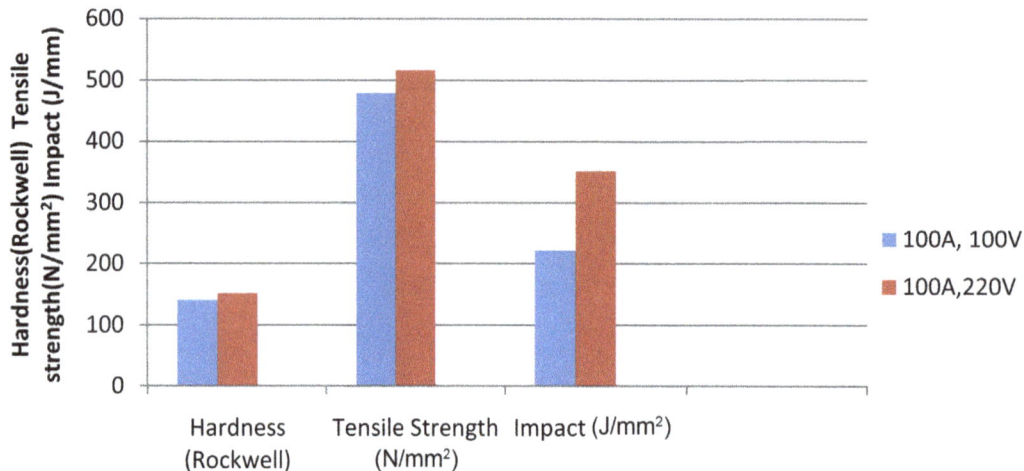

**Figure 9.** Fixed voltage and current for different edge preparations with V groove edge.

ences in edge preparations caused slight disparity in the values of the mechanical properties of the two types of preparation under study. Tensile strength is higher in V-grooved edge in **Figure 9** than that of straight edge in **Figure 8** as the effect of bonding is dominant over the residual stress due to constant heat input into the weldment. Similarly, the impact strength (toughness) also improved slightly in V-grooved edge in the **Figure 9** than the corresponding impact in **Figure 8**.

In **Figure 10** and **Figure 11**, the same scenario is exhibited except that the electrode is of different compositions. It was discovered that the use of different electrode at constant current and voltage did not have any significant changes in the tensile strength and hardness of the welded steel. However there is a noticeable improvement in the impact strength in V-grooved edge in **Figure 11** than that of **Figure 10**.

**Figure 12** and **Figure 13** shows the mechanical properties of the welded steel at fixed current and voltage for different cooling medium with straight edge and V-grooved edge preparations respectively. The cooled in still air samples has a lower tensile strength, and higher hardness values as shown in **Figure 12**. This may be due to a slow rate of cooling in draft air than that of still air. Low cooling rate led to grain growth of the welded joint [3] [8] [9]. In **Figure 13**, it was observed that there was decrease in the hardness value of the sample cooled in the still air, and this can be attributed to decrease in the concentration of hydrogen on the joint surface that caused long time or delay in cooling rate.

**Figure 14** shows the comparison of the mechanical properties of samples welded with oxy-acetylene in straight edge and V-Beveled edge preparations. From this figure, the tensile strength and the hardness values of the two types of preparations were totally different. As compared to SMAW, the mechanical properties of OAW as reported in **Figure 10** is lower than that of SMAW shown in **Figure 4** and **Figure 5** respectively. This is due low power density of OAW process.

The results of microstructural examination revealed that all the welded joints contain pearlite and ferrite phases.

**Figure 15(a)** shows the structure of the Base Metal of the Low Carbon Steel used for this study. The structure consists of about 80% to 85% ferrite (white) and about 15% to 20% pearlite (black).

It was observed that the grain-refined area exhibits extremely small ferrite grains, clearly indicating that very fine ferrite and pearlite is formed due to the heating and cooling cycles of the SMAW process [7].

**Figure 15(a)** and **Figure 15(b)** show the microstructures of the Weld Zone of samples welded at Constant Voltage and varied Current setting, representing a grain-coarsened area, which exhibits predominantly White Ferrite (WF) and pearlite.

**Figure 16(a)** and **Figure 16(b)** show the structure of the samples welded at constant current, and varied Voltage setting. It was observed that a large amount of pearlite is present in the ferrite matrix. As the current was increased, the pearlites become finely distributed within the coarse matrix with an increase in the proportion of the ferrite to pearlite.

**Figure 17(a)** and **Figure 17(b)** compare the prepared straight edge and the prepared V-grooved edge and it was discovered that ferrite phase were more than pearlite phase in their microstructure which also indicates

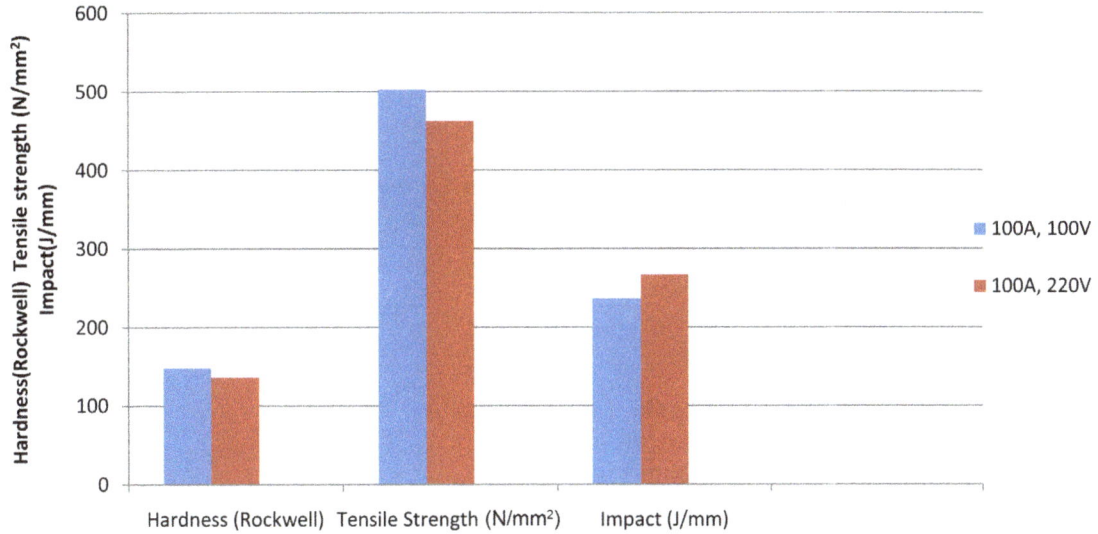

**Figure 10.** Fixed voltage and current for different types of electrode with straight edge.

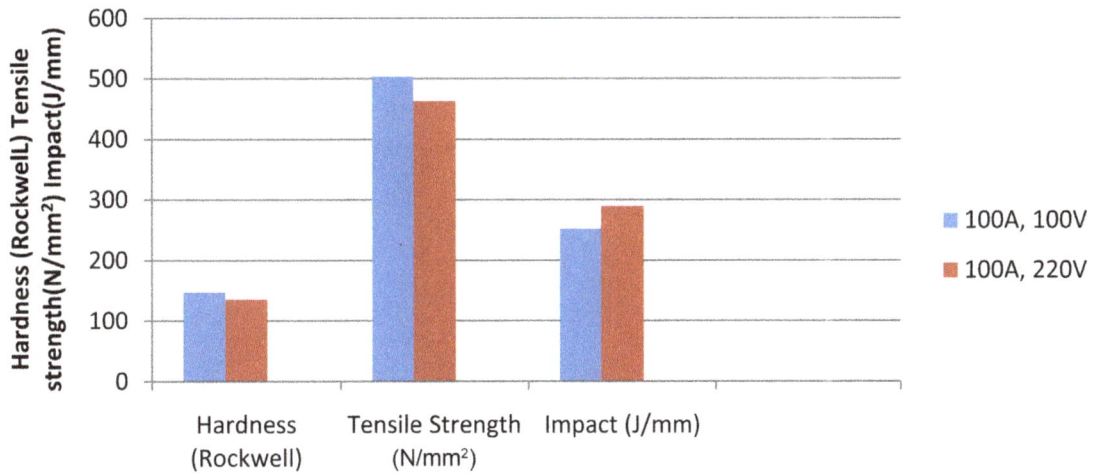

**Figure 11.** Fixed voltage and current for different types of electrode with V groove edge.

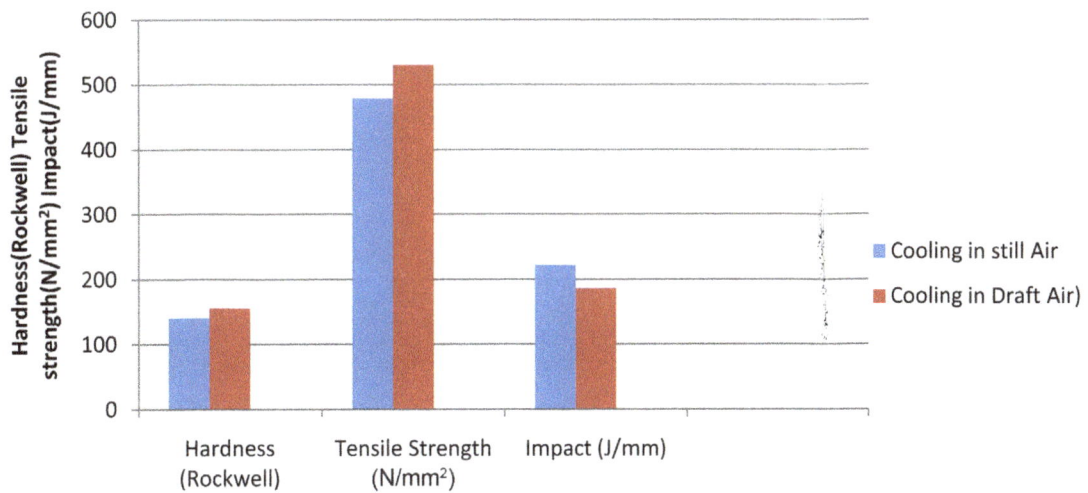

**Figure 12.** Fixed voltage and current for different cooling medium with straight edge.

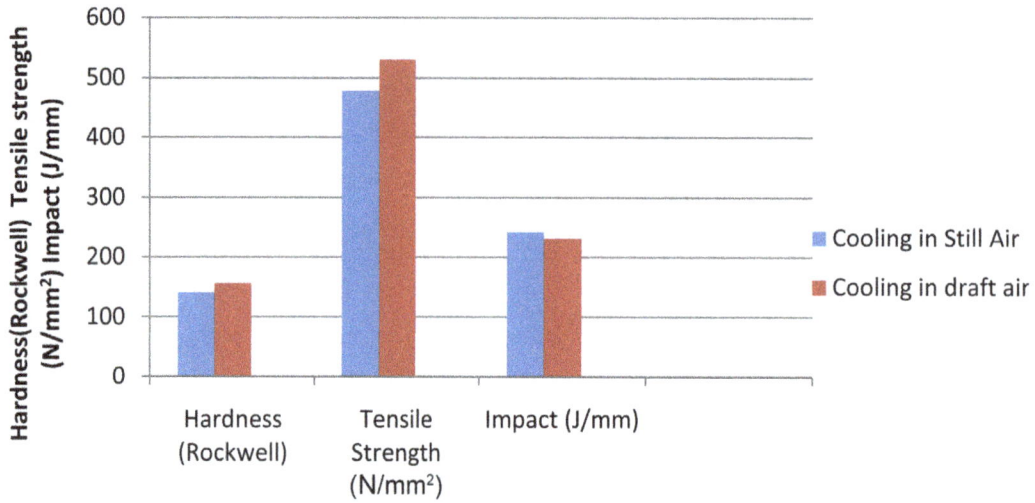

**Figure 13.** Fixed voltage and current for different cooling medium with V groove edge.

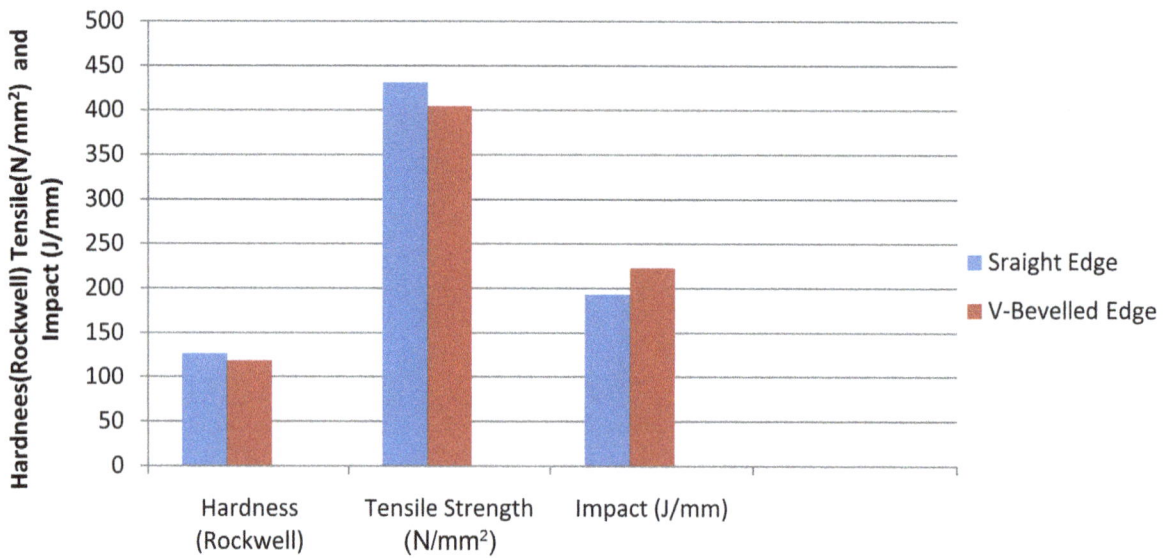

**Figure 14.** Welding with oxyacetylene flame with straight edge and V-bevelled edge.

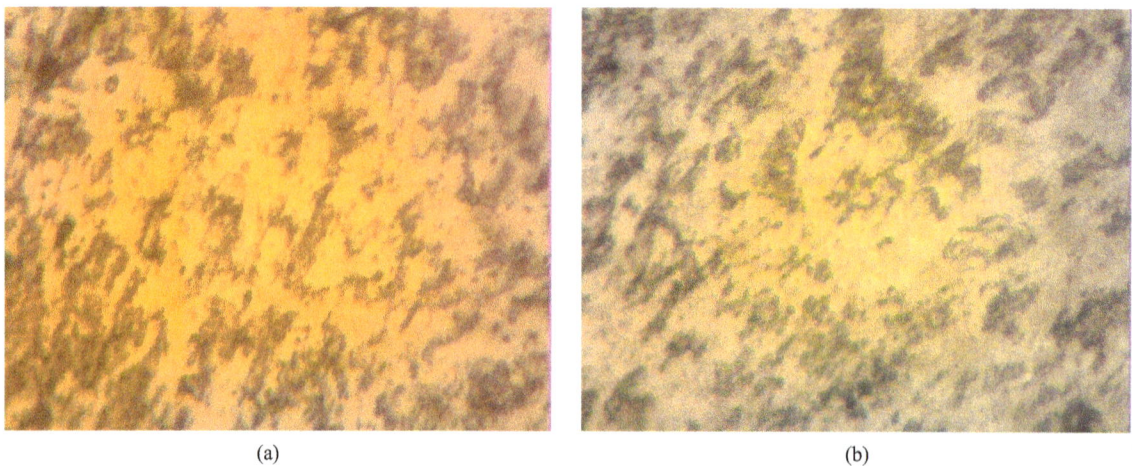

**Figure 15.** (a) Fixed voltage (220 V) and varying current; (b) Base metal with straight edge.

(a)                                                                (b)

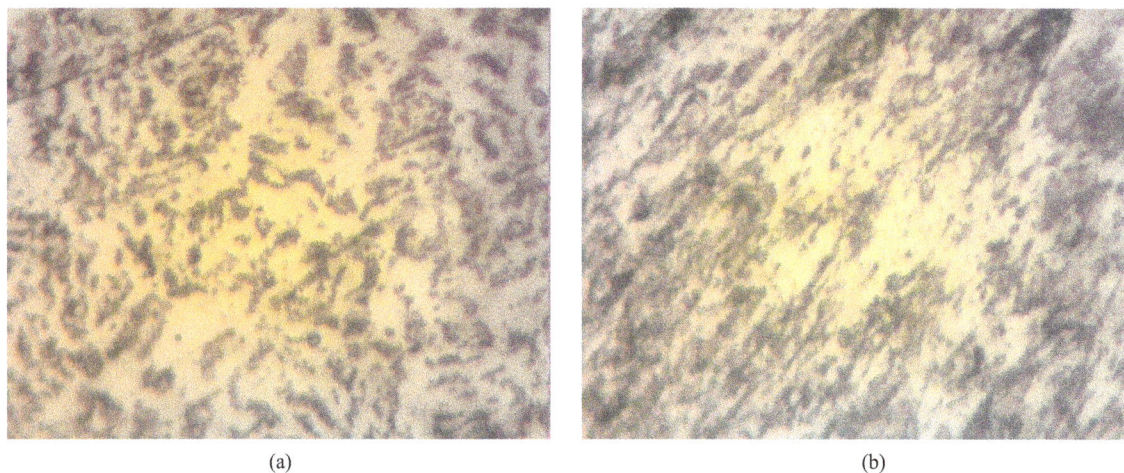

**Figure 16.** (a) Fixed current (100 A) varying voltage with straight edge; (b) Fixed voltage (220 V) and varying current with V groove edge.

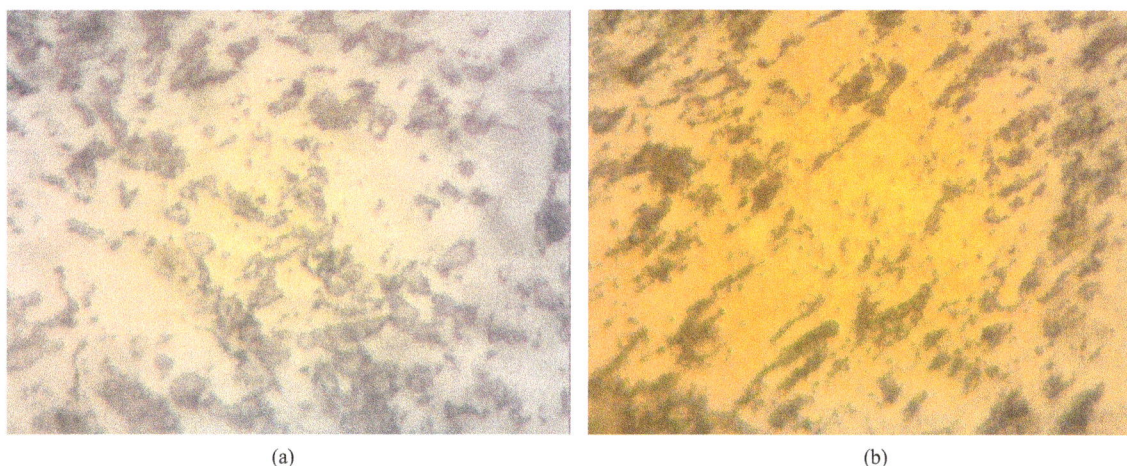

(a)                                                                (b)

**Figure 17.** (a) Fixed voltage and current for edge preparations with straight edge; (b) Fixed current (100 A) and varying voltage with V groove edge.

higher ductility in **Figure 17(a)**; this is in agreement with the work of [5] [6].

Examination of the Welded Zones in **Figure 18(a)** and **Figure 18(b)** also revealed the effect of electrode size on the microstructure. The observed structure is of coarse pearlite in ferrite matrix which became more coarse as the grain size increased. This observation explained the decrease in the hardness and tensile strength values as the heat energy input was increased.

Microstructure for samples welded with fixed voltage and current setting for different cooling medium were shown in **Figure 19(a)** and **Figure 19(b)**. These plates revealed the effects of cooling rate after welding on the properties of the steel. Cooling in still air produces coarse grain structure of both ferrite and pearlite while cooling in draft air produces fine grain size. This is due to slow cooling of the weldment in still air [3].

**Figure 20(a)** and **Figure 20(b)** revealed fine-grain microstructure of weldment of sample with Oxyacetylene welding procedure. Fine grain ferrite and pearlite were observed in this area as compared to the base metal microstructure (**Figure 19(a)**). However, as observed from **Figure 21(a)** and **Figure 21(b)** better grain refinement were noticed than in **Figures 20(a)** and **Figure 20(b)**. This clearly indicates that during the heating and cooling cycles, a very fine grain structure is formed in the FZ. This area is generally termed as the grain-refined area.

From the observation of the plates, the microstructures with higher concentration of the pearlite phase will have more hardness and strength but with lower ductility. Ferrite crystals were formed as a result of transformation in low-carbon steel upon cooling from the austenitic state. It is not possible to state the relationship mathematically, but since the ferrite grains are nucleated and grow; their size will depend on time and temperature,

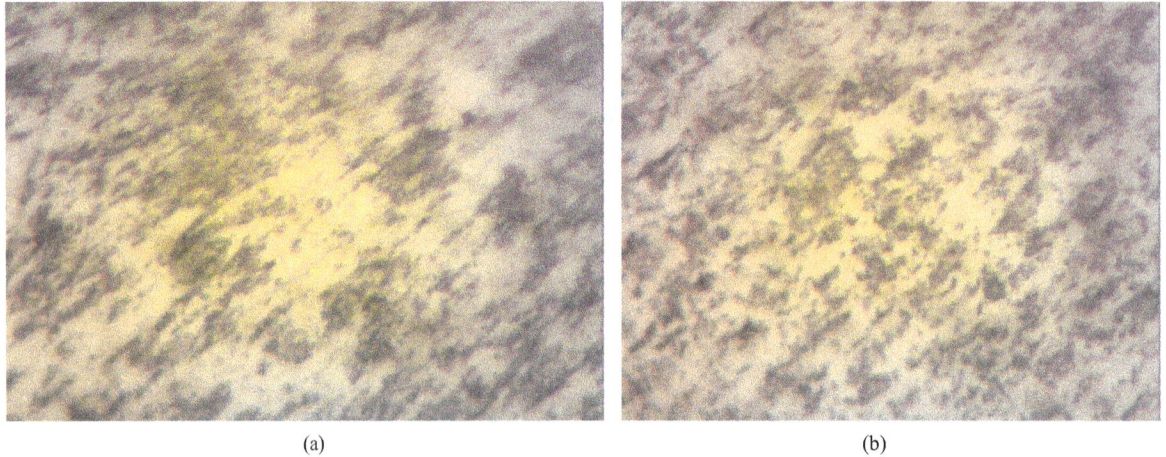

**Figure 18.** (a) Fixed voltage and current for different types of electrode with straight edge; (b) Fixed voltage and current for diff edge preparations with V groove edge.

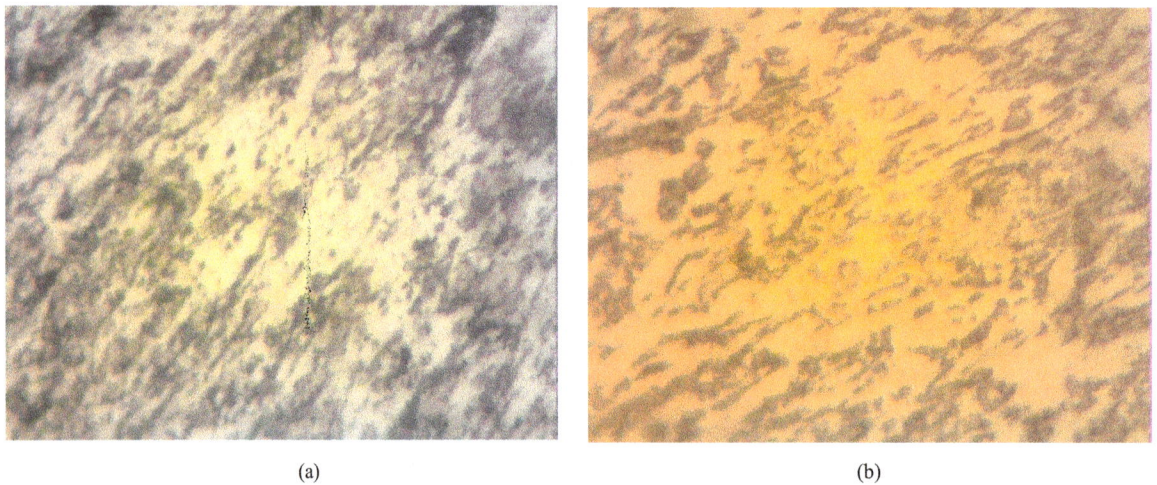

**Figure 19.** (a) Fixed voltage and current for medium with straight edge; (b) Fixed voltage and current for different types of electrode with V groove edge.

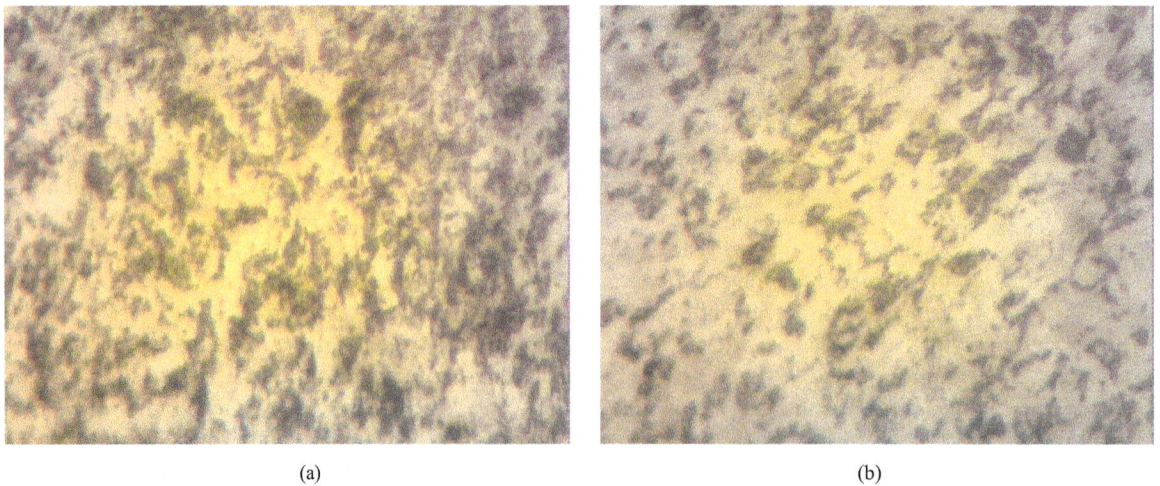

**Figure 20.** (a) Welding with oxyacetylene with straight edge; (b) Fixed voltage and current for cooling medium with V groove edge.

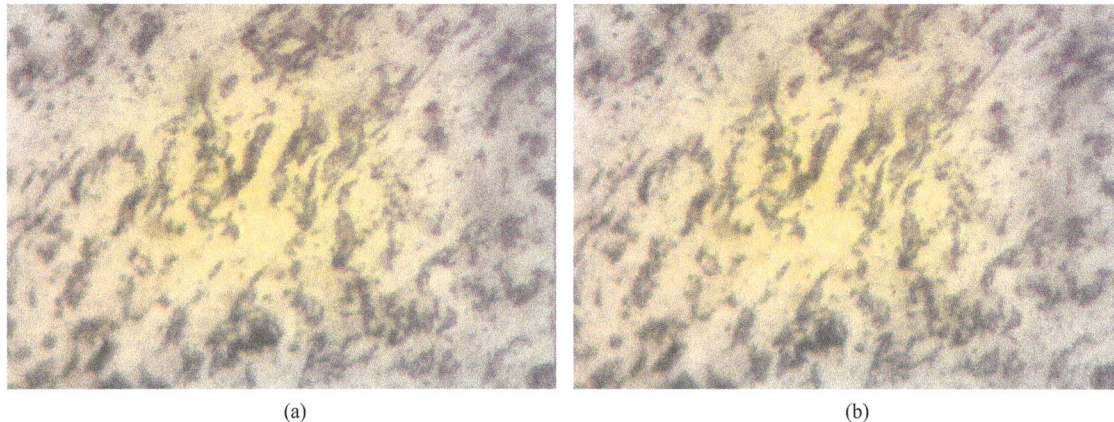

(a)    (b)

**Figure 21.** (a) Fixed voltage (220 V) and varying current with straight edge; (b) Welding with oxyacetylene flame with V groove edge.

especially on the cooling rate.

## 4. Conclusions

The following conclusion can be drawn from the present study:

- Adequate edge preparation of the weldment enhances the strength of the welded steel. This may be due to the high diffusion of the weld metal which in turn may enhance bonding of the two metal pieces.
- The hardness, tensile and impact strength of the welds are functions of the microstructure; therefore these properties are structural sensitive.
- It was observed that the strength and hardness of the joint increase with the decrease in heat input.
- The toughness increases while the hardness and tensile strength of the joint decrease.

## References

[1]   Kou, S. (2003) Welding Metallurgy. 2nd Edition, John Wiley& Sons, Inc., Hoboken, New Jersey, 17-20.

[2]   Puchoicela, J. (1998) Control of Distortion of Wed Steel Structures. *Welding Journal*, **77**, 49-52.

[3]   Ueji, R., Fujii, H., Cui, L., Nishiokioka, A., Kunishige, K. and Nogi, K. (2006) Friction Stir Welding of Ultrafine Grained Plain Low-Carbon Steel Formed by the Martensite Process. *Materials Science and Engineering*: A, **423**, 324-330. http://dx.doi.org/10.1016/j.msea.2006.02.038

[4]   Gery, H., Long, P. and Maropoulos, E. (2005) Effects of Welding Speed, Energy Input and Heat Source Distribution on Temperature Variations in but Joint Welding. *Journal of Material Processing Technology*, **167**, 393-401. http://dx.doi.org/10.1016/j.jmatprotec.2005.06.018

[5]   Muthupandi, V., Srinivasan, P., Bala, S.K. and Sundaresan, S. (2003) Effect of Weld Metal Chemistry and Heat Input on the Structure and Properties of Duplex Stainless Steel Welds. *Materials Science and Engineering*: A, **358**, 9-16. http://dx.doi.org/10.1016/S0921-5093(03)00077-7

[6]   Yan, J., Goa, M. and Zeng, X. (2010) Study on Microstructure and Mechanical Properties of 304 Stainless Steel Joints by TIG, Laser and Laser-TIG Hybrid Welding. *Optics and Lasers in Engineering*, **4**, 512-517. http://dx.doi.org/10.1016/j.optlaseng.2009.08.009

[7]   Monika, K., Bala, M.C., Nanda, P.K. and Prahalada, K.R. (2013) Effect of Heat Input on the Mechanical Properties of MIG Welded Dissimilar Joints. *International Journal of Engineering Research & Technology*, **2**.

[8]   Easterling, K.E. (1998) Modeling the Weld Thermal Cycle and Transformation Behavior in the Heat Affected Zone. In: Cerjak, H. and Easterling, K.E., Eds., *Mathematical Modeling of Weld Phenomenon*, The Institute of Materials.

[9]   Marashi, P., Pouranvari, M., Amirabdollahian, S. and Abedi, G. (2008) Microstructure and Failure Behavior of Dissimilar Metal Spot Welds between Low Carbon Steel, Galvanized and Austenistic Stainless Steels. *Materials Science and Engineering*: A, **420**, 175-180. http://dx.doi.org/10.1016/j.msea.2007.07.007

[10]  Lowe, T.C. and Zhu, Y.T. (2003) Commercialization of Nanostructured Metals Produced by Severe Plastic Deformation Processing. *Advanced Engineering Materials*, **5**, 373-378. http://dx.doi.org/10.1016/j.msea.2007.07.007

# 4

# Improving Laser Beam Welding Efficiency

**Mikhail Sokolov[1*], Antti Salminen[1,2]**

[1]Laboratory of Laser Materials Processing, Lappeenranta University of Technology, Lappeenranta, Finland
[2]Machine Technology Centre Turku Ltd, Turku, Finland
Email: *mikhail.sokolov@lut.fi

## Abstract

**Laser beam welding is becoming widely used in many industrial applications. This paper reviews recent research conducted on the performance, potential and problems of thick section butt joint laser welding. Common defects that occur in laser beam welding with high power laser welding are discussed and possible solutions proposed. Methods of welding process efficiency improvement are analyzed.**

## Keywords

**Laser Welding, Edge Preparation, Preheating, Vacuum**

## 1. Introduction

Recent developments in the field of high power fiber lasers (HPFL) have led to renewed interest in thick section laser welding. A number of studies have shown that HPFL welding offers the potential for high-speed processing of different metals. The potential of laser beam welding (LBW) of thick sections can be realized in many industrial applications like power plants [1] [2], ship building [3] [4] and pipelines [5]-[7].

### Keyhole Welding

Welding processes imply the following three-steps: melting the metal to form a weld pool on the site of the future joint, permitting the weld pool to grow to the desired size, and maintaining of weld pool stability until solidification. Welding processes may be achieved using different energy sources: from gas flame and electronic arc to electron or laser beam and ultrasound. Of the energy sources used, the laser beam is notable for having the highest power density currently available to industry (up to $10^9$ W/cm$^2$) that is focusable on a small spot (down to 0.1 mm) [8]. The absorption of such energy leads not only to material melting but also to evaporation of the material at the point of contact, forming a cylindrical hole in the material which may extend through the entire plate thickness. Over time, the cavity becomes deeper and forms a canal filled with evaporated material along

---

*Corresponding author.

the direction of the incoming laser beam. The canal, or keyhole, is prevented from closing by the interaction forces of the vapor pressure, the hydrostatic stability of the molten material, and surface tension forces. Five forces should be considered when analyzing the establishment and stability of the keyhole [9]:

1. Pr: recoil pressure, arising from particle evaporation at the surface;
2. Pv: vapor pressure caused by metal vaporization;
3. Pb: beam pressure, from impinging photons and electrons;
4. Pg: gravitation pressure, in the case of deep keyholes;
5. Ps: surface tension pressure, surface of the keyhole acts like a cylindrical elastic membrane and collapses the keyhole when the energy density falls below a critical level.

As the laser focusing point moves along the joint, a hole is traversed through the material, with the molten walls sealing up behind it [10]. The process and the forces are illustrated in **Figure 1**.

The laser beam is reflected multiple times on the walls of the keyhole and the molten material of the keyhole walls absorb nearly all the power of the beam. With high welding speed, the process gives a parallel sided fusion zone, narrow weld and high welding efficiency-less energy is spent on unnecessary heat (compared to traditional welding processes). The weld depth may be ten or more times greater than the weld width. The high accuracy and reduced distortion of keyhole welding have made this process common in many applications requiring deeper welds [11].

LBW provides increased processing speed, less post-processing and reduced consumption of materials, leading to increased productivity. The drawbacks of this welding technique are high costs of laser apparatus, difficulty in melting highly reflective or highly thermal-conductive metals, small gap tolerance, and a tendency to the formation of welding defects [12].

Methods to increase the efficiency of the LBW process are an important component of modern deep penetration laser welding research. The purpose of this paper is to review recent research into possible techniques for improving LBW process efficiency.

## 2. Laser and Process Efficiencies

Throughout this paper, the term "efficiency" will refer to LBW process efficiency. However, it is important to first define the term "efficiency" from the perspective of a laser system.

### 2.1. Beam Quality

Applications like thick section welding require high beam quality. Higher beam quality allows focusing of smaller point diameters to achieve very narrow and deep welds. As all modern high power lasers used in thick welding use fibers to deliver the laser beam, the beam quality can be described by Beam Parameter Product (BPP). BPP can be defined as in Equation (1). Beam geometry is shown in **Figure 2**.

Where:

$$BPP = \frac{\omega_0 \theta}{4} \tag{1}$$

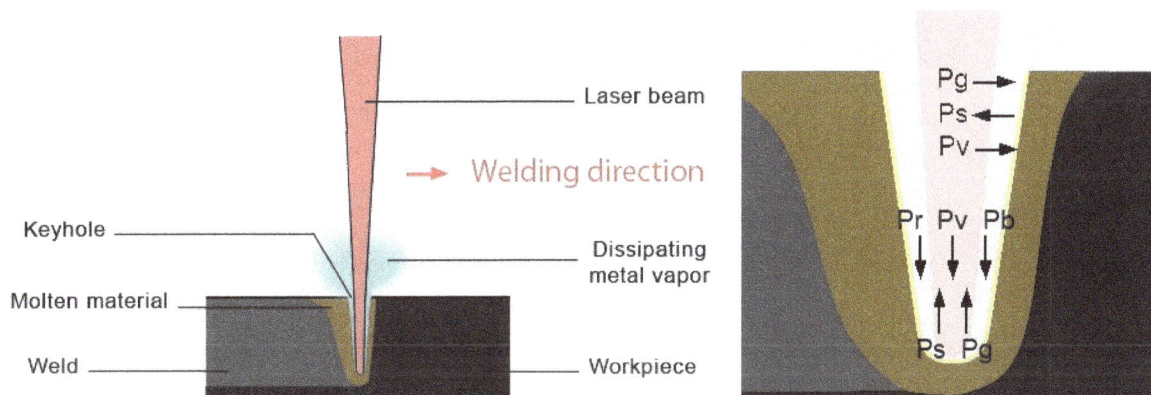

**Figure 1.** Keyhole welding process and balance of forces of a non-fully penetrating keyhole [9].

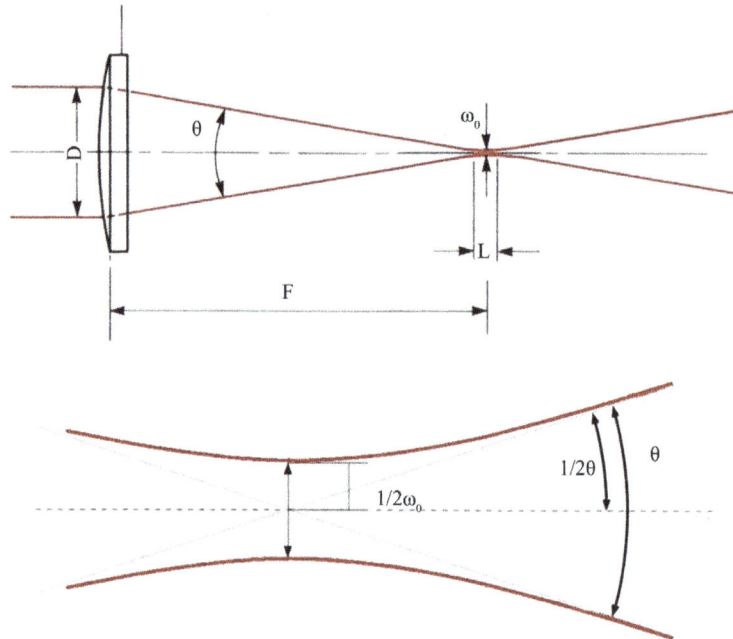

**Figure 2.** Beam geometry [8].

- $D$ : beam diameter incident on optic, mm;
- $\theta$ : beam convergence angle, degrees;
- $F$ : focal length, mm;
- $L$ : depth of focus, where minimum waist diameter does not increase beyond 5%, mm;
- $\omega_0$ : minimum waist diameter (focus spot size), mm;
- $\lambda$ : wavelength, mm.

In case of using HPFL, the focus spot size depends on the fiber core diameter and the magnification of the optical system. The higher the BBP the smaller spot size can be achieved and the smaller fiber core can be used. Therefore the smaller focus point will be on the workpiece.

## 2.2. Laser Electro-Optical Efficiency

The laser beam is a tool with properties that make it well suited for welding. However, the efficiencies of the conversion of electricity to optical power are limited by the distances where radiated transmission is possible [12] [13].

Dahmen *et al.* [12] presented an overview of the efficiencies of laser beam sources used in LBW dependent on the beam parameter product, as shown in **Figure 3**.

Lamp-pumped solid state lasers (LPSSL), for example Nd:YAG, show the lowest efficiency of 4%. Diode-pumped solid state lasers (DPSSL) show an efficiency of 15%. New high power fiber lasers (HPFL) and disc lasers achieve efficiencies of up to 25%. In recent years, due to rapid development of these lasers, the penetration depth and welding speed has been significantly improved [13] [14]. The further analysis of welding process efficiency will refer to HPFL and disc lasers.

## 2.3. Welding Process Efficiencies

There are two dimensionless process efficiencies can be measured as a function of controllable processing variables:
1. $\eta_a$ : laser energy transfer efficiency;
2. $\eta_m$ : melting efficiency.

The laser energy transfer efficiency is used to describe the ratio of energy that is absorbed by the workpiece over the incident laser energy [15]:

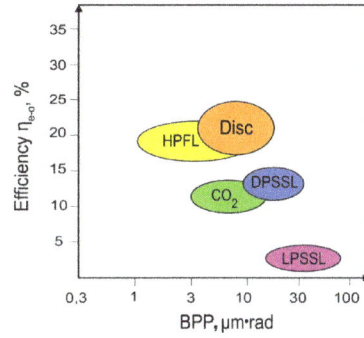

**Figure 3.** Electro-optical efficiencies of lasers [12].

$$\eta_w = \frac{E_a}{P_L \cdot t} \tag{2}$$

where
- $E_a$ : total energy absorbed by the workpiece, J;
- $P_L$ : laser output power, W or J/s;
- $t$ : laser time, s.

The laser energy transfer efficiency is always less than unity because not all of the energy generated by the laser is absorbed by the workpiece.

The second measurable process efficiency is the melting efficiency which is used to describe the amount of energy that is used to create a molten pool from the energy delivered to and absorbed by the workpiece [10]:

$$\eta_m = \frac{V_W \cdot T \cdot \omega_0 \cdot \Delta H_m}{P_L \cdot t} \tag{3}$$

where
- $V_W$ : welding speed, m/s;
- $T$ : material thickness, m;
- $P_L$ : laser power, W or J/s;
- $t$ : laser time, s;
- $\omega_0$ : minimum waist diameter, mm;
- $\Delta H_m$ : thermal content of the metal at the melt temperature (J).

LBW processing parameters and a number of other factors have a strong effect on the process efficiencies and, therefore, on weld penetration and geometry: laser power, beam diameter, welding speed and focal point position; material physical properties such as laser beam reflectivity, thermal diffusivity, surface tension, content of volatile elements and edge surface roughness; environment conditions such as air, atmosphere pressure, shielding gas type, shielding gas flow rate, laser inducted plasma and plume [16]-[18].

A review of all the possible factors affecting the process efficiency is far beyond the scope of this research. The aim of the paper is to review to most effective solutions thus far reported for improvement of LBW process efficiency.

## 2.4. Common Defects in LBW

Weld quality is an important factor that should be taken into account in analysis of LBW process efficiency. Appropriate choice of process parameters is not a function for maximization of welding process efficiencies, but a search for an acceptable balance between contending factors of physical and metallurgical effects. When this balance is achieved, a high quality weld is accomplished. The reliability of the welded component decreases when there are imperfections or defects in the weld. According to ISO 13919-1 [19] there are three quality levels for weld imperfections: B—stringent; C—intermediate; D—moderate. A list of the most common imperfections found in LBW is given in **Table 1** and examples of some defects are illustrated in **Figure 4**.

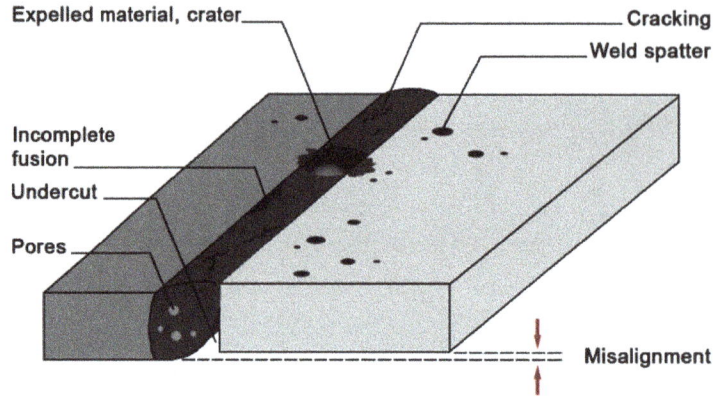

**Figure 4.** External and internal defects that can occur in butt-joint laser welding [22].

**Table 1.** Common imperfections in high power laser welding [19]-[21].

| Imperfection | Imperfection size limits | | | Illustration | Reasons |
|---|---|---|---|---|---|
| | Lack of penetration for full penetration welds. | | | | |
| Incomplete penetration | D. $h_1 \leq 0.15$ T max. 1 mm | C. not permitted | B. not permitted | | Low laser power; Bad edge surface preparation |
| Incompletely filled groove & excessive penetration | D. $h_2 \leq 0.15$ T max. 2 mm | C. $h_2 \leq 0.1$ T max. 1.5 mm | B. $h_2 \leq 0.05$ T max. 1 mm | | High laser power; Low welding speed; Metallurgical instability during welding |
| | D. $h_3 \leq 0.2 + 0.3$ T max. 5 mm | C. $h_3 \leq 0.2 + 0.2$ T max. 5 mm | B. $h_3 \leq 0.2 + 0.15$ T max. 5 mm | | |
| | Maximum dimensions for a single pore. | | | | |
| Porosity and gas pores | D. 1 or $h_4 \leq 0.5$ T max. 5 mm | C. 1 or $h_4 \leq 0.4$ T max. 3 mm | B. 1 or $h_4 \leq 0.3$ T max. 2 mm | | Incomplete cleaning of the edge surface; High gas saturation of metal; High welding speed |
| | All types of cracks except micro cracks (less than 1 mm² crack area) | | | | |
| Cracks | D. not permitted | C. not permitted | B. not permitted | | High internal stress; High amount of diffusive hydrogen in the weld metal; Brittle structure in the HAZ |

In addition to the defects listed in **Table 1**, it is important to mention the effect of the specific thermal cycle of thick section high power laser welding. This thermal cycle provides an extremely high rate weld cooling, which causes, in case of welding structural steels, hardening of the material in the HAZ. Around the fusion zone, where the cooling rate is highest, a large increase in the hardness is found [23]. Thick section laser beam welding has lower line energy (down to 5 J/mm) compared with traditional arc welding processes or laser arc hybrid welding, and the highest cooling rates (up to 3000 degrees/second) [8].

Efficiency increasing solutions in the research focus not only on increasing the process efficiency, but also on reducing the probability of defects occurrence.

## 2.5. Absorption as the Key to Welding Efficiency

The absorptance is the fraction of the incident laser light which is absorbed by the workpiece. The absorption level is not a steady factor, but a function of laser and metal properties, listed in **Table 2**.

There are several different mechanisms of beam absorption by a workpiece. In LBW, where laser energy can deeply penetrate the material via the keyhole, the energy absorption process is described by two mechanisms: Fresnel absorption and inverse Bremsstrahlung absorption. The first mechanism refers to the beam absorption at the solid/liquid surface of the material on the keyhole wall and the second one takes place in the partially ionized plume of vapor in the keyhole. These mechanisms are well described in literature and related articles [8]-[11], [25]-[27]. A number of comparative studies and simulation investigations of Fresnel absorption and multiple reflections in the keyhole have been carried out in recent years. Jin (2008) [28] analyzed factors affecting the keyhole formation and distribution of the laser intensity in the keyhole and noticed that laser energy is not uniformly distributed on the walls of the keyhole, but is concentrated in the region near the front keyhole wall. Kaplan (2012) [29] [30] provides an in-depth analysis of the modulation of the absorption mechanism and influence of surface roughness, wavelength and process parameters on the absorption level.

From Equation (2) it is clear that welding efficiency $\eta_W$ can be enhanced if the absorption can be increased. As the possibility of varying the laser parameters is usually limited in workshop conditions, this paper will focus on the material properties listed in **Table 2**. Based on the literature, absorption improvement methods that can be divided into three groups [8]-[11] [31] [32]:
1. Edge surface modifications;
2. Preheating techniques;
3. Ambient conditions modifications.

## 3. Methods to Improve Efficiency

### 3.1. Surface Modifications

The tendency of increased absorption with increased surface roughness in case of using $CO_2$ laser for low-alloyed steel processing was noted by Arata and Myamoto (1972) [33]. However, this factor has an impact on the absorption level only at the beginning of the welding process; after the stabilization, absorptionno longer depends on the optical properties of the surface.

Analysis of LBW of 4 mm thickness stainless steel with different edge surface roughness with 2.5 kW $CO_2$ laser was performed by Covelli (1988) [34]. The tests showed that the properties of the welds were not affected by the surface roughness.

**Table 2.** Properties influencing absorption [24].

| Laser Properties | Material Properties | Process Parameters |
|---|---|---|
| Power density | Composition | Welding speed |
| Wavelength | Temperature | Focal point position |
| Polarization | Roughness | Focal point size |
| Angle of incidence | Surface quality (oxide layers, dust, etc.) | Shielding gas |

Recent research shows that with use of HPFL, the edge surface roughness has a large effect on absorption, due to the multiple reflection undulations. Bergström *et al.* (2007) [35] recorded, by reflectance measurements, atrend of increasing absorptance for increasing roughness above Sa 1.5 μm for stainless steels and above 6 μm for low-alloyed steels. Sokolov *et al.* [36] [37], using calorimeter absorbed energy measurements, observed a correlation between edge surface roughness and absorption in welding structural steel in a butt-joint setup, as shown on **Figure 5**. A number of defects were recorded at edge surface roughness exceeding 6.3 μm.

Modifying surface roughness to improve energy absorption and welding efficiency should not incur additional costs as many manufacturing methods are available and an appropriate method can be selected in accordance with the desired surface roughness, as listed in **Table 3**.

## 3.2. Preheating Techniques

Preheating of the workpiece improves the LBW process efficiency by modification of the heat conduction characteristics inside the metal. As the temperature of the metal rises, there will be an increase in the photon population, causing more photon-electron energy exchange, as the electrons are more likely to interact with the material rather than oscillate and re-radiate. This phenomenon causes a fall in reflectivity and an increase in the absorptivity with the rise in temperature in the metal and therefore an increase in weld width and depth [27] [39].

Preheating is used to increase the weld width and depth as well as to prevent a number of defects. Changes in material properties due to high cooling rates (2000 - 3000 degrees/second) result in high hardness levelsand a risk

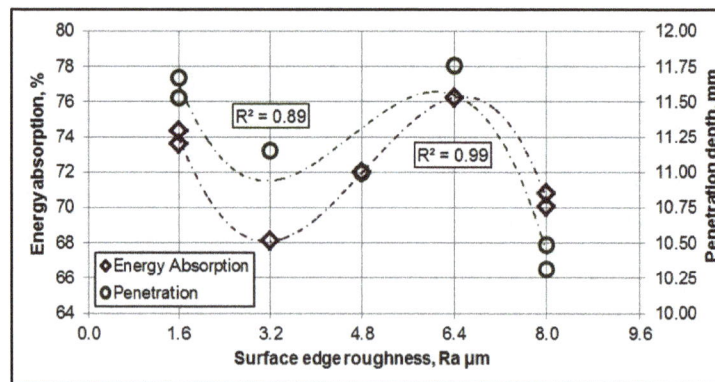

**Figure 5.** Energy absorption level and penetration depth at different roughness levels, HPFL, low-alloyed steel St 3, $t = 20$ mm , $P_L = 10$ kW , $V_W = 0.8$ m/min , $fpp = -4$ mm , $f_F = 250$ mm , $d_{OF} = 420$ μm ; Shielding. Ar 20 l/min [36].

**Table 3.** Example of surface roughness created using common cutting methods [38].

| Cutting method | Kerf surface roughness, Ra μm | | | | | | | |
|---|---|---|---|---|---|---|---|---|
| | 50 | 25 | 12.5 | 6.3 | 3.2 | 1.6 | 0.8 | 0.4 |
| Sawing | G | B | B | B | B | G | | |
| Milling | | G | B | B | G | G | G | |
| Flame cutting | G | B | B | | | | | |
| EB cutting | | | | B | B | B | G | |
| Laser cutting | | | G | B | B | G | | |
| Abrasive waterjet | | | | G | G | B | B | G |

The BLACK sections refer to common and GREEN sections refer to less frequent.

of cold and hot cracking. The critical cooling time between 800° and 500° is very short, so even steel with carbon content lower than 0.2% tends to form martensitic microstructure in the weld and run a high risk of crack formation [23] [40] [41].

The objective of preheating is to reduce the temperature gradient within the critical temperature range. The optimum preheating temperature depends on many factors: base metal composition, welding speed, workpiece thickness, external temperature, pressure and other environment parameters. In most applications preheating temperature and time are chosen such that the temperature of the workpiece is high enough so that the critical cooling temperature for the specified alloy is not reached. Too high preheating temperature has an adverse effect due to significant changes in the material composition in the HAZ [8] [42]-[44].

Preheating may be performed with the use of additional or splitted laser beams, electric heaters or flame torches—all these methods are proven to help prevent hot cracks and reduce hardness. However, the drawback of the above-mentioned techniques is the method of energy input: heat energy has to be transferred via the top surface of the weld, while hot-sensitive areas are situated several millimeters below the surface. In this situation, overheating of the top surface is very possible [45] [46].

Use of multi-beam welding techniques or dynamic beam forming to lower cooling rates, reduce volume percent martensite and therefore reduce hardness has been investigated in the last decade. A number of studies indicate a positive solution to the cracking problem in LBW [47]-[49]. Katayama, 2011 [50] utilized a twin beam method, using two laser sources as shown in **Figure 6**,whereas other authors (Xie, 2002 [51], Blackburn, 2010 [52]) utilized one laser source with a dual focus forming module or beam splitting system as shown in **Figure 7**.

In dynamic beam forming techniques mirrors deflect the focused laser beam and a mirror rotation of a few degrees causes laser beam to move several centimeters as shown on **Figure 8**. Oscillating mirrors, operating HPFL beams, are made of coated copper, or coated aluminum with water-cooling systems and are used in case of a small gap and less accurate geometry tolerances in butt-joint setups and allows widening of the weld, changing the focal point size or pattern. Vänskä and Salminen, 2012 [53] reported a successful adoption of beam forming techniques in tube-to-tube joints. Kraetzsch *et al.*, 2011 [54] pointed to ways of utilizing the laser beam oscillation for controlling weld seam geometry, meld pool turbulence and solidification behavior in welding of dissimilar materials.

In recent years, LBW has been assisted by induction heating. In induction preheating, low and medium frequency induction coils are heating the whole workpiece or move with the laser beam to achieve a constant effect of the critical cooling zones. For a butt-joint configuration a combination of two-armed induction coils were used by Göbel and Brener [46] (2006) to heat both sides of the weld, as shown in **Figure 9**.

The temperature field tailoring and reduction if tensile stresses possible with induction assisted LBW have

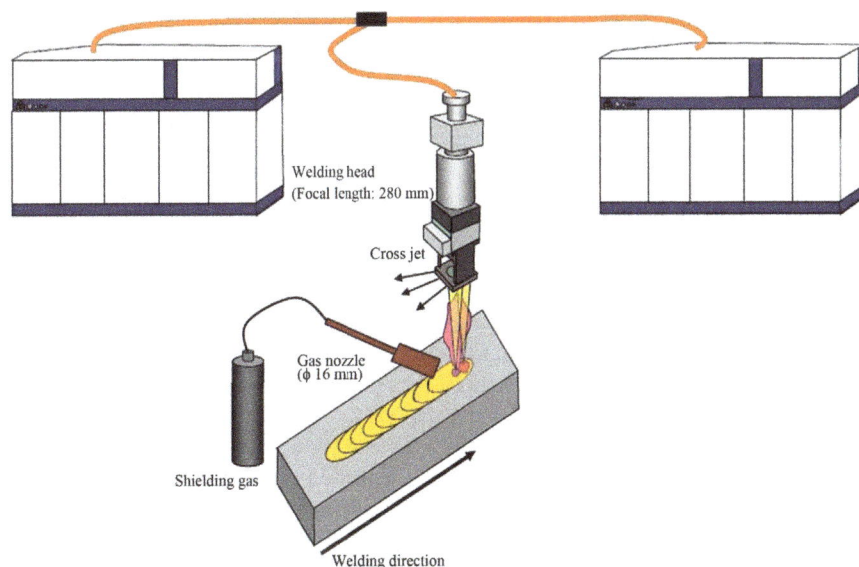

**Figure 6.** Twin of tandem beam system schematics [50].

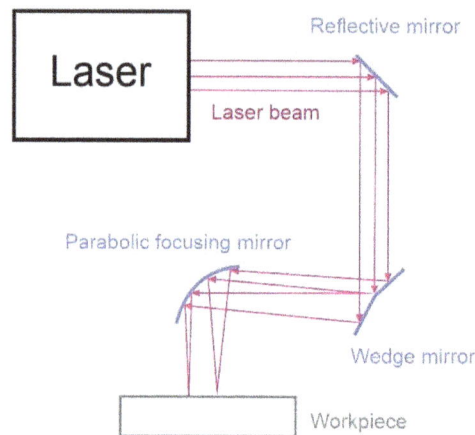

**Figure 7.** Beam splitting system schematics [51].

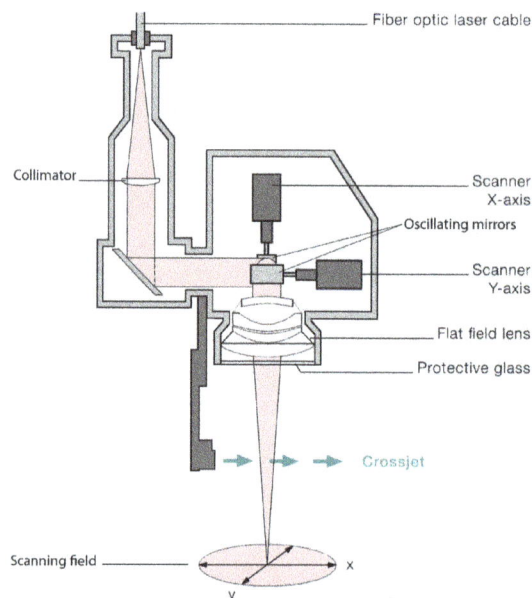

**Figure 8.** Dynamic beam forming techniques principle [22].

proved useful in avoiding hot cracks and other common defects, as well as being a relatively cost effective method compared to laser based beam splitting or scanning techniques [41] [46] [55].

## 3.3. Ambient Conditions Modifications

Katayama *et al.* 2001 [31] highlighted the potential for reduced pressure laser welding with a high-power. In later experiments with stainless steel, Katayama *et al.* 2011 [56] with a laser power up to 26 kW achieved an acceptable quality weld of 75 mm penetration depth with a 1 m/min speed single pass LBW at 1 kPa pressure. To achieve the desired pressure level, a vacuum chamber was sealed up, and the pressure was lowered by three rotary pumps. This method has certain limitations by size of the workpiece and use of the method in industry, as the welding process time have to be increased significantly.

Blackburn *et al.* (2013) [32] using a 5 kW Yb-fibre laser at reduced pressure achieved 11 mm penetration at a welding speed of 1 m/min using a sliding vacuum seal, illustrated in **Figure 10**.

Yang *et al.* (2013) [57] used more simple application of vacuum-assisted welding for zinc-coated steels: negative pressure zone is created directly on the top of the molten pool during the laser welding process by removal of the laser-induced plasma and plume enhances via copper tube installed on the top of the welding as shown in **Figure 11**.

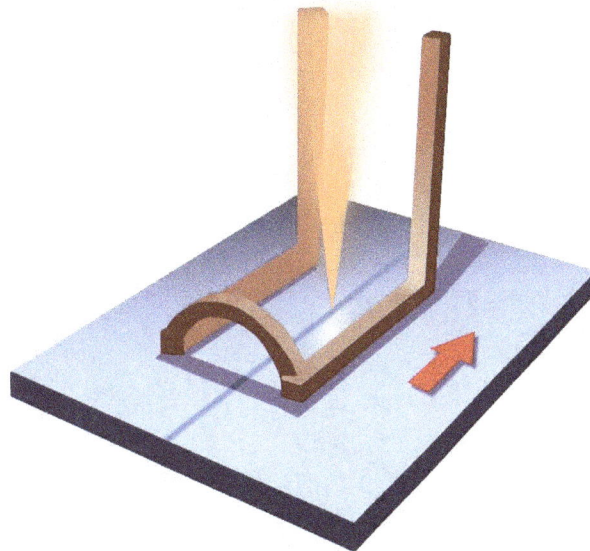

**Figure 9.** Inductive heating around the weld in butt-joint setup [46].

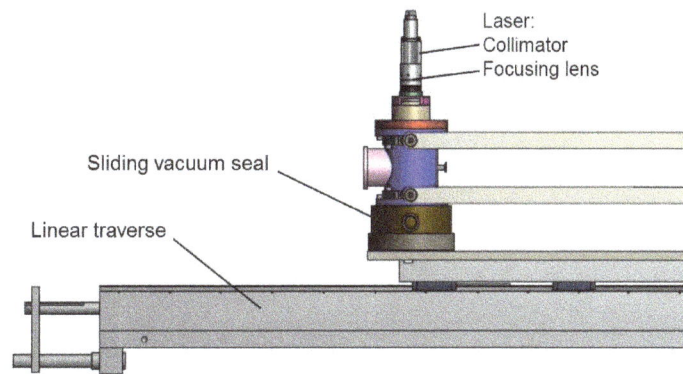

**Figure 10.** Schematic of a prototype sliding seal vacuum chamber, developed by TWI [32].

Significant increase in penetration depth (up to 200%) [32] as well as quality improvement compared with welding under atmospheric pressure [56]. Although this technology requires additional equipment and increase the preparation time, further research in this field would be of great help in improvement of LBW efficiency.

## 4. Conclusions

The present study was designed to review up-to-date methods for LBW of steels with HPFLs. The following conclusions can be drawn from the study:

1. LBW efficiency improvement can be achieved by increase in absorption.
2. There are three basic ways to improve LBW efficiency: by modifying the workpiece through surface preparation; by modifying the process with preheating or by lowering the pressure in the welding zone.
3. In thick section LBW at high power levels (≥10 kW) absorption has a significant dependence on the edge surface roughness. The influence of the roughness level has a tendency to increase with increasing laser power. Use of manufacturing methods that produce edge surfaces of a pre-determined roughness level is recommended.
4. Preheating techniques are preferable not only because of the resulting increase in process efficiency, but also because of reduction in the occurrence risk of defects.
5. Vacuum-assisted welding provides significant increase in penetration depth and weld quality.

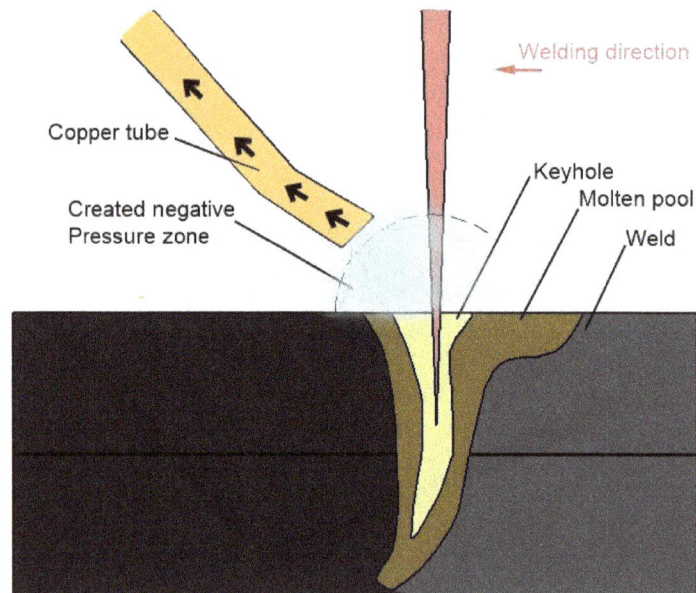

**Figure 11.** Schematic of the negative pressure zone above the molten pool [57].

6. Several described above methods may be used together to promote an additional increase in the absorption and, therefore, efficiency. The relationships and causalities of these factors require further investigation.

## Acknowledgements

This article was written as part of the Finnish Metals and Engineering Competence Cluster (FIMECC)'s Innovation & Network program.

## References

[1]   Shimokusu, Y., Fukumoto, S., Nayama, M., Ishide, T. and Tsubota, S. (2001) Application of High Power YAG Laser Welding to Stainless Steel Tanks, Mitsubishi Heavy Industries, Ltd. *Technical Review*, **38**, 118-121.

[2]   Ishide, T., Tsubota, S., Nayama, M., Shimokusu, Y., Nagashima, T. and Okimura, K. (2000) 10-kW-Class YAG Laser Application for Heavy Components. In: Chen, X.L., Fujioka, T., Matsunawa, A., Eds., *Advanced High-Power Lasers and Applications*, Society of Photo Optical, 543-550.

[3]   Roland, F., Reinert, T. and Pethan, G. (2003) Laser Welding in Shipbuilding—An Overview of the Activities at Meyer Werft. *Welding Research Abroad*, **49**, 39-51.

[4]   Tsirkas, S.A., Papanikos, P., Pericleous, K., Strusevich, N., Boitout, F. and Bergheau, J.M. (2003) Evaluation of Distortions in Laser Welded Shipbuilding Parts Using Local-Global Finite Element Approach. *Science and Technology of Welding & Joining*, **8**, 79-88. http://dx.doi.org/10.1179/136217103225010899

[5]   Moore, P.L., Howse, D.S. and Wallach, E.R. (2004) Microstructures and Properties of Laser/Arc Hybrid Welds and Autogenous Laser Welds in Pipeline Steels. *Science and Technology of Welding & Joining*, **9**, 314-322. http://dx.doi.org/10.1179/136217104225021652

[6]   Yapp, D. and Blackman, S.A. (2004) Recent Developments in High Productivity Pipeline Welding. *Journal of Brazilian Society of Mechanical Science and Engineering*, **26**, 89-97. http://dx.doi.org/10.1590/S1678-58782004000100015

[7]   Hecht, J. (2009) Fiber Lasers Ramp up the Power. *Laser Focus World*, **45**, 53-57.

[8]   Dawes, C. (1992) Laser Welding: A Practical Guide. Redwood Press Ltd, Cambridge. http://dx.doi.org/10.1533/9781845698843

[9]   Lancaster, J.F. (1984) The Physics of Welding. Pergamon Press, Oxford.

[10]  Duley, W.W. (1998) Laser Welding. John Wiley & Sons Inc., New York.

[11]  Ion, J.C. (2005) Laser Processing of Engineering Materials. Biddles Ltd., Norfolk.

[12]  Dahmen, M., Güdükkurt, O. and Kaierle, S. (2010) The Ecological Footprint of Laser Beam Welding. *Physics Proce-*

*dia*, **5**, 19-28. http://dx.doi.org/10.1016/j.phpro.2010.08.025

[13] Barnes, N.P. (2007) Solid-State Lasers from an Efficiency Perspective. *IEEE Journal of Selected Topics in Quantum Electronics*, **13**, 435-447.

[14] Zhang, X., Ashida, E., Tarasawa, S., Anma, Y., Okada, M., Katayama, S. and Mizutani, M. (2011) Welding of Thick Stainless Steel Plates up to 50 mm with High Brightness Lasers. *Journal of Laser Applications*, **23**, Article No. 022002. http://dx.doi.org/10.2351/1.3567961

[15] Unocic, R.R. and DuPont, J.N. (2004) Process Efficiency Measurements in the Laser Engineered Net Shaping Process. *Metallurgical and Material Transactions B*, **35**, 143-152.

[16] Katayama, S., Kawahito, Y. and Mizutani, M. (2010) Elucidation of Laser Welding Phenomena and Factors Affecting Weld Penetration and Welding Defects. *Physics Procedia*, **5**, 9-17. http://dx.doi.org/10.1016/j.phpro.2010.08.024

[17] Salminen, A., Piili, H. and Purtonen, T. (2010) The Characteristics of High Power Fibre Laser Welding. *Proceeding of the Institution of Mechanical Engineers Part C: Journal of Mechanical Engineering Science*, **224**, 1019-1029. http://dx.doi.org/10.1243/09544062JMES1762

[18] Zhang, M., Chen, G., Zhou, Y. and Liao, S. (2013) Optimization of Deep Penetration Laser Welding of Thick Stainless Steel with a 10 kW Fiber Laser. *Materials & Design*, **53**, 568-576.

[19] ISO 13919-1 (1996) Welding. Electrons and Laser Beam Welded Joints. Guidance on Quality Levels for Imperfections. Part 1: Steel.

[20] EN 10025-1 (2004) Hot Rolled Products of Structural Steels. Part 1: General Technical Delivery Conditions.

[21] EN 10025-2 (2004) Hot Rolled Products of Structural Steels. Part 2: Technical Delivery Condition for Non-Alloy Structural Steels.

[22] TRUMPF Group (2007) The Laser As a Tool. Vogel Buchverlag, Würzburg.

[23] Sokolov, M., Salminen, A., Kuznetsov, M. and Tsibulskiy, I. (2011) Laser Welding and Weld Hardness Analysis of Thick Section S355 Structural Steel. *Materials & Design*, **32**, 5127-5131. http://dx.doi.org/10.1016/j.matdes.2011.05.053

[24] Bergström, D. (2005) The Absorptance of Metallic Alloys to Nd:YAG and Nd:YLF Laser Light. Doctoral Dissertation, Luleå University of Technology, Luleå.

[25] Li, L. (2008) Lasers in Technology, Physical Methods. *Instruments and Measurements*, **4**, 159-173.

[26] Hagen, E. and Rubens, H. (1904) Emissivity and Electrical Conductivity of Alloys. *Deutsche Physikalische Gesellschaft*, **6**, 128-136.

[27] Steen, W.M. (2003) Laser Material Processing. 3rd Edition, Springer-Verlag, London. http://dx.doi.org/10.1007/978-1-4471-3752-8

[28] Jin, X. (2008) A Three-Dimensional Model of Multiple Reflections for High-Speed Deep Penetration Laser Welding Based on an Actual Keyhole. *Optics and Lasers in Engineering*, **46**, 83-93. http://dx.doi.org/10.1016/j.optlaseng.2007.05.009

[29] Kaplan, A.F.H. (2012) Fresnel Absorption of 1 μm- and 10 μm-Laser Beams at the Keyhole Wall during Laser Beam Welding: Comparison between Smooth and Wavy Surfaces. *Applied Surface Science*, **258**, 3354-3363. http://dx.doi.org/10.1016/j.apsusc.2011.08.086

[30] Kaplan, A.F.H. (2012) Local Absorptivity Modulation of a 1 μm-Laser Beam through Surface Waviness. *Applied Surface Science*, **258**, 9732-9736. http://dx.doi.org/10.1016/j.apsusc.2012.06.020

[31] Katayama, S., Kobayashi, Y., Mizutani, M. and Matsunawa, A. (2001) Effect of Vacuum on Penetration and Defects in Laser Welding. *Journal of Laser Applications*, **13**, 187. http://dx.doi.org/10.2351/1.1404413

[32] Blackburn, J., Allen, C., Smith, S. and Hilton, P. (2013) Thick-Section Laser Welding. *Proceeding of LAMP2013, the 6th International Congress on Laser Advanced Material Processing*, Niigata, 23-26 July 2013.

[33] Arata, Y. and Miyamoto, I. (1972) Some Fundamental Properties of High Power Laser Beam as a Heat Source (Report 2). *Transactions of the Japan Welding Society*, **3**, 152-162.

[34] Covelli, L., Jovane, F., De Iorio, L. and Tagliaferri, V. (1988) Laser Welding of Stainless Steel: Influence of the Edges Morphology. *CIRP Annals, Manufacturing Technology*, **37**, 545-548.

[35] Bergström, D., Powell, J. and Kaplan, A.F.H. (2007) The Absorptance of Steels to Nd:YLF and Nd:YAG Laser Light at Room Temperature. *Applied Surface Science*, **253**, 5017-5028. http://dx.doi.org/10.1016/j.apsusc.2006.11.018

[36] Sokolov, M., Salminen, A., Somonov, V. and Kaplan, A.F. (2012) Laser Welding of Structural Steels: Influence of the Edge Roughness Level. *Optics & Laser Technology*, **44**, 2064-2071. http://dx.doi.org/10.1016/j.optlastec.2012.03.025

[37] Sokolov, M. and Salminen, A. (2012) Experimental Investigation of the Influence of Edge Morphology in High Power Fiber Laser Welding. *Physics Procedia*, **39**, 33-42. http://dx.doi.org/10.1016/j.phpro.2012.10.011

[38] Degarmo, E.P., Black, J.T. and Kosher, R.A. (2003) Materials and Processes in Manufacturing. 9th Edition, Wiley, Hoboken.

[39] Peretz, R. (1988) The Preheating Temperature Parameter for Deep Penetration Welding with High Energy Focused Beams. *Optics and Lasers in Engineering*, **9**, 23-34. http://dx.doi.org/10.1016/0143-8166(88)90026-7

[40] Akesson, B. and Karlsson, L. (1976) Prevention of Hot Cracking of Butt Welds in Steel Panels by Controlled Additional Heating of the Panels. *Welding Research International*, **6**, 35-52.

[41] Rosenfeld, R., Herzog, D. and Haferkamp, H. (2009) Process Combination of Laser Welding and Induction Hardening. *Proceeding of the 5th International WLT-Conference on Lasers in Manufacturing*, Munich, 15-18 June 2009, 71-75.

[42] Miranda, R., Costa, A., Quintino, L., Yapp, D. and Iordachescu, D. (2009) Characterization of Fiber Laser Welds in X100 Pipeline Steel. *Materials & Design*, **30**, 2701-2707. http://dx.doi.org/10.1016/j.matdes.2008.09.042

[43] Hu, L.H., Huang, J., Li, Z.G. and Wu, Y.X. (2011) Effects of Preheating Temperature on Cold Cracks, Microstructures and Properties of High Power Laser Hybrid Welded 10Ni3CrMoV Steel. *Materials & Design*, **32**, 1931-1939. http://dx.doi.org/10.1016/j.matdes.2010.12.007

[44] Böllinghaus, T. and Herold, H. (2005) Hot Cracking Phenomena in Welds. Springer-Verlag Berlin Heidelberg, Berlin. http://dx.doi.org/10.1007/b139103

[45] Yang, S. and Kovacevic, R. (2009) Laser Welding of Galvanized DP980 Steel Assisted by the GTAW Preheating in a Gap-Free Lap Joint Configuration. *Journal of Laser Applications*, **21**, 139-148. http://dx.doi.org/10.2351/1.3184432

[46] Göbel, G. and Brenner, B. (2006) Avoiding Hot Cracking by Induction Based Change of Thermal Strains during Laser Welding. *Proceedings of 25th International Congress on Applications of Lasers & Electro Optics ICALEO*, 34-43.

[47] Iqbal, S. and Gualini, M. (2010) Dual Beam Method for Laser Welding of Galvanized Steel: Experimentation and Prospects. *Optics & Laser Technology*, **42**, 93-98. http://dx.doi.org/10.1016/j.optlastec.2009.05.009

[48] Ploshikhin, V., Prikhodovsky, A., Makhutin, M., Zoch, H.W., Heimerdinger, C. and Palm, F. (2004) Multi-Beam Welding: Advanced Technique for Crack-Free Laser Welding. *Proceedings of 4th International Conference Laser Assisted Net Shape Engineering LANE*, Erlangen, 21-24 September 2004, 131-136.

[49] Liu, Y.N. and Kannatey-Asibu Jr., E. (1997) Experimental Study of Dual-Beam Laser Welding of AISI 4140 Steel. *Welding Journal*, **76**, 342-348.

[50] Katayma, S., Hirayama, M., Mizutani, M. and Kawahito, Y. (2011) Deep Penetration Welds and Welding Phenomena with Combined Disk Lasers. *Proceedings of 30th International Congress on Applications of Lasers & Electro Optics ICALEO*, Orlando, 23-27 October 2011, 661-668.

[51] Xie, J. (2002) Dual Beam Laser Welding. *Welding Journal*, **81**, 223-230.

[52] Blackburn, J., Allen, C., Smith, S., Punshon, C. and Hilton, P. (2010) Dual Focus Nd: YAG Laser Welding of Titanium Alloys. *Proceedings of the 36th International MATADOR Conference*, Manchester, 14-16 July 2010, 279-282. http://dx.doi.org/10.1007/978-1-84996-432-6_64

[53] Vänskä, M. and Salminen, A. (2012) Laser Welding of Stainless Steel Self-Steering Tube-to-Tube Joints with Oscillating Mirror. *Proceedings of the Institute of Mechanical Engineering, Part B: Journal of Engineering Manufacture*, **226**, 632-640. http://dx.doi.org/10.1177/0954405411425114

[54] Kraetzsch, M., Standfuss, J., Klotzbach, A., Kaspar, J., Brenner, B. and Beyer, E. (2011) Laser Beam Welding with High-Frequency Beam Oscillation: Welding of Dissimilar Materials with Brilliant Fiber Lasers. *Physics Procedia*, **12**, 142-149. http://dx.doi.org/10.1016/j.phpro.2011.03.018

[55] Brenner, B., Standfuss, J., Wetzig, A., Fux, V., Lepski, D. and Beyer, E. (1998) New Developments in Induction Assisted Laser Materials Processing. *European Conference on Laser Treatment of Materials*, Hanover, 22-23 September 1998, 57-66.

[56] Katayama, S., Yohei, A., Mizutani, M. and Kawahito, Y. (2011) Development of Deep Penetration Welding Technology with High Brightness Laser under Vacuum. *Physics Procedia*, **12**, 75-80. http://dx.doi.org/10.1016/j.phpro.2011.03.010

[57] Yang, S., Wang, J., Carlson, B. and Zhang, J. (2013) Vacuum-Assisted Laser Welding of Zinc-Coated Steels in a Gap-Free Lap Joint Configuration. *Welding Journal*, **92**, 197-204.

# 5

# Investigation of the Weldability of Austanitic Stainless Steel

E. M. Anawa, M. F. Bograrah, S. B. Salem

Industrial and Manufacturing Department, Faculty of Engineering, University of Benghazi, Benghazi, Libya
Email: ezzeddin.anawa @uob.edu.ly

## Abstract

This work concerns with the study of weldability of austenitic stainless steel 316 by using automatic tungsten gas shielded arc welding under various welding conditions under which it is designed to weld the samples. Results have been studied using impact and tensile strength tastings of the prepared welding joints using statistical approach. Results obtained showed that as gas flow rate of ($CO_2$) increased the impact energy is increased, while increasing of welding current caused increasing of impact energy up to (120 ampere) then decreased. The tensile strength test results showed that as welding current is increased the tensile fracture load is decreased while increasing gas flow rate caused an increase in tensile fracture load up to 12 L/min then reduced. Microstructure examination of the weld zones did support the explanation of the variation of weld joint mechanical properties.

## Keywords

Weldability, Austenitic Stainless, Impact Energy, Tensile Strength

## 1. Introduction

The austenitic stainless steels, because of their high chromium and nickel content, are the most corrosion resistant of the stainless group providing unusually fine mechanical properties. Austenitic is the most widely used type of stainless steel. It has a nickel content of at least of 7%, which makes the steel structure fully austenitic and gives it ductility, a large scale of service temperature, non-magnetic properties and good weld ability. The range of applications of austenitic stainless steel includes house wares, containers, industrial piping and vessels, architectural facades and constructional structures [1]. Murat *et al.* [2] have studied on the resistance spot weldability of galvanized interstitial free steel sheets with austenitic stainless steel sheets. In microhardness measurements, the maximum hardness values were in the middle of the weld nugget. Emin Bayraktar *et al.* [3] have contributed their research on the selection of optimal welding conditions and developed new grade steels for automotive applications. The study is based on impact tensile testing to spot welded sheets. The effect of nucleus size on mechanical properties in electrical resistance spot welding of chromide micro alloyed steel sheets was investigated by Aslanlar [4]. Effects of Laser Welding Conditions on Toughness of Dissimilar (austenitic stainless steel 316 with low carbon steel F/A) Welded Components were investigated by E. M. Anawa and A.G. Olabi [5] using Taguchi approach to optimize the dissimilar F/A joints in terms of its mechanical properties.

Bouyousfi *et al.* [6] have studied the effect of process parameters (arc intensity, welding duration and applied load) on the mechanical characteristics of the weld joint of austenitic stainless steel 304L. The results showed that the applied load seems to be the control factor of the mechanical characteristics of weld joint compared to the welding duration and the current intensity. Nizamettin K [7] has focused his study on the influence of welding parameters on the joint strength of resistance spot-welded titanium sheets. The results indicated that increasing current time and electrode force increased the tensile shear strength and the joint obtained under the argon atmosphere gave better strength. Hardness measurement results showed that welding nugget gave the highest hardness. L. Suresh Kumar *et al.* [1] experimentally investigated welding aspects of AISI 304 and 316 by taguchi technique for the process of TIG & MIG welding. They used the TIG and MIG process to find out the characteristics of the metal after it is welded. The voltage is taken constant and various characteristics such as strength, hardness, ductility, grain structure, modulus of elasticity, tensile strength breaking point, HAZ are observed in two processes and analyzed and finally concluded. V. Shankar *et al.* [8] investigated the solidification cracking which is a significant problem during the welding of austenitic stainless steels, particularly in fully austenitic and stabilized compositions. They studied the solidification cracking in austenitic stainless steels with particular emphasis on nitrogen-alloyed and stabilized stainless steels. A statistical design of experiment (DOE) was used to optimise selected laser beam welding parameters (laser power, welding speed, and focus length). Optimization of tensile strength of ferritic/austenitic laser welded components was studied by E.M. Anawa and A.G. Olabi [9]. The experimental results indicate that the F/A laser welded joints are improved effectively by optimizing the input parameters using the Taguchi approach. The aim was to optimize the maximum ultimate tensile strength of F/A welded components, by minimizing the laser power and maximizing welding speed in order to optimize the cost and increase the production rate.

This work concerns with the study of weldability of austenitic stainless steel 316 by using automatic tungsten gas shielded arc welding under various welding conditions under which it is designed to welded the samples. The mechanical properties such as tensile strength and notch impact strength are tested and investigated.

## 2. Experimental Work

### 2.1. Base Metal Selection

The base metal used in the present work was stainless steel type (Austenitic 316) the dimensions are: (1250 × 1000 × 2 mm (l × w × h)), the chemical compositions and the mechanical properties are shown in **Table 1** and **Table 2** respectively.

**Table 1.** Chemical composition of base metal.

| Element | Cr | Ni | Mn | Si | C | N | Nb | Ti | Fe |
|---------|-----|-----|-----|------|-----------|-----------|------|------|---------|
| Weight % | 16 - 18 | 10 - 14 | 1 - 2 | 0.08 | 0.02 - 0.08 | 0 - 0.015 | <0.2 | <0.2 | balance |

**Table 2.** Mechanical properties of base metal.

| Mechanical Properties | |
|---|---|
| Density (×1000 Kg/m) | 7.8 |
| Poisson's Ratio | 0.27 - 0.30 |
| Elastic Modulus (GPa) | 200 |
| Tensile Strength (MPa) | 515 |
| Yield Strength (MPa) | 205 |
| Elongation (%) | 40 |
| Reduction in Area (%) | 50 |

## 2.2. Welding Process

The welding process was carried out by using TIG technique for welding process, this technique is widely used for different kinds of welding processes. The welding machine type is (PHOENIX 500) this machine can be used for multi welding processes such as: (TIG, MIG) technique, TIG welding technique was selected for this study to produce the joint for selected welding parameters. The welding process was carried out without filler metal.

## 2.3. Samples Preparation

The selected material was cut by using an electrical shear to (20) samples, which are exhibited in **Figure 1**. Than the welding process was applied to join samples at different welding conditions which are presented in **Table 3**, using butt joint design, exhibited in **Figure 2**.

## 2.4. Weld Joints Testing

### 2.4.1. Non Destructive Testing (Visual Inspection)
In an attempt to find any flout in the welded joints such as (cracks, pores, etc.) and the welding penetration. The applied visual inspection was shows that acceptable quality of joints.

### 2.4.2. Destructive Testing
- **Tensile Test**

Tensile tests were carried out at room temperature (25°C), for all the prepared tensile samples. The machine test and the diagram for tensile samples are shown in **Figure 3** and **Figure 4**.

**Tensile Samples Preparations**

After finished the welding process the samples intended to be used for the tensile test were prepared, they were cut by using a hacksaw and no high temperature was generated that keeps the metal properties away from phase transformations, then the shaping machine was used to obtain a V groove at the welded area, to prepare a Notched Tensile Strength (NTS) as shown in **Figure 4** [10]. For each welding condition five samples were prepared for tensile test.

- **Impact Test**

This test was carried out in (Mechanical Engineering Department) at room temperature (25°C), for all the impact samples. The impact testing machine diagram and the impact samples are shown in **Figure 5** and **Figure 6**. For each welding condition five samples were prepared for impact test.

**Table 3.** Experimental welding conditions.

| | Sample No. | 1 | 2 | 3 | 4 | 5 | 6 | 7 | 8 | 9 | 10 |
|---|---|---|---|---|---|---|---|---|---|---|---|
| Welding Conditions | GFR, (L/min) | 10 | 10 | 10 | 10 | 10 | 7 | 8 | 10 | 12 | 14 |
| | C, (A) | 80 | 100 | 120 | 160 | 180 | 120 | 120 | 120 | 120 | 120 |

*GFR: Gas Flow Rate, C: Current

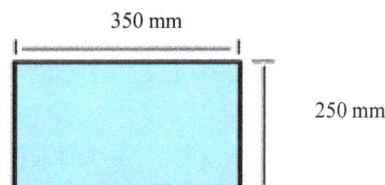

350 mm

250 mm

**Figure 1.** Material after cutting process.

**Figure 2.** The welded sample.

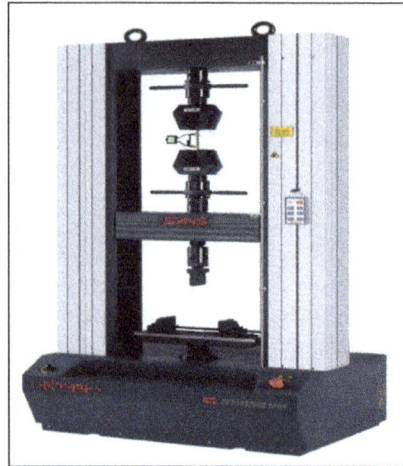

**Figure 3.** Tensile testing machine.

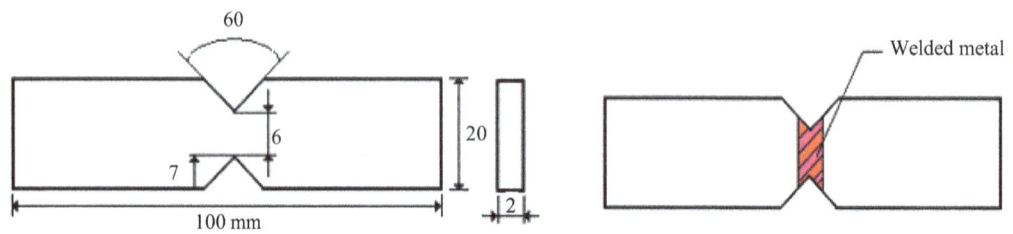

**Figure 4.** Schematic diagram for notched tensile strength (NTS).

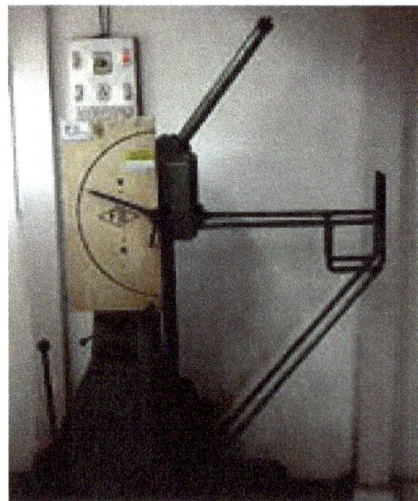

**Figure 5.** Impact testing machine.

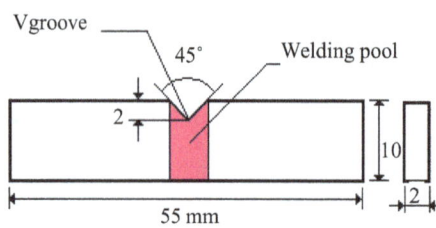

**Figure 6.** Schematic diagram of impact testing sample.

### Impact Samples the Preparation

After finishing the welding process the samples for the impact test were prepared, they were cut by using a hacksaw and no heat was introduced to keep the metal properties away from phase transformations, then the shaping machine was used to obtain a V groove in the welded area **Figure 6**.

### 2.4.3. Microstructure Testing

Microstructure study was carried out in Mechanical Engineering Department. A small sample was cut from each welded plate to study the microstructure. Emery papers grades (80, 500, and 800) were used for grinding the samples before polishing. For etching stainless steel, a (1 g) meta sulphate by meta sodium and (20 ml) of HCL and (100 ml) of distilled water was applied as an etchant. Microscopic examination was carried out using the **ZIESS** type optical microscope in Mechanical Department at Material laboratory at Benghazi University.

## 3. Results and Analysis

Following welding process and preparing of samples, the testing process was performed and the results obtained are exhibited in **Tables 4-7** and represents in **Figures 7-10**.

The result analyses for the effect of gas flow and welding current on the mechanical properties are as follows:

**Table 4.** Results of impact test.

| Current (A) | Gas Flow (10 L/min) | | | | | |
|---|---|---|---|---|---|---|
| | 1 | 2 | 3 | 4 | 5 | Average |
| 80 | 22 | 17 | 16 | 14 | 26 | 19 |
| 100 | 29 | 30 | 26 | 24 | 24 | 26.6 |
| 120 | 30 | 30 | 22 | 24 | 28 | 26.8 |
| 160 | 32 | 30 | 24 | 20 | 26 | 26.4 |
| 180 | 24 | 30 | 22 | 26 | 24 | 25.2 |

**Table 5.** Impact testing result.

| Gas Flow (L/min) | Current (120 A) | | | | | |
|---|---|---|---|---|---|---|
| | 1 | 2 | 3 | 4 | 5 | Average |
| 7 | 26 | 24 | 24 | 20 | 26 | 24 |
| 8 | 24 | 24 | 30 | 24 | 26 | 25.6 |
| 10 | 30 | 26 | 24 | 24 | 24 | 25.6 |
| 12 | 24 | 30 | 30 | 22 | 22 | 25.6 |
| 14 | 30 | 32 | 30 | 20 | 24 | 27.2 |

**Table 6.** Readings of tensile test.

| Current (A) | Gas Flow (×10 L/min) | | | | | |
|---|---|---|---|---|---|---|
| | 1 | 2 | 3 | 4 | 5 | Average |
| 80 | 900 | 800 | 900 | 900 | 850 | 870 |
| 100 | 900 | 1000 | 900 | 900 | 900 | 920 |
| 120 | 900 | 920 | 800 | 820 | 850 | 858 |
| 160 | 950 | 800 | 850 | 900 | 820 | 864 |
| 180 | 800 | 800 | 900 | 800 | 800 | 820 |

**Table 7.** Readings of tensile test.

| Gas Flow (L/min) | Current (120 A) | | | | | |
|---|---|---|---|---|---|---|
| | 1 | 2 | 3 | 4 | 5 | Average |
| 7 | 900 | 1000 | 900 | 940 | 970 | 942 |
| 8 | 830 | 980 | 900 | 850 | 820 | 876 |
| 10 | 860 | 900 | 950 | 900 | 970 | 916 |
| 12 | 950 | 1000 | 950 | 900 | 940 | 948 |
| 14 | 940 | 910 | 900 | 960 | 920 | 926 |

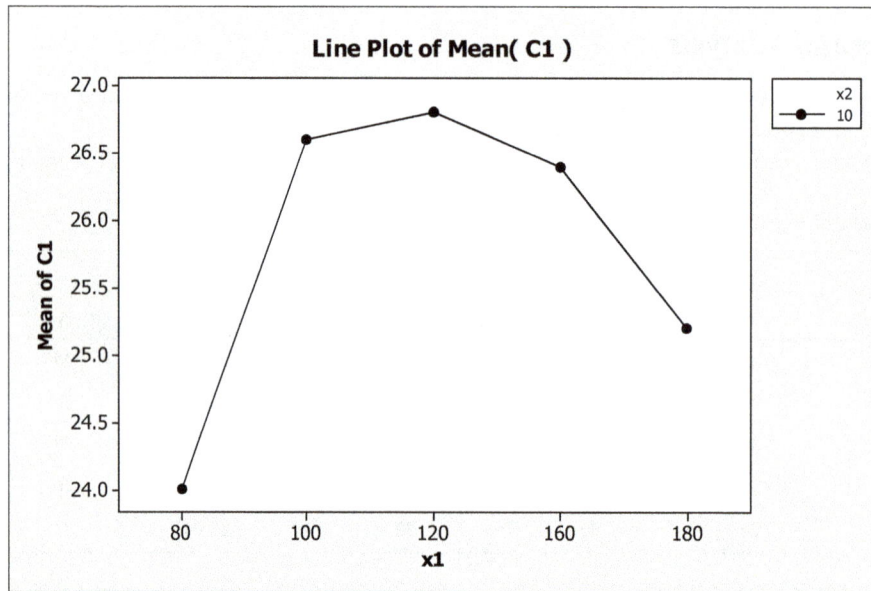

Where: The **X** axis is presenting the gas flows. The **Y** axis is presenting the mean of readings of impact tests.

**Figure 7.** Plot of impact test.

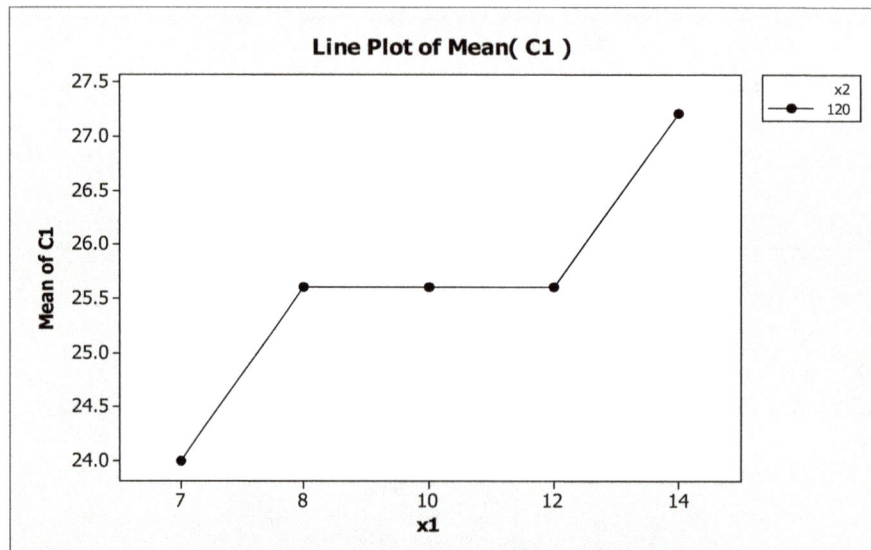

Where: The **X** axis is presenting the gas flows. The **Y** axis is presenting the mean of readings of impact tests.

**Figure 8.** A plot of impact test.

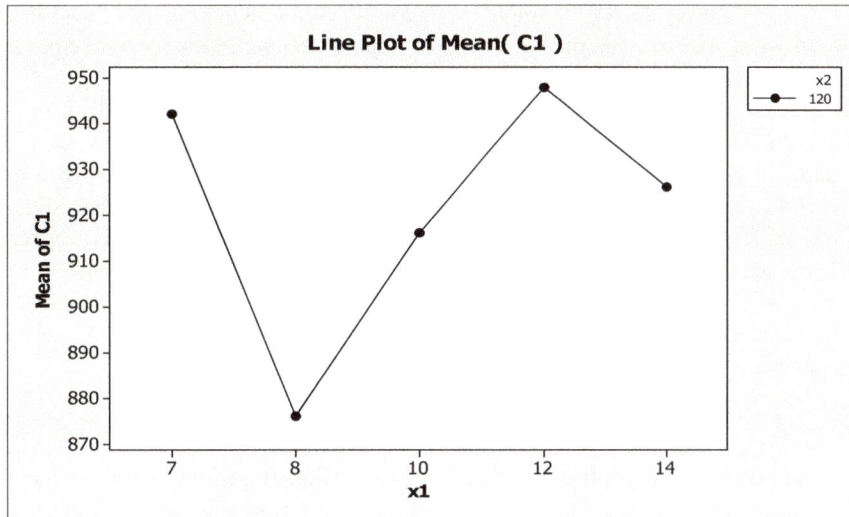

Where: The X axis is presents the changing in the gas flow. The Y axis is the mean of tensile strength.

**Figure 9.** A plot of tensile test result.

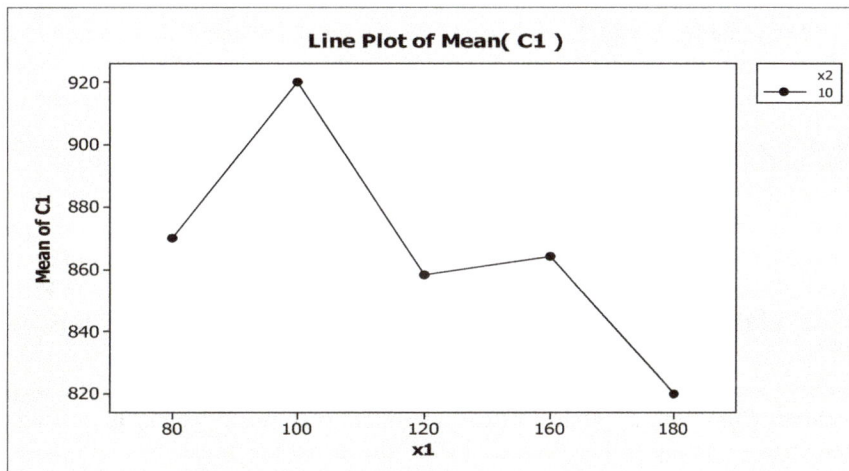

Where: The X axis is presents the changing in the gas flow. The Y axis is the mean of tensile strength.

**Figure 10.** A plot of tensile test result.

## 3.1. Impact Test

### 3.1.1. Analysis of Impact Energy Results for Fixed Gas Flow Rate and Variable Electrical Welding Current.

The welded samples were tested by charpy impact. An average of five impact specimens values were determined and tabulated in **Table 4**. These results are represented in a plot of impact energy with the variable welding current as shown in **Figure 7**.

**Table 4** shows that the maximum value of absorbed energy achieved was **26.8 J** at **120** A and the minimum value was **19 J** at **80** A. The range between the maximum and minimum absorbed energy was **7.8 J**. The results represented in **Figure 7** shows that the effect of changing of welding current at a fixed gas flow rate, it is clear that increasing of welding current results to increasing in absorbed impact energy and toughness of the welded joint of 316 austenitic stainless steel.

### 3.1.2. Analysis Result of the Fixed Welding Current and Changing of Gas Flow Rate

The effect of gas flow rate on the impact energy of the welded joints of studied stainless steel was also investigated. **Table 5** presents the results for the impact tests for the various shielding gas flow rates at a fixed welding current of 120 Amperes. The average of five values of impact strength specimens were plotted against the various gas flow rates as presented in **Figure 8**.

From **Table 5** the maximum value was achieved for impact test was **27.2 J** at **14** L/min and the minimum value was **24 J** at **7** L/min. The range between the maximum and minimum value was **3.2 J**. **Figure 8** shows the effect of changing of gas flow rate at fixe welding current, and it is clear that increasing of gas flow rate caused increasing of absorbed energy which is reflecting increase in toughness of 316 stainless steel welded joints. The increasing of welding current and gas flow rate results an increasing of absorbed impact energy.

In **Figure 7**, an increasing welding current (as it is increased to **120** A) the absorbed impact energy is increased while at low current (**80** A) the impact energy was decreased. In **Figure 8** where at increased gas flow rate (as it is increased up to **14** L/min) the absorbed impact energy is increased while at low gas flow rate (7 L/min) the impact energy was decreased.

## 3.2. Tensile Test

### 3.2.1. Analysis of Fixed Gas Flow Rate and Changing of Welding Electrical Current

Tensile testing was performed also for the study the effects of gas flow rates and welding currents on the tensile fracture load. For gas flow rate of **10** L/min, welding currents of 80, 100, 120, 160, and 180 amps were applied and five tensile samples were tested for each welding condition. **Table 6** presents the received testing results. To demonstrate the tensile strength behavior, the results were plotted as shown in **Figure 9**.

**Table 6** exhibits that the maximum tensile strength value was **920 MPa** achieved at **100** A and the minimum value was **820 MPa** achieved at **180** A. The range between the maximum and minimum tensile strength is **100 MPa**. **Figure 9** represents the effect of changing of welding current on tensile strength. It is clear that increasing of electrical current caused decreasing tensile strength and hence the ductility of the welded joints.

### 3.2.2. Analysis of the Fixed Electrical Current and Changing of Gas Flow Rate

The effect of gas flow rate was also studied on the tensile strength behavior of stainless steel. **Table 7** presents the results for tensile strength at a fixed welding current of 120 A but for various shielding gas flow rates. The average of five tensile testing samples was plotted against the shielding gas flow rate in **Figure 10**.

From the **Table 7** the maximum tensile strength test value achieved was **948 MPa** at **12** L/min and the minimum value was **876 MPa** at **8** L/min, the range between the maximum and minimum tensile strength value is **72 MPa**.

**Figure 10** shows the effect of changing of gas flow rate at fixe welding current and increasing of gas flow rate results an increase of tensile strength and hence the ductility of welded joints. Increasing of welding current, the tensile strength and ductility will be decreased while the increasing of gas flow rate caused increasing of tensile strength and ductility of welded joints.

In **Figure 9** where for increased welding current as it is increased (**100** A) the tensile strength is increased while at low welding current (**80** A) the tensile strength was decreased. In **Figure 10** whereas the gas flow is increased (**12** L/min), the absorbed tensile strength is increased while at low gas flow (**8** L/min) the tensile strength was decreased.

## 3.3. Solidification and Microstructure Study

### Solidification in the Fusion Zone

A review of solidification cracking in austenitic stainless steel welds shows that the problem is more prevalent in fully austenitic and stabilized stainless steels. Solidification mode is a major determinant of cracking susceptibility; ensuring an FA or F mode ensures the best resistance to cracking. An important aspect of weld solidification is the effect of solidification kinetics on the phases formed. The eutectic reaction $L \leftrightarrow \sigma + \gamma$ in stainless steels is not typical of the classical eutectic in the sense that the composition difference between $\sigma$ and $\gamma$ phases is minor. The relatively minor difference in thermodynamic stability of the two phases in the vicinity of the eutectic permits non-equilibrium solidification to a metastable phase obtained by extrapolation of the equilibrium phase boundaries [11]. Thus, the weld microstructure in stainless steels depends significantly on kinetic factors such as cooling rate and epitaxy, apart from equilibrium stability considerations. Under the rapid cooling and fast growth rates aided by epitaxy, the weld structure could solidify far away from equilibrium [12].

As Pb is insoluble in molten steel, and austenite/ferrite has a low capacity for dissolving S and P, all of these elements are vigorously segregated in the liquid during solidification. The resulting high impurity concentrations in the last liquid to solidify in the interdendritic regions have much lower melting points than those of the primary solidifying phase. The melting point of Pb is only 327°C - 502°C and the melting point of the sulphides (MnS, FeS, CrS) is about 1100°C - 1200°C, *i.e.* much lower than that of Fe (1538°C). If sufficiently high stresses are generated before final solidification, the boundaries with segregated Pb and sulfides may separate to form solidification cracks in the fusion zone, which providentially was not observed in this experimental study. **Figure 11(a)**, **Figure 11(b)** shows the base metal (BM)/HAZ of AISI316.

## 4. Microstructure in the HAZ

Owing to the epitaxial nature of solidification, the grain boundary in the HAZ can link up with the solidification grain boundary in the fusion zone. Segregation of S, Pb, Mn and P during solidification means that these elements are able to diffuse into the HAZ from the fusion zone along the grain boundaries. The impurities and dissolved elements diffuse more rapidly along the grain boundaries than through the crystal lattice, which results in a local depression of the melting temperature. As a consequence, the grain boundary may melt during welding thermal cycles, but the local stress is insufficiently high to impose the melted grain boundary to separate in the HAZ. **Figure 12** exhibits redistribution of elements in the melted zone and HAZ of a butt weld joining AISI316.

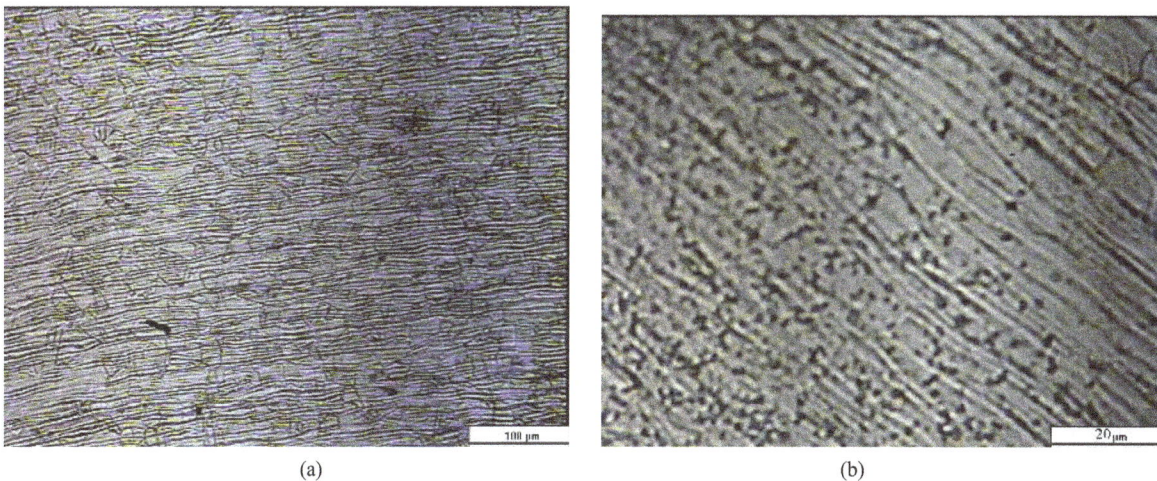

(a)                                                    (b)

**Figure 11.** (a): SST 316 base metal. (b): SST 316 HAZ.

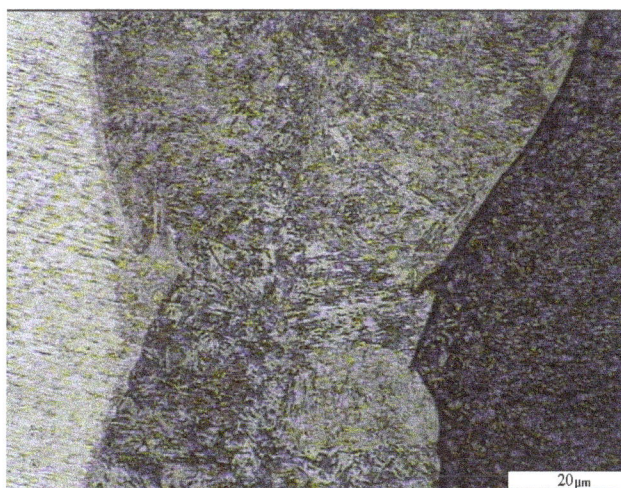

**Figure 12.** The redistribution of elements in the fusion zone of a butt weld joining AISI316.

No cold cracking was observed during the study. The cooling rate which is estimated to be very high is responsible for the martensite structure formation. The martensitic structure in WZ and HAZ is the main reason for controlling the mechanical properties of the welded joints. The martensitic structure improves tensile strength but harms impact strength.

## 5. Conclusions

Referring to the previous discussion and the results obtained the following are concluded:

**1)** The welding parameters studied in this manuscript had high effect in the quality of welding and mechanical properties of the joints.

**2)** The electrical current welding parameter had a greatest effect in the weld quality. It is inversely proportional with tensile strength and directly proportional with impact strength.

**3)** The increasing electrical current resulted in the increasing impact energy and toughness but also caused decrease for tensile strength and ductility. But the increasing gas flow rate resulted in increasing impact and tensile strength which was an indication of improvement of toughness and ductility properties.

**4)** The gas flow rate had strong effective on weld quality of austenitic stainless steel 316. It is directly proportional with impact and tensile strength.

## Acknowledgements

Benghazi University, Faculty of Engineering (Industrial and Mechanical Departments) is gratefully acknowledged for the support of this research. Technical support from General Pipe Company is also gratefully acknowledged.

## References

[1]   Suresh Kumar, L., Verma, S.M., Radhakrishna Prasad, P., Kiran Kumar P. and Siva Shanker, T. (2011) Experimental Investigation for Welding Aspects of AISI 304 & 316 by Taguchi Technique for the Process of TIG & MIG Welding. *International Journal of Engineering Trends and Technology*, **2**, 28-33.

[2]   Murat, V. and Ahmet, A. (2004) On the Resistance Spot Weldability of Galvanized Interstitial Free Steel Sheets with Austenitic Stainless Steel Sheets. *Journal of Materials Processing Technology*, **153-154**, 1-6. http://dx.doi.org/10.1016/j.jmatprotec.2004.04.063

[3]   Emin, B., Dominique, K. and Marc, G. (2004) Application of Impact Tensile Testing to Spot Welded Sheets. *Journal of Materials Processing Technology*, **153-154**, 80-86. http://dx.doi.org/10.1016/j.jmatprotec.2004.04.020

[4]   Aslanlar, S. (2006) The Effect of Nucleus Size on Mechanical Properties in Electrical Resistance Spot Welding of Sheets Used in Automotive Industry. *Materials and Design*, **27**, 125-131. http://dx.doi.org/10.1016/j.matdes.2004.09.025

[5]   Anawa, E.M. and Olabi, A.G. (2006) Effects of Laser Welding Conditions on Toughness of Dissimilar Welded Components. *Applied Mechanics and Materials*, **5-6**, 375-380. http://dx.doi.org/10.4028/www.scientific.net/AMM.5-6.375

[6]   Bouyousfi, B., Sahraoui, T., Guessasma, S. and Tahar Chaouch, K. (2007) Effect of Process Parameters on the Physical Characteristics of Spot Weld Joints. *Materials and Design*, **28**, 414-419. http://dx.doi.org/10.1016/j.matdes.2005.09.020

[7]   Nizamettin, K. (2007) The Influence of Welding Parameters on the Joint Strength of Resistance Spot-Welded Titanium Sheets. *Materials and Design*, **28**, 420-427.

[8]   Shankar, V., Gill, T., Mannan, S. and Sundaresan, S. (2003) Solidification Cracking in Austenitic Stainless Steel Welds. *Sadhana*, **28**, 359-382.

[9]   Anawa, E.M. and Olabi, A.G., Anawa, E.M. and Olabi, A.G. (2008) Optimization of Tensile Strength of Ferritic/Austenitic Laser-Welded Components. *Optics and Lasers in Engineering*, **46**, 571-577. http://dx.doi.org/10.1016/j.optlaseng.2008.04.014

[10]  Hassan, E.M. (2008) Feasibility and Optimization of Dissimilar Laser Welding Components. Ph.D. Thesis, Dublin City University, Dublin.

[11]  Kelly, T.F., Cohen, M. and Vandersande, J.B. (1984) Rapid Solidification of a Droplet-Processed Stainless Steel. *Metallurgical Transactions A*, **15**, 819-833. http://dx.doi.org/10.1007/BF02644556

[12]  Balluffi, R. (1982) Grain Boundary Diffusion Mechanism IN Metal's. *Metallurgical Transactions A*, **13A**, 2069-2095. http://dx.doi.org/10.1007/BF02648378

# 6

# Study of Deformation Coating for Sheets by Using Tensile Test

**Milan Dvořák[1], Emil Schwarzer[1], Miloš Klíma[2]**

[1]BUT, Faculty of Mechanical Engineering, IMT, Brno, Czech Republic
[2]MU, Faculty of Science, PF MU, Brno, Czech Republic
Email: dvorak.m@fme.vutbr.cz, emil.schwarzer@seznam.cz, klima@sci.muni.cz

## Abstract

This article focuses on the study of the defined values of tensile strain and the effect of low temperature plasma adhesion selected coatings on steel samples using a tensile testing flat test bars. Samples were made by machining and welding technologies. The flat test bars were tested by pulling on a test rig UPC 1200. Part of the samples was treated on the surface prior to coating by a tensile test, second base coat and with a final coat continuous multi plasma system. The selected test samples were determined from the tensile test of the material characteristics apparent from the tensile diagrams. The examined samples were fitted top and base coat. Another group was the KTL basis. The presented graphs show the dependence of the strength on elongation of a sample according to DIN EN ISO 6892-2. The samples were then examined under a stereo microscope SCHUT brand, type SSM-E in the laboratory to conduct coating on a steel sheet at the moment of total violation sectional samples. The base layer, in which the temperature ranges from 160°C - 180°C, was applied by electrophoresis method.

## Keywords

Tensile Test, Sheet with Coating, Adhesion of Coating, Multi-Jet Plasma System

## 1. Introduction

Distortion between the base steel material and its coating, causing defects occurring during sheet forming, such as sheet surface waviness, warpage, breaking, metal release, inadequate structure of the surface or poor choice of combinations of basic materials and coatings [1] [2]. Current requirements for surface treatment technology is to preserve the integrity of the surface layer and bonds between the surface and the base material after the drawing respectively bending [3].

Protective coatings play an irreplaceable role at a very wide field of construction materials used [4]-[6]. In the area of testing and assessment of adhesion of protective coatings for sheets is characterized by great variety of methods and procedures [7]-[10].

Were investigated values of the material characteristics such as: such as e.g. the yield strength $R_{p0,2}$, tensile strength $R_m$ is, an elongation $A_{50}$, $A_g$ apparent from the tensile diagrams. The thickness of the base coat at the sample ranges from 15 - 30 microns. Sheets supplied with a certificate validated and chemical composition have been previously degreased and then scrubbed with a special deburring machine. Designation of material is 1.0322 (DX56D). This is unalloyed quality deep-drawing steel thickness of 1 mm. In **Table 1** results of adhesion, samples were sorted from best to worst.

## 2. Experiments

The aim of the experiment was submission of selected specimens with the proposed dimensions (**Figure 1**) tensile test. Tensile test of samples was performed on a UPC 1200.

Samples were technologically prepared with using by welding technology with support $CO_2$ and dimensionally adapted for tensile test. Selected samples before tensile test were photographically documented, see **Figure 2**.

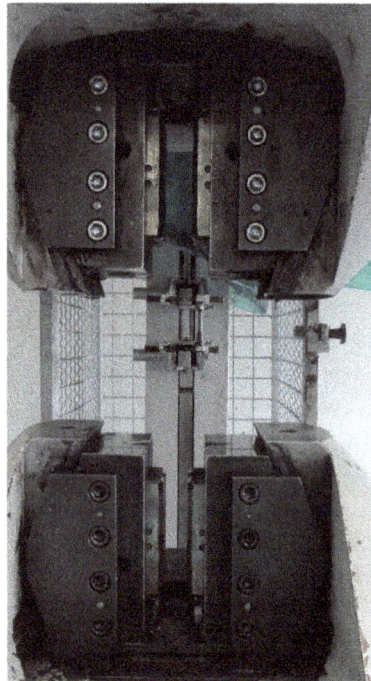

**Figure 1.** Detailed view on the clamped sample in clamping jaws device.

**Table 1.** Selected samples with plasma treatment and applying different plasma.

| Samples with plasma treatment | | |
|---|---|---|
| BP11 | $A_r + N_2$ | (BP + PL + Z) |
| BP16 | $A_r + H_2O$ | (BP + PL + Z) |
| BP7 | $A_r + O_2$ | (BP + PL + Z) |
| TC6 | $A_r$ | (TC + PL) |
| TC9 | $A_r + O_2$ | (TC + PL) |
| Z12 | $A_r + N_2$ | (Z + PL + TC) |

Before own experiments were divided to four groups with different modes of composition of the plasma.

# 3. Results

## 3.1. Measured Values for Selected Samples from the Tensile Tests

From tensile tests were processed results of experimental work by a tensile test in the form of tabular and graphical outputs, see **Table 2**.

The test specimens, see **Figure 2** and the width [mm] and thickness b [mm] was subjected to a tensile test. The results are shown in **Table 2**.

The most variable results for Young's modulus E [MPa]. Why is this large margin given, has not been found and neither was the subject of this experiment.

Values contractual yield strength $R_p$ [MPa] and the tensile strength $R_m$ [MPa] correspond to the material 1.0322 (DX56D). Like the $A_{50}$ elongation and contraction Z [%]. The plastic elongation $A_g$ ranged from 9.98 to 21.85 [%].

It was interesting to observe the time to break the sample t [s], which is also significantly different from the method of application of various plasma [11].

Also, the force F [kN] for all samples was very similar with the exception for sample BP16, which was significantly lower and the voltage for rise of the separated parts of the sample $R_{m1}$ [MPa] was at sample BP16

**Figure 2.** Overview of samples which were subjected to tensile test.

**Table 2.** Measured values for selected samples after tensile test.

| Samples | a [mm] | b [mm] | E [MPa] | $R_p$ [MPa] | $R_m$ [MPa] | $A_g$ [%] | $A_{50}$ [%] | Z [%] | t [s] | s [mm] | F [kN] | dL [mm] | $R_{m1}$ [MPa] |
|---|---|---|---|---|---|---|---|---|---|---|---|---|---|
| **BP11** | 13.10 | 1.0 | 68.52 | 129.71 | 259 | 21.85 | 42.18 | 43.01 | 31.44 | 17.11 | 3.27 | 14.66 | 249.71 |
| **BP16** | 12.70 | 1.0 | 74.79 | 135.87 | 258 | 11.93 | 37.70 | 34.19 | 35.91 | 17.21 | 1.49 | 10.35 | 117.45 |
| **BP7** | 13.20 | 1.0 | 22.02 | 143.91 | 253 | 9.98 | 29.44 | 39.20 | 37.23 | 19.12 | 2.74 | 10.15 | 207.82 |
| **TC6** | 13.80 | 1.0 | 38.38 | 136.70 | 256 | 15.44 | 42.04 | 44.18 | 20.05 | 11.96 | 3.45 | 9.26 | 249.65 |
| **TC9** | 12.38 | 1.0 | 68.91 | 138.32 | 263 | 18.22 | 27.37 | 42.48 | 18.21 | 9.74 | 3.19 | 7.28 | 257.74 |
| **Z12** | 13.20 | 1.0 | 64.51 | 147.45 | 265 | 14.73 | 32.23 | 43.0 | 18.09 | 7.72 | 3.42 | 5.76 | 259.0 |

different and significantly lower.

Track s [mm] and extension dL [mm] were evaluated by a computer program on the device UPC 1200.

## 3.2. Graphic Results from the Tensile Test

Results of tensile test are elaborated in **Table 2** and results of tension are shown in **Figures 3-8**.

All these samples were fixed, as shown in **Figure 9**, where the fixed sample BP7.

## 3.3. Selected Samples after Tensile Test under the Microscope

Samples BP11, BP16, BP7, TC6, TC9, and Z12 were examined under a stereo-microscope under magnification

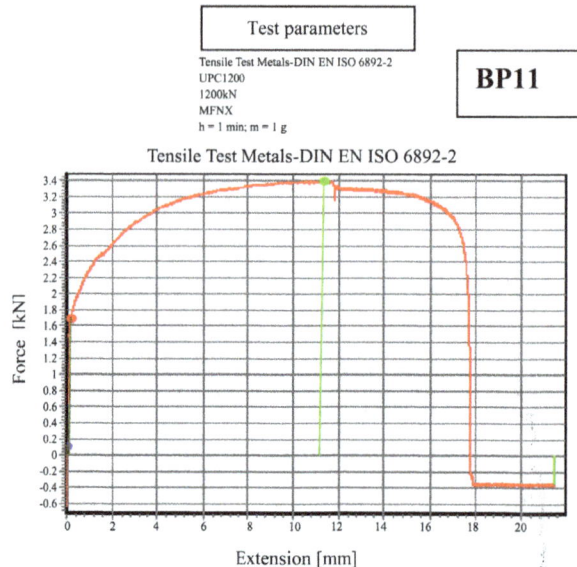

**Figure 3.** Progress of force versus elongation and values of the tensile test, the sample BP11.

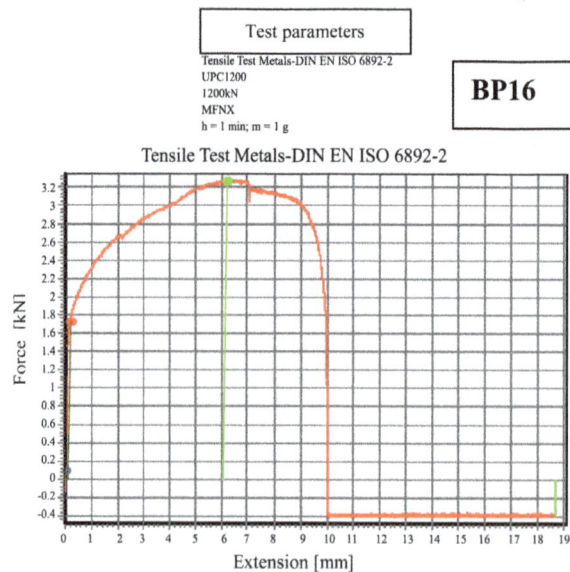

**Figure 4.** Progress of force versus elongation and values of the tensile test, the sample BP16.

Test parameters

Tensile Test Metals-DIN EN ISO 6892-2
UPC1200
1200kN
MFNX
h = 1 min; m = 1 g

BP7

Tensile Test Metals-DIN EN ISO 6892-2

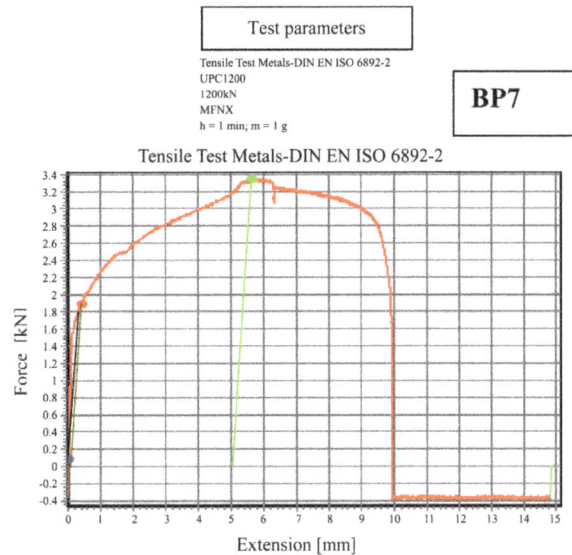

**Figure 5.** Progress of force versus elongation and values of the tensile test, the sample BP7.

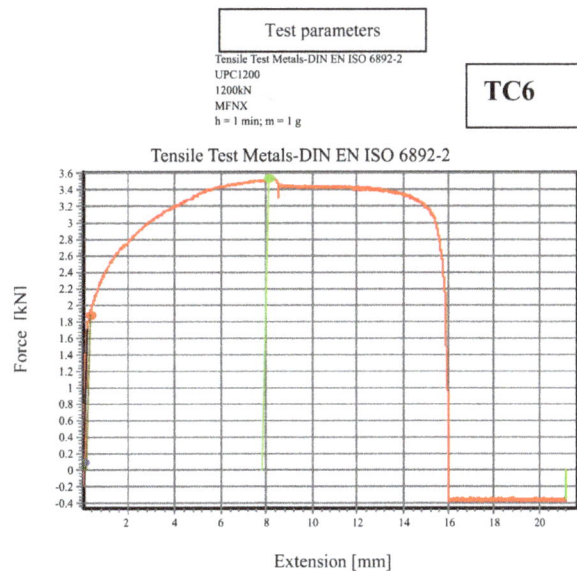

Test parameters

Tensile Test Metals-DIN EN ISO 6892-2
UPC1200
1200kN
MFNX
h = 1 min; m = 1 g

TC6

Tensile Test Metals-DIN EN ISO 6892-2

**Figure 6.** Progress of force versus elongation and values of the tensile test, the sample TC6.

10×. These scales in **Figures 10-15** are in mm.

The samples were examined under a stereo microscope SCHUT brand, type SSM-E in the laboratory of the Department of Engineering Technology BUT in Brno to conduct coating on a steel sheet at the moment of total violation sectional samples.

## 3.4. The Principle of High-Frequency Hollow Cathode

The basis of nozzles was used dielectric capillary of quartz glass, which flows argon, including any impurities. The plasma ejected from the cavity and from the plasma nozzle orifice into the external environment, where it acts on the test specimens coated steel sheet. Discharge is over the whole length of its actively generated plasma channel.

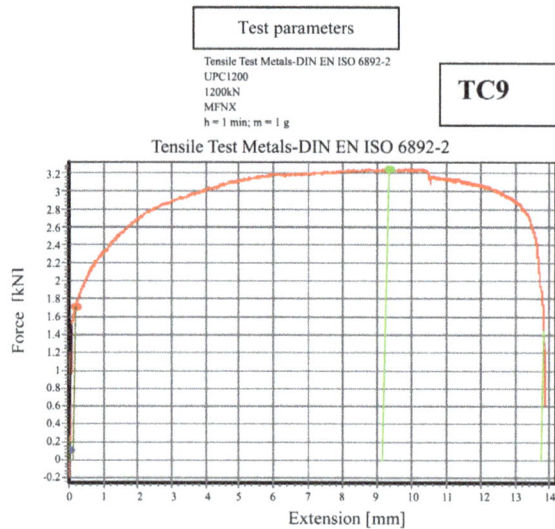

**Figure 7.** Progress of force versus elongation and values of the tensile test, the sample TC9.

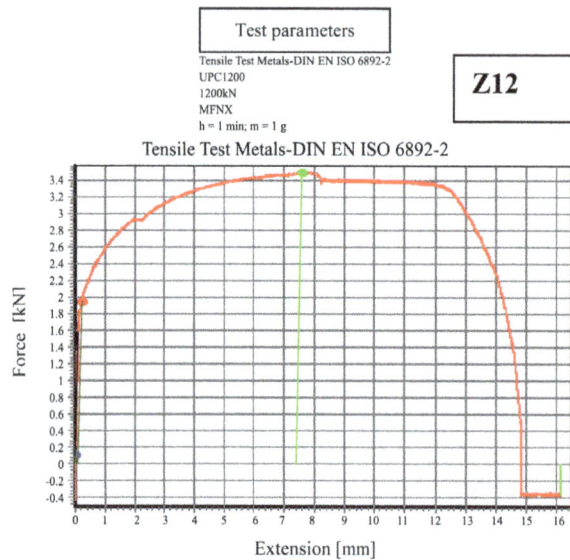

**Figure 8.** Progress of force versus elongation and values of the tensile test, the sample Z12.

Power absorbed at plasma channel multi-jet device was used according to the selected working conditions in the range of $10^2$ - $10^3$ W·cm$^{-2}$. Unlike electron beam welding where the power density at the spot welding up to $10^9$ W·cm$^{-2}$ [12].

The thermal effects on the surface of samples can be range from 30°C to 1600°C while maintaining substantially no-isometric character of discharge (energetic particles at a temperature up to 10,000 K). Based on these characteristics of the plasma nozzle system provides a high reaction mixture with a high efficiency for the chemical and physical modification of the material surface [13].

Suitable array of nozzles to linear or other units allows cutting of larger areas of test samples respectively semi-finished steel in industrial practice.

## 3.5. Composition of the Material Used in KTL

KTL material contains positively charged paint particles (the largest particles leftmost in **Figure 1**) which have

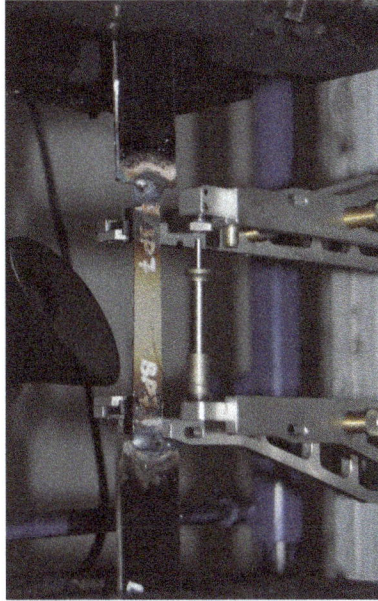

**Figure 9.** Demonstration fixing of sample BP7 for experiment.

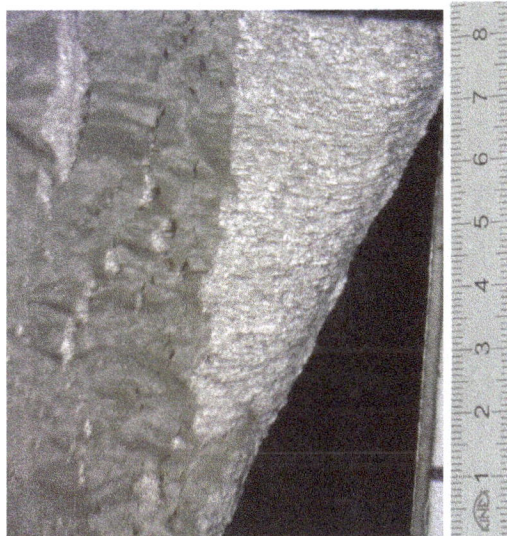

**Figure 10.** Sample BP11.

incorporated therein a pigment and a binder (synthetic resin). Another component (see the right side of **Figure 16**) is negatively charged residues org. $COO^-$ acids (e.g. acetic acid anions). It is essentially the electro-neutral colloidal solution, wherein the colloidal particles on the resin $(CH_3)$ a large positive charge wrapping anions are simple organic acids. The resin is a high molecular weight tertiary amine which in an acidic aqueous environment creates the following structure.

Addition of the organic acid thus formed colloidal solution whose resin part is charged positively and the acidic portion is negatively charged [4]. Thus formed bath after connecting a DC current to the cathode (part of the body to which you want to apply KTL coating) and firmly installed anode enables electro-chemical processes leading to the exclusion of KTL layer.

## 3.6. Treatment of the Surface by Plasma Jets

At the base steel sheet and the coat should be seen as a system. When technological forming operations occur,

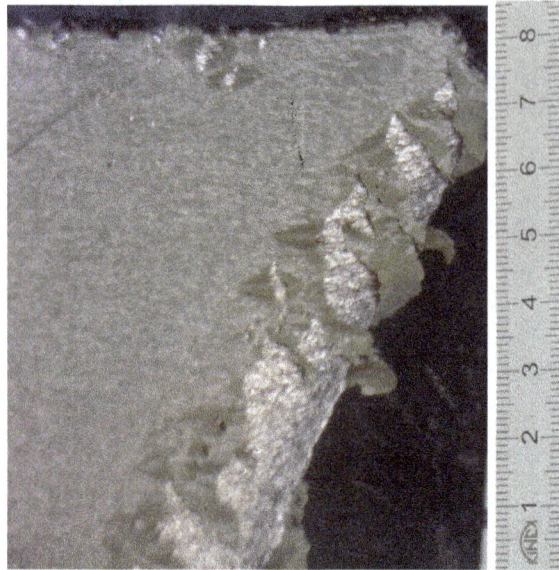

**Figure 11.** Sample BP16.

**Figure 12.** Sample BP7.

however, in distortion of these pairs of common bonds [14] [15]. Disruption of these bonds may in some cases lead to disruption of the integrity of the coating and its subsequent flaking.

Distortion between the base steel material and its coating can cause defects occurring during sheet forming (metal surface waviness, warpage, breaking, metal release, inadequate structure of the surface) or a combination of a bad choice base material and coating.

Goal is thus to maintain the integrity of the coating, *i.e.* all bonds between the coating and the base material [5] [6]. This can be achieved by optimizing the forming operations, *i.e.* optimization of process parameters and further improving the coating technology [16]-[18].

Plasma-chemical equipment with conveyor is on **Figure 17**.

## 4. Conclusion

This paper presents another basic criterion tensile test to detect adhesion of organic coatings in one or more layers

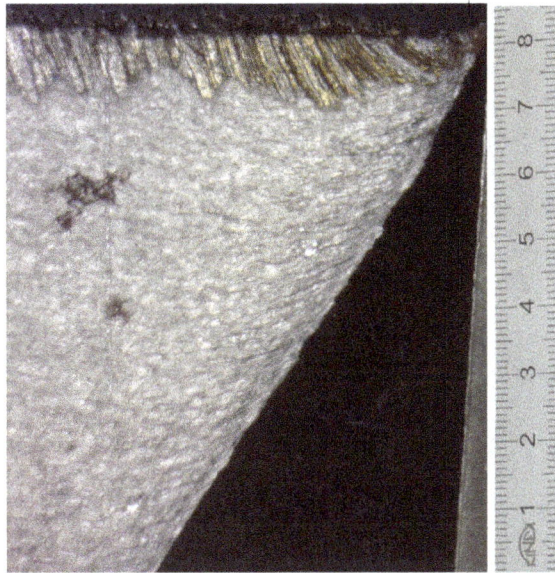

**Figure 13.** Sample TC6.

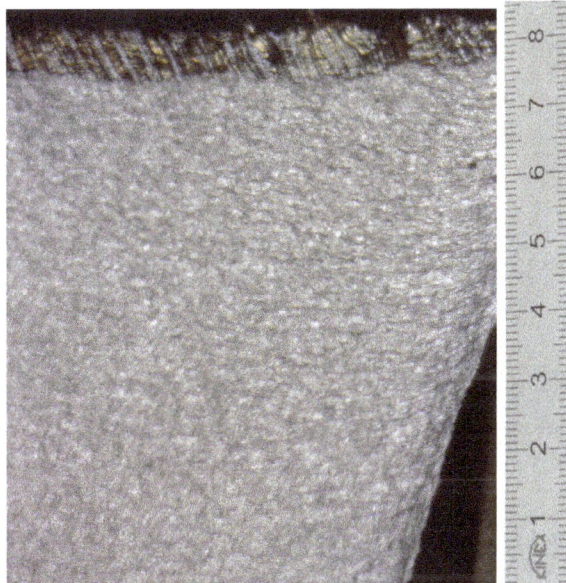

**Figure 14.** Sample TC9.

in the interaction with the metal base.

Experiments have shown the influence of the tensile stress on adhesion coatings. Functional cataphortic coloring bath consists of the following basic components: resin (binder), paste (pigment), which determines the color shade.

Common colors are gray, black eventually white or beige. Further additives (solvents, pH regulator). All these components are homogeneously mixed in demi-water.

By employing various types of gases in the multi-jet plasma system on selected samples of steel with organic coatings were found to what combination of gases is most suitable in terms of adhesion and of course where and in which the interface is coated using multi-jet plasma system optimum.

Experimental results show that using a multi-jet plasma system with optimal composition of low temperature plasma samples are more resistant from the viewpoint of adhesion of the selected organic coating selected flat steel samples structural steel, nominal thickness 1 mm.

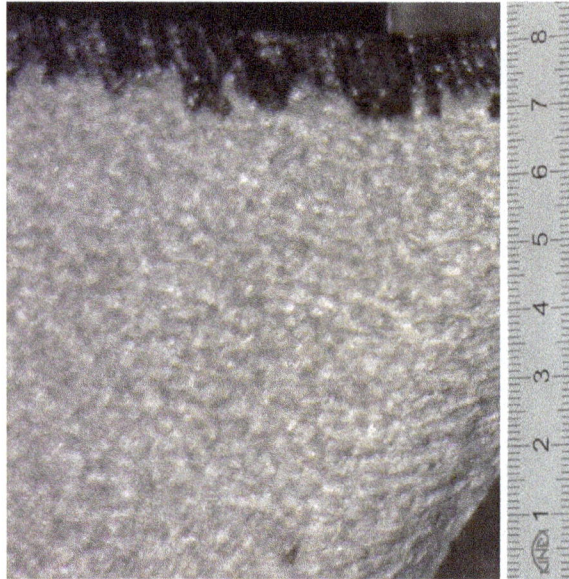

**Figure 15.** Sample Z12.

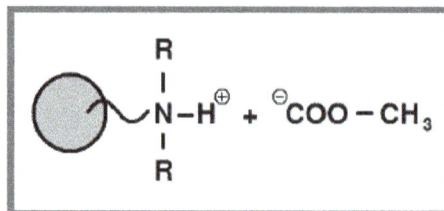

**Figure 16.** Molecular structure of colloid solution.

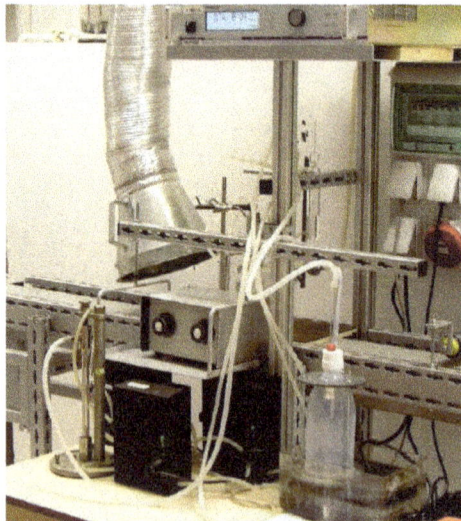

**Figure 17.** Plasma-chemical equipment with conveyor, laboratory MU Brno.

Experiments with selected samples were evaluated with the result that the sample should the best BP11 resistant coating from the viewpoint of adhesion. This coating was treated with the following procedure application:

The sheet material 1.0322 (DX56D), which was properly degreased, subsequently the sheet surface was treated by multi-jet plasma system $A_r + N_2$ mixtures, that have proven to improve adhesion of a tensile test and subsequently applied the base coat (KTL).

## Note

The article is supported by project Technical University in Brno, Faculty of Mechanical Engineering: BUT FME-S-12-5 from 2012 and VAV 13313.

## References

[1] Hermann, F. and Schiller, M. (2007) Testing of Paints and Protective Coatings. Pardubice-Green Suburb: SYNPO PLC.

[2] Test of Resistance of Coatings (2004) Measuring Equipment, Thickness, Hardness, Gloss Meters, Thermometers. Pro-Inex Instruments, Ltd., Ostrava.

[3] Dvořák, M., et al. (2001) Technology II. Academic Publishing CERM, Ltd., Brno, 236-238.

[4] ČSN EN ISO 1519 (2002) Paints Substance—Bend Test (Cylindrical Mandrel). Czech Standards Institute.

[5] ČSN EN ISO 7438 (2005) Metallic Materials Bend Test. Czech Standards Institute, 11-12.

[6] ČSN ISO 24213 (2009) Metallic Materials—Sheets and Belt: Evaluation Method of Suspension for Flexural Bending. Czech Standards Institute, 14-15.

[7] Blanks, T. (1985) Metals Handbook: Mechanical Testing. American Society for Metals, 837-842.

[8] Čada, R. (1998) Surface Formability of Metallic Materials. Technical University of Ostrava, Ostrava, 90-92.

[9] ČSN EN ISO 20482 (2004) Metallic Materials-Sheet and Belts-Bulge Tests According to Erichsen. Czech Standards Institute, Prague.

[10] ČSN EN 13144 (2003) Metallic and Other Inorganic Coatings: Method for Quantitative Measurement of Adhesion for Tensile Test. Czech Standards Institute, 12-13.

[11] Dvořák, M. and Schwarzer, E. (2013) Study of Formability of Coated Sheets from the Plasma Chemical Pretreatment of Surfaces. International Journal of Engineering and Innovative Technology (IJEIT), 3, 356-360.

[12] Patent EP 1077021, US 6,525,481 (2005) Method of Making a Physically and Chemically active Environment by Means of Plasma Jet and the Related Plasma JET. Masaryk University, Brno, 5-6.

[13] Krejčík, V. (1988) Surface Treatment of Metals II. Publishing House of Technical Literature, Prague.

[14] Kraus, V. (2000) Surface Modification. University of West Bohemia, Plzeň, 218-220.

[15] Kreibich, V. (1996) Theory and Technology of Surface Treatment. Publishing House of CVUT, Prague, 89-92.

[16] Forejt, M. and Píška, M. (2006) Theory of Machining, Molding and Tool. CERM, Academic Publishing, Ltd., Brno, 225-226.

[17] Hušek, M. and Dvořák, M. (2010) Test of Adherence Multifunctional Coating on the Sheet Using a Graduated Bending Jig. Engineering Technology, 15, 15-20.

[18] Dvořák, M. and Schwarzer, E. (2012) New Methods Testing of Adhesion of the Coating to Sheet Metal by Bending. Journal of Surface Engineered Materials and Advanced Technology (JSEMAT), 2, 61-64.

# 7

# Optimization of Gas Metal Arc Welding Process Parameters Using Standard Deviation (SDV) and Multi-Objective Optimization on the Basis of Ratio Analysis (MOORA)

**Joseph Achebo[1]\*, William Ejenavi Odinikuku[2]**

[1]Department of Production Engineering, University of Benin, Benin City, Nigeria
[2]Department of Mechanical Engineering, Petroleum Training Institute, Effurun, Nigeria
Email: \*josephachebo@yahoo.co.uk

## Abstract

Welding technology is very vital for the industrial development and technological advancement of any country. In this regard achieving good quality machine manufactured products cannot be over emphasized. Since welding is a very reliable method of joining metals together permanently, several methodologies have been adopted to improve the quality of weldments, such as the neural network, fuzzy logic, surface response methodology, full factorial method, and so on. In this case, the multi-objective optimization on the basis of ratio analysis (MOORA) is applied. MOORA is used to solve multi-criteria (objective) optimization problem in welding. MOORA in combination with standard deviation (SDV) was used for the optimization process. SDV was used to determine the weights that were used for normalizing the responses obtained from the mechanical test results. From applying the SDV-MOORA method, it was found that welding current of 350 A, welding voltage of 22 V, an electrode diameter of 3.2 mm and welding speed of 100 mm/s produced the weldment with the best mechanical properties. The mechanical properties compare very well with those obtained from other literature. It is, therefore, concluded that the SDV-MOORA method has successfully optimized the welding process parameters used in this study.

## Keywords

SDV, MOORA, Welding Process, Bead Geometry, Mechanical Property

---

\*Corresponding author.

# 1. Introduction

The failure of structural materials especially steel is of great concern globally. In Nigeria, steel pipes used for the transportation of water, oil, gas etc. are joined together by welding. Welding is designed to permanently join pieces of materials producing weldments which significantly enhance the rigidity, stability, reliability, and integrity of structural materials. If the quality of the weldment is poor, the weldment will fail either by breaking off due to its brittle nature or by corrosion. It has been proven by several researchers that the choice of welding input process parameters can alter the quality of the weldment. Therefore, optimizing these process parameters to obtain the best weld quality and multi-response properties cannot be over emphasized. The choice of the appropriate optimization tool is an ongoing research process. Researchers are using different methods to obtain the most economic input process parameters. Today, there is no known particular optimization method used for optimizing these input process parameters. Instead, several optimization methods are used and the one that has produced the most acceptable results is selected. MOORA is one of such techniques that have not been fully utilized in optimizing welding input process parameters and multi-response properties.

The international engineering and welding community is keenly interested in investing in research and development geared towards finding optimized methods for obtaining weldments of acceptable quality. Several expert methods such as artificial neural network, fuzzy logic, finite element, genetic algorithm etc. have been applied for optimizing process parameters, since it has been found that applying the most appropriate process parameters has a huge impact on the eventual quality of each weldment. It has become imperative that new methods should be explored for the purpose of obtaining better weldments to meet even more specific engineering requirements.

In this study, the Multi-Objective Optimization on the basis of Ratio Analysis (MOORA) is used to optimize the process parameters. Görener et al. [1] were of the opinion that the MOORA method was first used by Brauers [2]. The MOORA method is a relatively new multi-criteria decision-making method which is based on the ratio system as well as dimemsionless measurement [3] [4].

Gadakh et al. [5] were of the opinion that multi-objective optimization (or programming), also known as multi criteria or multi-attribute optimization, was the process of simultaneously optimizing two or more conflicting attributes (objectives) subject to certain constraints. Hwang and Yoon [6] said that multi-criteria decision-making was applied to decisions among available classified alternatives by multiple attributes. Mandal and Sarkar [7] wrote that MOORA was the process of simultaneously optimizing two or more conflicting attributes (objectives) subject to certain constraints.

Stanujkic et al. [8] were of the opinion that multiple-criteria decision-making (MCDM) could be generally described as the process of selecting one from a set of available alternatives, or ranking alternatives, based on a set of criteria, which usually had a different significance. The multi-objective optimization by ratio analysis (MOORA) which was a part of MCDM was first introduced by Brauers and Zavadskas [9].

Some researchers have used MOORA for solving product or system optimization problems. Chakraborty [10] applied the MOORA method for decision-making in a manufacturing environment. Karande and Chakraborty [11] applied the MOORA method for the selection of materials. Chaturvedi and Sharma [12] optimized CNC wire cut EDM for OHNS steel using MOORA methodology. Görener et al. [1] applied the MOORA method for selecting where a bank should be located. Gadakh et al. [5] optimized welding process parameters using the MOORA method.

Brauers and Zavadskas [13] expanded the scope of MOORA to be known as MULTI-MOORA. Ozcelik et al. [14] in their paper wrote that Brauers and Zavadskas [13] developed the equation for the full multiplicative form of MOORA known as MULTI-MOORA method. Farzamnia and Babolghani [15] applied the group decision making process for material supplier selection in a supply chain using MULTI-MOORA technique under fuzzy environment.

In this study, the SDV-MOORA method was used to optimize the welding process parameters used for gas metal arc welding of mild steel plates. SDV was the standard deviation method used for determining the weight attached to each mechanical property. The results obtained from the optimized process parameters would be compared with those obtained using other optimization methods.

# 2. Materials and Methods

## 2.1. Materials

Five large weld deposits were made for each application of the sixteen process parameters on 4 mm mild steel plates. Each of these weld deposits are sectioned into three parts. One part was used to determine the Bead geometry (see **Figure 1**), the other part was used to conduct the Charpy V-Notch (CVN) Impact test. The CVN specimen is shown in **Figure 2**, while the remaining part was machined into the tensile specimen (see **Figure 3**) for conducting the tensile test.

   Five tensile specimens are prepared using a CNC Lathe machine. Tensile tests are carried out in 100 kN computer controlled Universal Testing Machine as used by Prasad *et al.* [16]. The specimens were loaded at a rate of 1.8 kN/min as per ASTM specifications, so that these tensile specimens can undergo the deformation process. From the stress strain curve obtained, the ultimate tensile strength (UTS) of the weld joints is evaluated and the average of the five test results was recorded.

## 2.2. Method

The method adopted by El-Santawy and Ahmed [18] was used in this study. The methodology is expressed as contained herein under:

### 2.2.1. Weight Allocation via Standard Deviation

Standard deviation is applied to this study for unbiased allocation of weights. The importance of weights in solving Multi-Criteria Decision Making (MCDM) problems cannot be over emphasized. To determine the standard deviation, the range standardization was done using Equation (1) to transform different scales and units among various criteria into common measurable units in order to compute their weights.

$$X'_{ij} = \frac{X_{ij} - \min_{1<j<n} X_{ij}}{\max_{1<j<n} X_{ij} - \min_{1<j<n} X_{ij}} \tag{1}$$

where $\max X_{ij}$, $\min X_{ij}$ are the maximum and minimum values of the criterion ($j$) respectively. The Standard deviation ($SDV$) is calculated for every criterion using Equation (2)

$$SDV_j = \sqrt{\frac{1}{m}\sum_{i=1}^{m}\left(X_{ij} - \overline{X'_j}\right)^2} \tag{2}$$

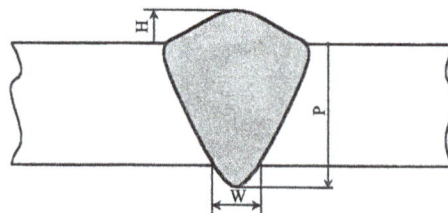

**Figure 1.** Bead geometry [17].

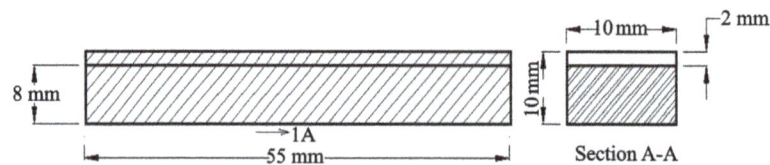

**Figure 2.** Schematic diagram of charpy V notch impact test specimen.

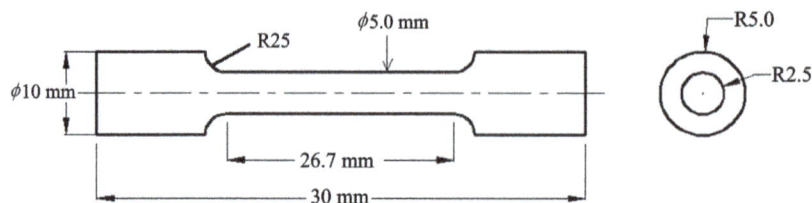

**Figure 3.** Schematic diagram of tensile specimen.

where $\overline{X'_j}$ is the mean of the values of the $j^{th}$ criterion after normalization and $j = 1, 2, \cdots, n$. After calculating for $SDV$ for all criteria, the next step is to determine the weights, $W_j$ of all the criteria considered.

$$W_j = \frac{SDV_J}{\sum\limits_{j=1}^{n} SDV_J} \tag{3}$$

where $j = 1, 2, \cdots, n$.

## 2.2.2. Application of MOORA

The Multi-Objective Optimization on the basis of Ratio Analysis (MOORA) method starts with a decision matrix as expressed by Equation (4)

$$D = \begin{array}{c} \\ A_1 \\ A_2 \\ A_3 \\ \vdots \\ A_n \end{array} \begin{vmatrix} C_1 & C_2 & C_3 & \cdots & C_n \\ X_{11} & X_{12} & X_{13} & \cdots & X_{1n} \\ X_{21} & X_{22} & X_{23} & \cdots & X_{2n} \\ X_{31} & X_{32} & X_{33} & \cdots & X_{3n} \\ \vdots & \vdots & \vdots & \vdots & \ddots & \vdots \\ X_{m1} & X_{m2} & X_{m3} & \cdots & X_{mn} \end{vmatrix} \tag{4}$$

The procedure for using MOORA for ranking alternatives is described here under;
Step 1: Compute the normalized decision matrix by vector method as defined by Equation (5)

$$X'_{ij} = \frac{X_{ij}}{\sqrt{\sum\limits_{i=1}^{m} X_{ij}^2}} \tag{5}$$

where $i = 1, \cdots, m$; $j = 1, \cdots, n$
Step 2: Calculate the composite score as expressed in Equation (6)

$$Z_i = \sum\limits_{j=1}^{b} X'_{ij} - \sum\limits_{j=b+1}^{n} X'_{ij}; \text{ where } i = 1, \cdots, m \tag{6}$$

where $\sum\limits_{j=1}^{b} X'_{ij}$ and $\sum\limits_{j=1}^{n} X'_{ij}$ are the benefit and non benefit (cost) criteria, respectively. If there are some attributes more important than the others, the composite score becomes as expressed in Equation (7)

$$Z_i = \sum\limits_{j=1}^{b} W_j X'_{ij} - \sum\limits_{j=b+1}^{n} W_j X'_{ij}, i = 1, \cdots, m \tag{7}$$

where $W_j$ is the weight of $j^{th}$ criterion
Step 3: Rank the alternatives in descending order.
**Figure 4** shows the sequence of operations that was performed in the framework of multicriteria decision support system

## 3. Results and Discussion

### 3.1. Results

The matrix design used to prepare the layout for the welding procedure is presented in **Table 1**.
**Table 2** shows the welding process parameters and their levels
**Table 3** shows the decision matrix used for categorizing the weld mechanical properties and bead geometry
Where UTS is the Ultimate Tensile Strength, CVN is the Charpy V-Notch Impact Energy, BP is the Bead Penetration, BH is the Bead Height and BW is the Bead Width.

### 3.1.1. Weight Allocation

In this study, the weight allocation for each of the output parameters, that is, the weld mechanical properties and the bead geometry were determined. In determining the weight, the range of standardized decision matrix is determined using Equation (1). **Table 4** shows the summary of the range of standardized decision matrix.

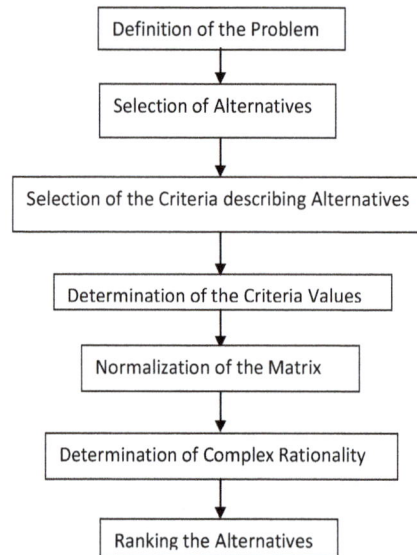

**Figure 4.** Sequence of operations performed in the framework of multi-criteria decision support system [19].

**Table 1.** Matrix design.

|   | A | B | C | D |
|---|---|---|---|---|
| 0 | 1 | 2 | 3 | 4 |
| + | + | - | - | - |
| + | + | + | - | - |
| + | + | + | + | - |
| + | + | + | + | + |
| + | - | + | + | + |
| + | + | - | + | + |
| + | - | + | - | + |
| + | + | - | + | - |
| + | + | + | - | + |
| + | - | + | + | - |
| + | - | - | + | + |
| + | + | - | - | + |
| + | - | + | - | - |
| + | - | - | + | - |
| + | - | - | - | + |
| + | - | - | - | - |

**Table 2.** Process parameters and their levels.

| Process Parameters | Unit | Levels | |
|---|---|---|---|
|  |  | Low | High |
| Current | A | 280 | 350 |
| Voltage | V | 22 | 38 |
| Electrode diameter | mm | 1.6 | 3.2 |
| Welding speed | mm/s | 100 | 135 |

**Table 3.** Decision matrix.

| Sample Number | Maximum | | | Minimum | |
|---|---|---|---|---|---|
| | UTS (MPa) | CVN (J) | BP (mm) | BH (mm) | BW (mm) |
| 1 | 420 | 110 | 2.04 | 2.25 | 10.82 |
| 2 | 500 | 100 | 1.12 | 2.85 | 5.14 |
| 3 | 380 | 80 | 2.58 | 3.10 | 7.22 |
| 4 | 320 | 90 | 1.03 | 2.51 | 11.42 |
| 5 | 410 | 60 | 1.45 | 3.72 | 5.35 |
| 6 | 220 | 100 | 1.05 | 2.05 | 8.83 |
| 7 | 280 | 55 | 2.01 | 2.15 | 10.72 |
| 8 | 510 | 115 | 3.50 | 3.88 | 4.50 |
| 9 | 480 | 85 | 3.78 | 2.85 | 6.85 |
| 10 | 320 | 60 | 2.15 | 2.15 | 11.20 |
| 11 | 250 | 95 | 1.90 | 2.98 | 12.40 |
| 12 | 310 | 83 | 2.42 | 2.06 | 9.80 |
| 13 | 520 | 100 | 3.82 | 2.97 | 4.18 |
| 14 | 430 | 70 | 2.25 | 3.08 | 8.32 |
| 15 | 270 | 60 | 1.65 | 2.15 | 10.74 |
| 16 | 290 | 80 | 1.88 | 2.70 | 12.88 |

**Table 4.** Summary of range of standardized decision matrix.

| Sample No. | UTS, MPa | CVN, J | BP, mm | BH, mm | BW, mm |
|---|---|---|---|---|---|
| 1 | 0.67 | 0.92 | 0.36 | 0.11 | 0.76 |
| 2 | 0.93 | 0.75 | 0.03 | 0.44 | 0.11 |
| 3 | 0.53 | 0.42 | 0.56 | 0.57 | 0.35 |
| 4 | 0.33 | 0.58 | 0 | 0.25 | 0.83 |
| 5 | 0.63 | 0.08 | 0.15 | 0.91 | 0.13 |
| 6 | 0 | 0.75 | 0.01 | 0 | 0.53 |
| 7 | 0.20 | 0 | 0.35 | 0.06 | 0.75 |
| 8 | 0.97 | 1 | 0.89 | 1 | 0.04 |
| 9 | 0.87 | 0.50 | 0.99 | 0.44 | 0.31 |
| 10 | 0.33 | 0.08 | 0.40 | 0.06 | 0.81 |
| 11 | 0.10 | 0.67 | 0.31 | 0.51 | 0.95 |
| 12 | 0.30 | 0.47 | 0.50 | 0.01 | 0.65 |
| 13 | 1 | 0.78 | 1 | 0.50 | 0 |
| 14 | 0.70 | 0.25 | 0.44 | 0.56 | 0.48 |
| 15 | 0.17 | 0.08 | 0.22 | 0.06 | 0.75 |
| 16 | 0.23 | 0.42 | 0.31 | 0.37 | 1 |

The next step is to determine the standard deviation and weights using Equation (2) and Equation (3) as shown in **Table 5**.

### 3.1.2. Application of MOORA

In applying the MOORA method, the first step was to square each value in **Table 1**, $X_{ij}^2$, this lead to the creation of **Table 6**.

Applying $X_{ij}' = \dfrac{X_{ij}}{\sqrt{\sum\limits_{i=1}^{m} X_{ij}^2}}$ for each column in **Table 3** and **Table 6**, **Table 7** was created therefrom.

**Table 5.** Weights assigned to criteria.

| Property | SDV$_j$ | W$_j$ |
|---|---|---|
| UTS | 0.57607 | 0.20578 |
| CVN | 0.55074 | 0.19674 |
| BP | 0.55655 | 0.19881 |
| BH | 0.53803 | 0.19220 |
| BW | 0.57800 | 0.20647 |

**Table 6.** The square value of $X_{ij}$.

| Sample No. | UTS, MPa | CVN, J | BP, mm | BH, mm | BW, mm |
|---|---|---|---|---|---|
| 1 | 176,400 | 12,100 | 4.1616 | 5.0625 | 117.0724 |
| 2 | 250,000 | 10,000 | 1.2544 | 8.1225 | 26.4196 |
| 3 | 144,400 | 6400 | 6.6564 | 9.6100 | 52.1284 |
| 4 | 102,400 | 8100 | 1.0609 | 6.3001 | 130.4164 |
| 5 | 168,100 | 3600 | 2.1025 | 13.8384 | 28.6225 |
| 6 | 48,400 | 10,000 | 1.1025 | 4.2025 | 77.9689 |
| 7 | 78,400 | 3025 | 4.0401 | 4.6225 | 114.9184 |
| 8 | 260,100 | 13,225 | 12.2500 | 15.0544 | 20.2500 |
| 9 | 230,400 | 7225 | 14.2884 | 8.1225 | 46.9225 |
| 10 | 102,400 | 3600 | 4.6225 | 4.6225 | 125.4400 |
| 11 | 62,500 | 9025 | 3.6100 | 8.8804 | 153.7600 |
| 12 | 96,100 | 6889 | 5.8564 | 4.2436 | 96.0400 |
| 13 | 270,400 | 10,404 | 14.5924 | 8.8209 | 17.4724 |
| 14 | 184,900 | 4900 | 5.0625 | 9.4864 | 69.2224 |
| 15 | 72,900 | 3600 | 2.7225 | 4.6225 | 115.3476 |
| 16 | 84,100 | 6400 | 3.5344 | 7.3984 | 165.8944 |
| $\sum_{i=1}^{m} X_{ij}^2$ | 2,331,900 | 118,493 | 86.9175 | 123.0101 | 1357.8959 |
| $\sqrt{\sum_{i=1}^{m} X_{ij}^2}$ | 1527.0560 | 344.2281 | 9.3230 | 11.0910 | 36.8496 |

**Table 7.** Normalized weld properties.

| Sample No. | UTS, MPa | CVN, J | BP, mm | BH, mm | BW, mm |
|---|---|---|---|---|---|
| 1 | 0.2750 | 0.3196 | 0.2188 | 0.2029 | 0.2936 |
| 2 | 0.3274 | 0.2905 | 0.1201 | 0.2570 | 0.1395 |
| 3 | 0.2488 | 0.2324 | 0.2767 | 0.2795 | 0.1959 |
| 4 | 0.2096 | 0.2615 | 0.1105 | 0.2263 | 0.3099 |
| 5 | 0.2685 | 0.1743 | 0.1555 | 0.3354 | 0.1452 |
| 6 | 0.1441 | 0.2905 | 0.1126 | 0.1848 | 0.2396 |
| 7 | 0.1834 | 0.1598 | 0.2156 | 0.1939 | 0.2909 |
| 8 | 0.3340 | 0.3341 | 0.3754 | 0.3498 | 0.1221 |
| 9 | 0.3143 | 0.2469 | 0.4054 | 0.2570 | 0.1859 |
| 10 | 0.2096 | 0.1743 | 0.2306 | 0.1939 | 0.3039 |
| 11 | 0.1637 | 0.2760 | 0.2038 | 0.2687 | 0.3365 |
| 12 | 0.2030 | 0.2411 | 0.2596 | 0.1857 | 0.2660 |
| 13 | 0.3405 | 0.2963 | 0.4097 | 0.2678 | 0.1134 |
| 14 | 0.2816 | 0.2034 | 0.2413 | 0.2777 | 0.2258 |
| 15 | 0.1768 | 0.1743 | 0.1770 | 0.1939 | 0.2915 |
| 16 | 0.1899 | 0.2324 | 0.2017 | 0.2452 | 0.3495 |
| Weight, $w_j$ | 0.20578 | 0.19674 | 0.19881 | 0.19220 | 0.20647 |

The next step is to multiply each parameter value in **Table 7**, with their corresponding weights. This action leads to the creation of **Table 8**.

This last step is to sum the parameters comprising of the higher the better (maximum) and the smaller the better (minimum) respectively. **Table 9** is created and ranked therefrom.

## 3.2. Discussion of Results

### 3.2.1. Categorization of Test Results

This study investigates the utilization of standard deviation and multi-objective optimization on the basis of ratio analysis (MOORA) tools in the selection of appropriate gas metal arc welding process parameters.

**Table 8.** Clustered weld properties and bead geometry according to criteria.

| Sample No. | Maximum | | | Minimum | |
|---|---|---|---|---|---|
| | UTS, MPa | CVN, J | BP, mm | BH, mm | BW, mm |
| 1 | 0.0566 | 0.0629 | 0.0435 | 0.0390 | 0.0606 |
| 2 | 0.0674 | 0.0572 | 0.0239 | 0.0494 | 0.0288 |
| 3 | 0.0512 | 0.0457 | 0.0550 | 0.0537 | 0.0404 |
| 4 | 0.0431 | 0.0514 | 0.0220 | 0.0435 | 0.0640 |
| 5 | 0.0553 | 0.0343 | 0.0309 | 0.0645 | 0.0300 |
| 6 | 0.0297 | 0.0572 | 0.0224 | 0.0355 | 0.0495 |
| 7 | 0.0377 | 0.0314 | 0.0429 | 0.0373 | 0.0601 |
| 8 | 0.0687 | 0.0657 | 0.0746 | 0.0672 | 0.0252 |
| 9 | 0.0647 | 0.0486 | 0.0806 | 0.0494 | 0.0384 |
| 10 | 0.0431 | 0.0343 | 0.0459 | 0.0373 | 0.0627 |
| 11 | 0.0337 | 0.0543 | 0.0405 | 0.0516 | 0.0695 |
| 12 | 0.0418 | 0.0474 | 0.0516 | 0.0357 | 0.0549 |
| 13 | 0.0701 | 0.0583 | 0.0815 | 0.0515 | 0.0234 |
| 14 | 0.0580 | 0.0400 | 0.0480 | 0.0534 | 0.0466 |
| 15 | 0.0364 | 0.0343 | 0.0352 | 0.0373 | 0.0602 |
| 16 | 0.0391 | 0.0457 | 0.0401 | 0.0471 | 0.0722 |

**Table 9.** Ranking step.

| Sample No. | $\sum$max | $\sum$min | $\sum$max$-\sum$min | Rank | |
|---|---|---|---|---|---|
| 1 | 0.1630 | 0.0996 | 0.0634 | 5 | |
| 2 | 0.1485 | 0.0782 | 0.0703 | 4 | |
| 3 | 0.1519 | 0.0941 | 0.0578 | 6 | |
| 4 | 0.1165 | 0.1075 | 0.0090 | 13 | |
| 5 | 0.1205 | 0.0945 | 0.0260 | 9 | |
| 6 | 0.1093 | 0.0850 | 0.0243 | 10 | |
| 7 | 0.1120 | 0.0974 | 0.0146 | 12 | |
| 8 | 0.2090 | 0.0924 | 0.1166 | 2 | |
| 9 | 0.1939 | 0.0878 | 0.1061 | 3 | |
| 10 | 0.1233 | 0.1000 | 0.0233 | 11 | |
| 11 | 0.1285 | 0.1211 | 0.0074 | 15 | |
| 12 | 0.1408 | 0.0906 | 0.0502 | 7 | |
| 13 | 0.2099 | 0.0749 | 0.1350 | 1 | Best |
| 14 | 0.1460 | 0.1000 | 0.0460 | 8 | |
| 15 | 0.1059 | 0.0975 | 0.0084 | 14 | |
| 16 | 0.1249 | 0.1193 | 0.0056 | 16 | |

In the first instance, a layout matrix design is established as contained in **Table 1**. **Table 2** contains the process parameters, which comprises of the current, voltage, electrode diameter, and welding speed. The process parameters are either in low level (−) or high level (+). The low and high levels of the process parameters in **Table 2** are placed in their various locations in **Table 1**, where current, voltage, electrode diameter and welding speed are denoted, as A, B, C and C respectively in **Table 1**.

From **Table 1**, it can be seen that there is a sixteen process parameter layout design. Each process parameter welding operation was used to make five weldments. The UTS, and CVN of the weldments were determined by conducting the tensile test, as well as, the charpy V-Notch impact test. The height, width, and penetration of the bead geometry of these weldments were measured and determined. **Table 3** classified these properties according to their quality values, which shows that the larger the UTS, CVN, and BP the better the quality of the weldment. Whereas, the smaller the BH and BW the better the quality of the weldment.

This is because, UTS defines the strength of the weldment. Therefore, the greater the strength of the weldment, the more the weldment possesses the capacity to carry loads. This quality actually extends the service life of the weldment. The CVN measures the energy required to absorb impact loads. The higher the CVN value, the greater the chances of the weldments to absorb any applied impact load. This on the one hand tends to increase the service life of the weldment. The weld bead penetration is an important factor considered in assessing the quality of weldments. The higher the weld penetration, the lower the weld undercuts, and the higher the weld joint reinforcements. This however increases the strength and quality of the weldment.

**Table 4** shows the sixteen standard deviations, determined for each of the mechanical properties, whereas, **Table 5** shows the overall standard deviations and the corresponding weights assigned to each mechanical property. **Tables 6-9** show the MOORA application process for determining the optimum welding process parameters.

### 3.2.2. Result Analysis

The UTS considered in this study is within the range of 220 MPa and 520 MPa. Appling MOORA the selected process parameters thereof produced a weldment with a UTS of 520 MPa. The CVN considered in this study is in the range of 55 J and 115 J. By Appling the MOORA a CVN of 100 J was obtained. This indicates that when the CVN value is above the threshold value of 100 J. This may negatively affect on the long term the service life of the weldment.

The BP considered in this study is within the range of 1.03 mm and 3.8 mm. By applying the MOORA technique, BP was found to be 3.83 mm. This indicates that the more the gaps between the weld joints are covered by the molten weld metal, the better, because the strength is increased, porosity is reduced to the barest minimum or eliminated, and the weld joints are held together to an acceptable level. This study shows that the joint gap of weldment 13 was fully covered by the molten weld metal.

On the other hand, the second category shows that the smaller the bead height and bead width the better the quality of the weldment. This corresponds with actual welding practice. The smaller the BH and BW are, the better the quality of the weldment will be. The range of BH considered in this study is 2.05 mm and 3.88 mm. By applying the MOORA method, the BH obtained was 2.97 mm. This indicates that BH of 2.05 mm was too small to be considered and a BH below 2.97 mm may not have enough weld metal to sustain the strength possessed by the weldment when loads are applied. Therefore, for this study, BH with a value of 2.97 mm is considered the threshold value when using the optimum process parameters. The BW considered in this study is within the range of 4.18 mm and 12.88 mm. By applying the MOORA method, BW of 4.18 mm was obtained. This indicates that BW values above 4.18 mm may contain too much weld metal. Too much weld metal adds to the weight of the weldment and this may not be good for the overall structure of the material.

For this study, weldment 13, is found to possess the best mechanical property. From **Table 1** and **Table 2**, the process parameters for weldment 13 correspond to a welding current of 350 A, a welding voltage of 22V, an electrode diameter of 3.2 mm and welding speed of 100 mm/s. The mechanical properties produced by the weldment made by these process parameters are UTS of 520 MPa, CVN of 100 J, BP of 3.8 mm, BH of 2.97 mm and BW of 4.18 mm.

The results from this study were compared with similar work found in literature. Such as Gunaraj and Murugan [20] who predicted and optimized the weld bead volume for submerged arc process and obtained bead penetration of between the range of 3.04 mm and 3.80 mm, and bead width was in the range of 7.9 mm and 9.1 mm. From this study, the BP matches that obtained by Gunaraj and Murugan [20] and also the BW obtained in

**Figure 5.** Microstructure of weldment from optimized process parameters.

this study is better than the one determined by Gunaraj and Murugan [20].

**Figure 5** shows the microstructure of the optimized weldment. From the microstructural view it can be seen that the black and white colours are heterogeneously prominent. The white grains are ferrite, while the black ones are the pearlite. However, pearlite contains ferrite and cementite. Cementite is considered to be very hard and dense. From **Figure 5**, it can be seen that obviously pearlite is more in proportion than ferrite. As a result of this, the strength, that is the UTS of the optimized weldment is expected to be high and the weldment ductile as the ferrite is considerably high. Since it is observed that the optimized process parameters gave a good level of weld metal penetration, the fusion between the parent material and the weld metal would also be high with very good machinability because of its ductility. This analysis reveals that the optimized weldment is of very good quality.

## 4. Conclusions

Mild steel plates were joined by applying specific process parameters to carry out the welding operation. The weld metals were machined into various test specimens. Mechanical properties, UTS, and CVN were determined using the test specimens. The mechanical properties were found within the category. The larger the test result was, the better the quality of the weldment would be, whereas the individual weldment was bisected and BP, BH and BW were measured.

BP falls into the larger the test result, the better the quality of the weldment whereas, BH and BW fall into the smaller the test results, the better the quality of the weldment or bead geometry. The MOORA technique was applied to optimally select the welding process parameters that produced the weldment with the best properties. However, standard deviation was used to determine the weights allocated to each value of the mechanical property utilized in the course of running the MOORA process.

This study summarily covers the application of MOORA method in the selection of optimized welding process parameters for welding mild steel plates using the gas metal arc welding techniques. This multi-objective optimization tool utilizes a ranking method for the process parameters selection process.

From the results obtained, it can be found that the selected optimized process parameters are within the range of optimized process parameters obtained in literature. It is hereby concluded that MOORA method has successfully optimized the process parameters considered in this study and that these optimized process parameters are compared well with those obtained by other investigators who apply other known optimization models. The microstructure of the weldment produced by the optimized process parameters was also investigated to confirm the quality of the weldment. The analysis of the microstructure reveals that the weldment produced by the optimized process parameters is of excellent weld quality.

## References

[1]   Görener, A., Dinçer, H. and Hacıoğlu, Ü. (2013) Application of Multi-Objective Optimization on the Basis of Ratio Analysis (MOORA) Method for Bank Branch Location Selection. *International Journal of Finance & Banking Studies*, **2**, 41-52.

[2]   Brauers, W.K. (2003) Optimization Methods for a Stakeholder Society, a Revolution in Economic Thinking by Multi-Objective Optimization, Series: Nonconvex Optimization and Its Applications, 73. Kluwer Academic Publishers, Boston, 342.

[3]   Brauers, W.K.M., Zavadskas, E.K., Turskis, Z. and Vilutiene, T. (2008) Multi-Objective Contractors's Ranking by Applying the MOORA Method. *Journal of Business Economics and Management*, **9**, 245-255.
http://dx.doi.org/10.3846/1611-1699.2008.9.245-255

[4]   Brauers, W.K.M., Ginevičius, R. and Podvezko, V. (2010) Regional Development in Lithuania Considering Multiple

Objectives by the MOORA Method. *Technological and Economic Development of Economy*, **16**, 613-640.
http://dx.doi.org/10.3846/tede.2010.38

[5]   Gadakh, V.S., Shinde, V.B. and Khemnar, N.S. (2013) Optimization of Welding Process Parameters Using MOORA Method. *The International Journal of Advanced Manufacturing Technology*, **69**, 2031-2039.
http://dx.doi.org/10.1007/s00170-013-5188-2

[6]   Hwang, C.L. and Yoon, K.P. (1995) Multiple Attribute Decision Making and Introduction. Sage Publication, London, 2.

[7]   Mandel, U.K. and Sarkar, B. (2012) Selection of Best Intelligent Manufacturing System (IMS) Under Fuzzy MOORA Conflicting MCDM Environment. *International Journal of Emerging Technology and Advanced Engineering*, **2**, 301-310.

[8]   Stanujkic, D. Dordevic, B. and Dordevic, M. (2013) Comparative Analysis of Some Prominent MCDM Method: A case of Ranking Serbian Banks. *Serbian Journal of Management*, **8**, 213-241. http://dx.doi.org/10.5937/sjm8-3774

[9]   Brauers, W.K.M. and Zavadskas E.K. (2006) The MOORA Method and Its Applications to Privatization in a Transition Economy. *Control and Cybernetics*, **35**, 445-469.

[10]  Chakraborty, S. (2010) Application of the MOORA Method for Decision Making in Manufacturing Environment. *International Journal of Advanced Manufacturing Technology*, **54**, 1155-1166.
http://dx.doi.org/10.1007/s00170-010-2972-0

[11]  Karande, P. and Chakraborty, S. (2012) Application of Multi-Objective Optimization on the Basis of Ratio Analysis (MOORA) Method for Materials Selection. *Materials and Design*, **37**, 317-324.
http://dx.doi.org/10.1016/j.matdes.2012.01.013

[12]  Chaturved, V. and Sharma, A.K. (2014) Parametric Optimization of CNC Wire Cut EDM for OHNS Steel Using MOORA Methodology. *Proceedings of the 10th IRF International Conference*, Bengaluru, 4 October 2014, 79-84.

[13]  Brauers, W.K.M. and Zavadskas, E.K. (2010) Project Management by Multimoora as an Instrument for Transition Economics. *Technological and Economic Development of Economy*, **16**, 5-24. http://dx.doi.org/10.3846/tede.2010.01

[14]  Ozcelik, G., Aydogan, E.K. and Gencer, C. (2014) A Hybrid Moora-Fuzzy Algorithm for Special Education and Rehabilitation Centre Selection. *Journal of Military and Information Science*, **2**, 53-62.
http://dx.doi.org/10.17858/jmisci.53708

[15]  Farzamnia, F. and Babolghani, M.B. (2014) Group Decision Making Process for Supplier Selection Using Multimoora Technique under Fuzzy Environment. *Kuwait Chapter of Arabian Journal of Business and Management Review*, **3**, 203-218.

[16]  Prasad, K.S., Rao, C.S. and Rao, D.N. (2011) Optimizing Pulsed Current Micro Plasma Arc Welding Parameters to Maximize Ultimate Tensile Strength of Inconel 625 Nickel Alloy Using Response Surface Method. *International Journal of Engineering Science and Technology*, **3**, 226-236.

[17]  Kim, D., Kang, M. and Rhee, S. (2005) Determination of Optimal Welding Conditions with a Controlled Random Search Procedure. *Welding Journal*, **84**, 125-130.

[18]  El-Santawy, M.F. and Ahmed, A.N. (2012) Analysis of Project Selection by Using SDV-MOORA Approach. *Life Science Journal*, **9**, 129-131.

[19]  Kalibatas, D. and Turskis, Z. (2008) Multicriteria Evaluation of Inner Climate by Using MOORA Method. *Information Technology and Control*, **37**, 79-83.

[20]  Gunaraj, V. and Murugan, N. (2000) Prediction and Optimization of Weld Bead Volume for the Submerged Arc Process—Part 2. *Welding Journal*, **79**, 331s-338s.

# 8

# Reduction of Undercuts in Fillet Welded Joints Using Taguchi Optimization Method

## Joseph Achebo[1*], Sule Salisu[2]

[1]Department of Production Engineering, University of Benin, Benin, Nigeria
[2]Department of Mechanical Engineering, Petroleum Training Institute, Effurun, Nigeria
Email: *josephachebo@yahoo.co.uk

## Abstract

This project work focuses on the reduction of weld undercuts using the Taguchi method. The phenomenon of weld undercuts constitutes a major problem for the welding industry. When undercuts occur, and particularly when such cuts are deep, it has a negative impact on the weld as it lowers the integrity and quality of the weldment. Therefore, efforts are made globally to reduce the depth of such weld undercuts to the barest minimum. Several optimization methods have been adopted; however, in this study, the Taguchi method is applied. "The smaller the better components" of the Taguchi method is applied. From the results obtained from applying this Taguchi method, the optimum process parameters obtained are $A_2$-$B_1$-$C_2$, which are a voltage of 20 V, a current of 180 A, and a welding speed of 130 mm/s, required to form an undercut of 0.03 mm. Whereas the existing process parameters used by the company are $A_1$-$B_3$-C, which make an undercut to a depth of 0.09 mm. It is concluded that the use of Taguchi method has been able to reduce the depth of undercut as shown in this study. A step-by-step approach is presented in the study.

## Keywords

Bead Geometry, Fillet Weld, Process Parameters, Taguchi Method, Weld Undercut

## 1. Introduction

Weld bead geometry is severely negatively affected by the occurrence of the undercut phenomenon. Undercuts can be described as the presence of grooves along the edges of the weldment, usually observed in the welding of unskilled welders. These welders are often satisfied with their metal material joints simply being held together

---

*Corresponding author.

by the solidified molten weld metal upon cooling, without considering whether or not there was an adequate molten weld metal penetration in the parent metal's joints gaps. Petershagen [1] described an undercut as an irregular groove along the toe of a weld. Undercuts can also be seen as unfilled grooves in the base metal at the edge of the weld. These undercuts could similarly be caused by inappropriate electrode angle, excessive current and unacceptable travel speed. Petershagen [1] also said that undercuts are inherent in the welding process and may occur in either the base or filler metal. He classified undercuts into three categories. Undercuts of category one are referred to as wide and curved. Undercuts of category two are narrow and undercuts of category three are considered micro-flaws, with depth less than 0.25 mm. Undercuts of category three are believed to be unavoidable during welding and impossible to detect by visual inspection. Xu *et al.* [2] also said that bead undercut defects not only affect the appearance of weld beads, but also cause a severe stress concentration at the weld edges, which has a great effect on the reliability of the weld joints. Whatever the category, insufficient penetration of molten weld metal which is a major cause of undercuts, lowers the strength of the weldments, and this has led to structural failures of engineering projects. Major structural failures could lead to significant safety hazards.

This study attempts to find the most economic solutions of applying models, tested in other areas, to weld technology. This is all in the bid to optimize gas metal arc welding process parameters required to reduce the formation of undercuts to the barest minimum. The integrity of the weldment could be improved by tweaking the welding process and removing practices which undermine the process. The top and undersides of a weldment are expected to meet specific standards. The top side which constitutes the bead is expected to have optimized bead geometry, where the bead height and width are preferred to be smaller in size. The undersides, where undercuts predominantly occur, usually appear in between parent metal joints at the point where the weldment ought to have leveled out with the parent metals.

When there is insufficient molten weld metal penetration and the point where the weld metal does not level out with the parent metal, the difference between the parent metal and the weld metal determines the size of the weld undercut. Garg *et al.* [3] were of the opinion that a weld undercut occurs when the weld reduces the cross-sectional thickness of the parent metal.

The heat affected zone (HAZ) includes the part of the parent metal which is intensively affected by the arc heat. Parts of this parent metal may melt and the melt would either flow into a lower part of the parent metal or vaporize into the atmosphere. This movement of melt could eventually cause the cross-sectional area of the parent metal to be reduced.

Alloying elements vaporization must also be taken into consideration. Alloying elements like zinc and magnesium are very volatile when exposed to intense heat. Therefore, if these alloying elements are exposed to prolonged heat application during welding, they have the tendency to vaporize. Garg *et al.* [3] asserted that undercuts could be caused by excessive current. Excessive current usually causes weld spatter; weld spatters are metal droplets expelled from the weld/weld pool that stick to the surrounding surfaces of the weldment.

The authors also said that insufficient deposition of filler metal during welding could also cause weld undercut. This insufficient filler metal deposition could be as a result of very low voltage and current. These very low levels of voltage and current may not generate enough arc heat to melt the filler metal. Xu *et al.* [2] has suggested that in gas metal arc welding process, by increasing the welding speed beyond a particular threshold value, the weld formation quality will worsen and bead undercut defects would occur. It has also been suggested that these undercuts can be filled or repaired by welding them with smaller electrodes.

In this study, "the smaller the better part" of the Taguchi method is applied to reduce the formation of weld undercuts to the barest minimum.

## 2. Materials and Methods

Five weld joints were made on a 60 mm × 40 mm × 10 mm mild steel plate. This process was repeated nine times producing a total of forty five weld joints. For each application of the nine process parameters, the five undercuts made in each weld joint were measured and the average recorded. In all, a total of nine undercuts were recorded as presented in **Table 1**. **Figure 1** shows the undercut scan gauge used for measuring the depth of undercut.

Spadea and Frank [5] cited in their work that Petershagen [1] survey also indicated that undercuts are generally characterized by their length, height, and width. In addition, the transverse orthogonal plane angel, $\rho$, and the notch root radius, $r$, were used to describe the geometry of an undercut as shown in **Figure 2**.

**Figure 1.** Wiki Scan weld undercut measurement gauge [4].

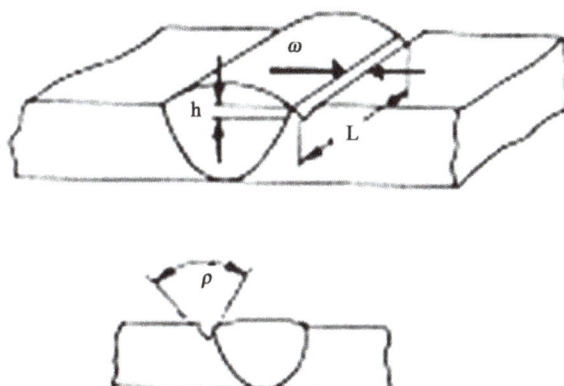

**Figure 2.** Undercut dimensions [1].

**Table 1.** Orthogonal matrix layout and undercut measurements of corresponding weldments.

| Experiment runs | voltage | current | Welding speed | Undercut sizes, mm |
|---|---|---|---|---|
| 1 | 1 | 1 | 1 | 0.08 |
| 2 | 1 | 2 | 2 | 0.05 |
| 3 | 1 | 3 | 3 | 0.12 |
| 4 | 2 | 1 | 3 | 0.03 |
| 5 | 2 | 2 | 1 | 0.15 |
| 6 | 2 | 3 | 2 | 0.09 |
| 7 | 3 | 1 | 2 | 0.11 |
| 8 | 3 | 2 | 3 | 0.07 |
| 9 | 3 | 3 | 1 | 0.06 |

The methods adopted in conducting this study are as follows:
- An $L_9$ Taguchi orthogonal matrix layout, shown in **Table 1** was chosen for this study;
- A welding operation was done by utilizing the range of process parameters as shown in **Table 2** and using a gas metal arc welding (GMAW) machine to join a pair of 60 mm × 40 mm × 10 mm mild steel plate;
- Undercuts in the weldments were measured and recorded;
- The Taguchi optimization process relating to the smaller the better was applied using Equation (1);
- The signal to noise ratio generated was clustered and arranged into their respective positions indicating their optimum process parameters as shown in **Table 3**;
- Signal to noise ratio vs process parameters graphs were drawn to show the effect of S/N ratios on the process parameters in relation to the optimum process parameters as shown in **Figures 3-5**;

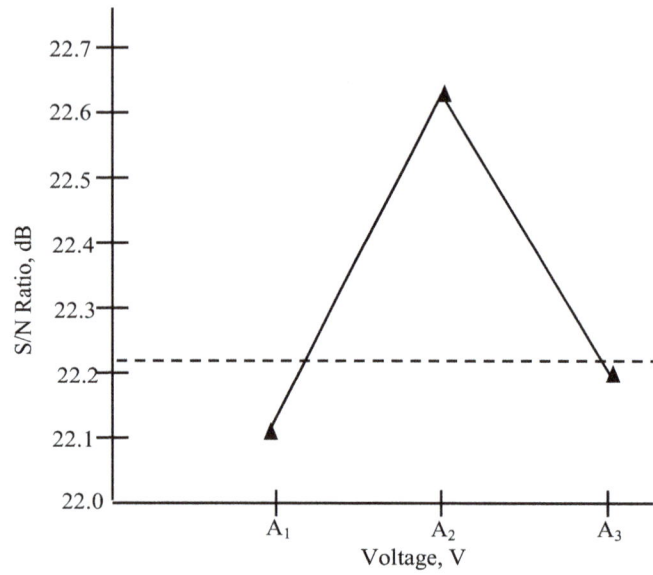

**Figure 3.** S/N ratio vs voltage.

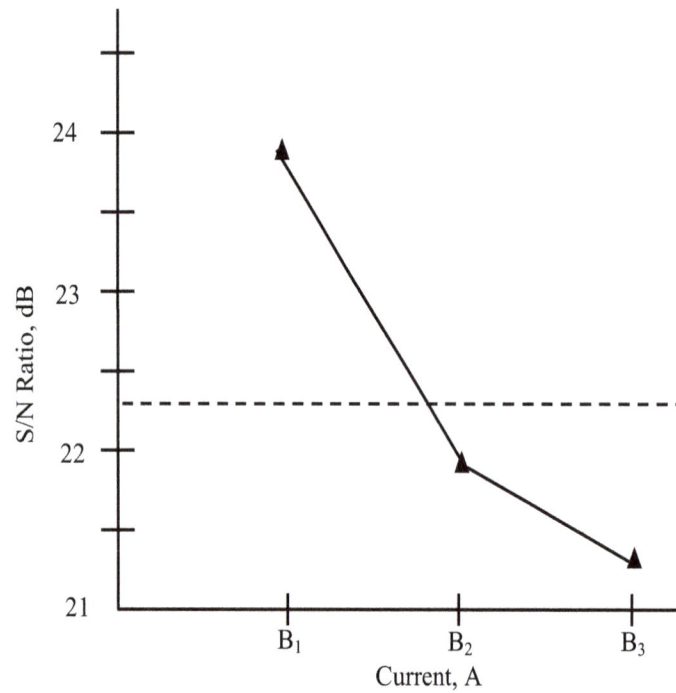

**Figure 4.** S/N ratio vs current.

**Table 2.** Range of process parameters.

| Factors | Levels | | |
|---|---|---|---|
| | 1 | 2 | 3 |
| A. Voltage, V | 18 | 20 | 22 |
| B. Current, A | 180 | 200 | 260 |
| C. Welding speed, mm/s | 85 | 105 | 130 |

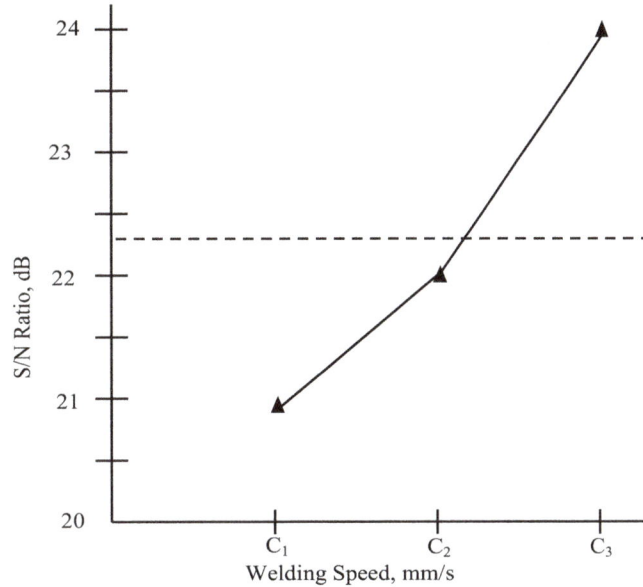

**Figure 5.** S/N ratio vs welding speed.

**Table 3.** Sum and average S/N ratio for each process parameter.

| A Level | A Voltage | | B Current | | C Welding Speed | | Total Average of S/N Ratio |
|---|---|---|---|---|---|---|---|
| | Sum of S/N ratio | Average of S/N ratio | Sum of S/N ratio | Average of S/N ratio | Sum of S/N ratio | Average of S/N ratio | |
| 1 | 66.3752 | 22.1251 | 71.5680 | 23.8560* | 62.8534 | 20.9511 | 22.3259 |
| 2 | 67.8510 | 22.6170* | 65.5968 | 21.8656 | 66.1080 | 22.0360 | |
| 3 | 66.7072 | 22.2357 | 63.7686 | 21.2562 | 71.9720 | 23.9907* | |

*signifies the optimum level based on the smaller-the-better criterion.

- The analysis of variance of the parameters shown in **Table 4** are calculated for to determine the contribution of each process parameter to the formation of undercuts;
- A confirmation test as shown in **Table 5** is done using Equation (4). This is done to find out if the optimized process parameters have better mechanical properties or signal to noise ratio than those obtained from using the existing process parameters.

## 3. Results

### 3.1. Presentation of Results

**Table 1** shows the orthogonal matrix layout and undercut measurement results.

**Table 2** shows the range of process parameters.

Applying the smaller the better methodology by using Equation (1).

Signal to Noise ratio for the smaller the better function is

$$\eta = -10\log_{10}\left(\frac{1}{n}\sum_{i=1}^{n}y_i^2\right) \tag{1}$$

where, $n$ is the sample size and $y$ is the mean weld undercut in mm.

The S/N ratio for the individual control factors as extracted from **Table 1**, are as calculated hereunder:

For voltage

$$S_{v1} = \eta_1 + \eta_2 + \eta_3, \ S_{v2} = \eta_4 + \eta_5 + \eta_6, \ S_{v3} = \eta_7 + \eta_8 + \eta_9$$

**Table 4.** Analysis of variance (ANOVA).

| Parameter | Process Parameter | Degree of Freedom | Sum of Squares | Variance | F-ratio | Contribution (%) |
|-----------|-------------------|-------------------|----------------|----------|---------|------------------|
| A | Voltage, V | 2 | 0.1332 | 0.0666 | 0.00096 | 0.09 |
| B | Current, A | 2 | 3.6973 | 1.8487 | 0.0268 | 2.53 |
| C | Welding Speed, mm/s | 2 | 4.7457 | 2.3729 | 0.0344 | 3.24 |
| Error | | 2 | 137.8180 | 68.9090 | - | 94.14 |
| Total | | 8 | 146.3942 | - | - | 100 |

**Table 5.** Confirmation test results.

| | Existing Process Parameters | Optimum Process Parameters | Improvement in S/N ratio |
|---|------------------------------|-----------------------------|---------------------------|
| Process Parameters | $A_1$-$B_3$-$C_1$ | $A_2$-$B_1$-$C_3$ | |
| Undercut Measurement (mm) | 0.09 | 0.03 | 0.06 |
| S/N dB | 23.4759 | 25.8119 | 2.336 |

Average S/N ratio for voltage

$$\frac{S_{v1}}{3}; \frac{S_{v2}}{3}; \text{ and } \frac{S_{v3}}{3}$$

For current,

$$S_{c1} = \eta_1 + \eta_4 + \eta_7, \, S_{c2} = \eta_2 + \eta_5 + \eta_8, \, S_{c3} = \eta_3 + \eta_6 + \eta_9$$

Average S/N ratio for current

$$\frac{S_{c1}}{3}; \frac{S_{c2}}{3}; \text{ and } \frac{S_{c3}}{3}$$

For welding speed,

$$S_{w1} = \eta_1 + \eta_5 + \eta_9, \, S_{w2} = \eta_2 + \eta_6 + \eta_7, \, S_{w3} = \eta_3 + \eta_4 + \eta_8$$

Average S/N ratio for welding speed

$$\frac{S_{w1}}{3}; \frac{S_{w2}}{3}; \text{ and } \frac{S_{w3}}{3}$$

The sum and average S/N ratios for the process parameters are shown in **Table 3**.

Athreya and Venkatesh [6] calculated the surface roughness of lathe facing operation by using the smaller the better type of Taguchi method and selected the factor levels corresponding to the highest S/N ratio as the optimum process parameters. From **Table 3**, the optimum level corresponding to the highest S/N Ratio is **$A_2$-$B_1$-$C_3$**.

**Figures 3-5** show the S/N ratio graph where the dashed line is the value of the total mean of the S/N ratios. The Figures clearly show the interactions between the levels of each process parameters.

**Analysis of Variance (ANOVA)**

**Table 4** shows the analysis of variance tabulation.

$$\text{Sum of square, } S_i^2 = \sum \left( y_i^2 \right) - \frac{\left( \sum y_i \right)^2}{n} \quad [7] \tag{1}$$

where $n$ is the number of test conducted

$$\% \text{ Contribution } = \frac{S_i^2}{S_T^2} \times 100\% \quad [7] \tag{2}$$

$$F\text{-ratio} = \frac{\text{Variance}_i}{\text{Variance}_{error}} \quad [7] \tag{3}$$

**Confirmation Test**

$$\text{Using} \quad \eta = \eta_m + \sum_{i=1}^{n}(\bar{\eta} - \eta_m) \tag{4}$$

where $\eta_m$ is the total mean of S/N ratio; $\bar{\eta}$ is the mean of S/N ratio at the optimal level and n is the number of main welding parameters.

The optimal parameters are $A_2$-$B_1$-$C_3$ and the corresponding S/N ratios are 22.6170, 23.8560 and 23.9907 respectively.

The total mean of S/N ratio, $\eta_m = 22.3259$.

Therefore, $\eta = 22.3259 + \left[(22.6170 - 22.3259) + (23.8560 - 22.3259) + (23.9907 - 22.3259)\right] = 25.8119$.

The existing process parameters in use for welding processes are $A_3$-$B_1$-$C_2$.

Its S/N ratio is calculated as follows:

$$\eta = 22.3259 + \left[(22.2357 - 22.3259) + (23.8560 - 22.3259) + (22.0360 - 22.3259)\right] = 23.4759$$

**Table 5** shows the confirmation test results.

## 3.2. Discussion of Results

This paper tends to use the Taguchi optimization tool relating to the smaller the better technique in reducing the depth of weld undercut made during the welding operations of weldments. The orthogonal matrix layout in **Table 1** was used to allocate the process parameters in **Table 2**, these process parameters were used to make weldments. Nine welding operations were carried out in this experiment. Each welding operation was conducted five times and the corresponding undercuts were measured. The average of these five weldments obtained from one welding operation was recorded. The values of these measurements are recorded in **Table 1**. The signal to noise ratio relating to the smaller the better technique was obtained for the values of the undercuts in **Table 1**. The average value of the calculated s/n ratios were recorded in **Table 3**.

From **Table 3**, the optimum process parameters were obtained utilizing the smaller the better technique of the Taguchi method. The obtained process parameters are $A_2$-$B_1$-$C_3$. This indicates that a welding process parameter consisting of 20 V, a current of 180 A and a welding speed of 130 min/s are required to make an undercut in the weldment to a value of 0.03 mm. The existing process parameter utilized by the company is $A_1$-$B_3$-$C_1$ which made an undercut with an average depth of 0.09 mm. By using the optimum welding process parameters, an improvement of an S/N ratio of 2.336 and 0.06 mm depth of undercut were made over the S/N ratio and depth of cut made by applying the existing process parameters.

The effects of the S/N ratios on the process parameters were also investigated and are expressed in **Figures 3-5**. **Figure 3**, shows the effect of S/N ratio on voltage, the dashed line is the value of the total mean of the S/N ratios and it is indicated by dote lines. From **Figure 3**, it is seen that a voltage of 20V that has an S/N ratio of 22.6170 db is above the total mean S/N ratio of 22.3259 db. Whereas, a voltage of 18V which has an S/N ratio of 22.1251 db and another voltage of 22 V with an S/N ratio of 22.2357 db, are both below the total mean S/N ratio. This indicates that using the voltage of 20V has the probability of reducing the noise level of the welding operation that can result in the production of weldment with the lowest undercut. **Figure 4**, shows the effect of the S/N ratio on the welding current. From **Figure 4**, it is seen that a current of 180 A with an S/N ratio of 23.8560 db was above the total mean S/N ratio, whereas, the 200 A current, having an S/N ratio of 21.8656 and a current of 260 A with an S/N ratio of 21.2562 db fall below the total mean value. The total mean value is a threshold value that determines the viability of the process parameter. This further shows that from the interactions of these components of the process parameters (in this case current), the 180 A current with an S/N ratio of 23.8560 db has the ability of reducing the noise, in this case, noise are likely factors that would adversely affect the welding operation capacity to a very insignificant value that can eventually produce a weldment with the lowest undercut.

**Figure 5** shows the effect of the S/N ratio on the welding speed. From **Figure 5**, it is observed that a welding speed of 85 mm/s having an S/N ratio of 20.9511 and another welding speed of 105 mm/s with an S/N ratio of

22.0360 db respectively were below the total mean S/N ratio. However, the welding speed of 130 mm/s having an S/N ratio of 23.9907 db is above the total mean S/N ratio This indicates that the third component of the welding speed parameter, which is 130 mm/s having an S/N ratio of 23.9907 is the optimum component of welding speed and this component has the ability to reduce noise to the barest minimum.

However, the extent of these process parameters consisting of voltage, current and welding speed in reducing the depth of the weld undercut were measured by determining the contribution made by each process parameter. The contribution made by these process parameters were determined by applying the analysis of variance approach as presented in **Table 4**.

From **Table 4**, it is observed that the welding speed has the greatest contribution of 3.24%, followed by current, which contributed 2.53% and voltage has the lowest contribution of 0.09%. This indicates that as the welding speed increase, which also reduces the welding time and with a significant amount of current, a weldment with insignificant or no undercut is likely to be produced. From this study, voltage does not significantly affect the formation of undercuts.

**Table 5** shows the confirmation test results. The importance of the confirmation test is to determine the effectiveness and reliability status of the optimum process parameters over the existing process parameters used by the welding and fabrication operators. From **Table 5**, it is observed that the existing process parameters of $A_1$-$B_3$-$C_1$, utilized by the company has an S/N ratio of 23.4759 db which produced a weldment with an average undercut measurement of 0.09 mm. The optimum process parameters of $A_2$-$B_1$-$C_3$ obtained by applying the Taguchi method has an S/N ratio of 25.8119 db which produced a weldment with an average undercut of 0.03mm, from the above, it can be deduced that by using the optimum process parameters there is a reduction of 0.06mm of undercut from 0.09 mm obtained in the weldment made by using the existing process parameters. This also gave an improvement in S/N ratio of 2.3360 db over that made by using the existing process parameters.

## 4. Conclusions

Weld undercuts are a very important factor considered in assessing the integrity of weldments. The higher the depth of the undercut, the lower the integrity of the weldment. Therefore, this study is designed to determine optimum process parameters that can reduce this undercut to the barest minimum. The application of the wrong process parameters can greatly affect the integrity of the weldment.

Taguchi orthogonal array, $L_9$, was used to design the various compositions or combinations of the process parameters. These process parameters were used to make weldments. The undercuts formed in these weldments were measured by using a Wiki Scan as a laser based tool, which does a 3D scan of joints and weld information on size, porosity and undercut.

The S/N ratios of the measured undercuts were determined and the optimum process parameters were obtained by applying "the smaller the better technique" of the Taguchi method. Also the contributions of each of the process parameters were determined to find out the extent to which each of the process parameters affected the formation of undercut in the corresponding weldment. It was found that welding speed has the greatest input in the reduction of weld undercuts.

## References

[1]   Petershagen, H. (1990) The Influence of Undercut on the Fatigue Strength of Welds: A Literature Survey. *Welding in the World*, **28**, 114-125.

[2]   Xu, W., Wu, C. and Zou, D. (2008) Predicting of Bead Undercut Defects in High Speed Gas Metal Arc Welding (GMAW). *Frontiers of Materials Science China*, **2**, 402-408. http://dx.doi.org/10.1007/s11706-008-0065-x

[3]   Garg, S., Kakkar, I., Pandey, A., Gupta, M. and Kishor, N. (2013) Effect of Different Coating Composition's Rutile-Type Welding Electrodes on Undercut Defect in Manual Electric Arc Welding. *International Journal of Mechanical Engineering and Research*, **3**, 381-388.

[4]   Servo-Robot (2015) Wiki-Scan Welding Inspection System. http://wikiscan.servo-robot.com/wp-content/uploads/2012/07/wikiscanbrochurefinal.pdf

[5]   Spadea, J.R. and Frank, K.H. (2002) Fatigue Strength of Fillet-Welded Transverse Stiffeners with Undercuts. Center for Transportation Research, The University of Texas at Austin, Austin, Report No. FHWA/TX-05/0-4178-1.

[6]   Athreya, S. and Venkatesh, Y.D. (2012) Application of Taguchi Method for Optimization of Process Parameters in Improving the Surface Roughness of Lathe Facing Operation. *International Refereed Journal of Engineering and*

*Science* (*IRJES*), **1**, 13-19.

[7]     Achebo, J.I. (2012) Pipe Joint Strength Design and Service Life of a Pseudo Homogeneous All Weld Metal under Continuum Flow. In: Rivero, M.G. and Mansillo, L.M., Eds., *Pipelines*: *Design, Applications and Safety*, *Chapter* 9, Nova Science Publishers, New York, 225-257.

# Zr-Ti-Ni-Cu Amorphous Brazing Fillers Applied to Brazing Titanium TA2 and Q235 Steel

Jie Cui[1], Qiuya Zhai[1*], Jinfeng Xu[1], Yahui Wang[2], Jianlin Ye[2]

[1]School of Materials Science and Engineering, Xi'an University of Technology, Xi'an, China
[2]Xi'an Unit Container Manufacturing Co., Ltd., Xi'an, China
Email: *qiuyazhai@xaut.edu.cn

## Abstract

Ti-Zr-Cu-Ni amorphous filler with good performance is suitable for joining TC and TB titanium alloy, but its melting temperature is higher than 882.5°C, the $\alpha \rightarrow \beta$ phase transition temperature of TA2, which makes the ductility of TA2 fall and the microstructure of the joint coarse. In this paper, Ti-Zr-Cu-Ni amorphous filler was redesigned and optimized by using orthogonal experiment to obtain three easy-to-use Zr-Ti-Ni-Cu amorphous fillers with low melting points and good plasticity. The fast cooling equipment was used to fabricate the brazing filler foils to implement the braze welding of TA2 and Q235 with high frequency inductance. The results indicate that all the brazing foils are amorphous structure with lower melting temperature, for example, Zr52Ti22Ni18Cu8 filler's is 538°C. The technical parameters in brazing welding are: welding temperature T = 800°C; heating electric current I = 25 A; heating time t = 15 s and holding time t = 15 s, in the case of these conditions, the jointing head shear strength of TA2/Zr52Ti24Ni13Cu11/Q235 is 139 MPa. Fracture is mainly located in the brazing seam. The white brittle intermetallic TiFe, TiFe2 and enhancement TiC spread in the center zone of brazing seam.

## Keywords

Titanium/Steel Brazing, Amorphous Brazing Filler, Joint Microstructure, Shear Strength

## 1. Introduction

Since the 1950s, titanium has gradually become an important metal with high specific strength, low density,

---

*Corresponding author.

good thermostability, tenacity, thermal conductivity and fatigue resistance but high price. Q235 mild steel is a common engineering material with good performance and low price. So, if these two materials can be connected together to be used, their merits can be expressed better, which has good practical worth and economical benefit [1]. However, there is big difference between the physical and chemical properties of titanium and steel, which makes it hard to connect these dissimilar metals. Many methods can be used to connect titanium and its alloy at present [2]. And brazing with simple technology, equipment and low welding temperature is the most appropriate for joining dissimilar metals. Titanium has active chemical property, so it must be brazed under vacuum or dry inert gas atmosphere.

At the moment, the brazing fillers applied to the brazing of titanium and its alloy can be divided into four kinds: Ag-based, Al-based, Pd-based and Ti-based fillers. Through rapid solidification, Ti based fillers can be made into amorphous brazing fillers which has uniform microstructure, little thickness, low welding temperature [3] and good brazing quality *et al*. Ti-Zr-Cu-Ni alloy is now considered to be the best amorphous filler for brazing titanium alloy, especially in high temperature and severe corrosion environments, but most of this kind of fillers are appropriate for TC and TB series of titanium alloys [4], rarely for the connection of commercial pure titanium TA2 and mild steel Q235. The melting temperatures of Ti-Zr-Cu-Ni brazing fillers is in a range from 840°C to 900°C lower than the phase inversion temperatures of most titanium alloys, such as the most widely used TC4, whose phase inversion temperature is a range of 980°C - 1000°C [5] [6]; however higher than the one of TA2, 882.5°C. During the process of heating, when welding temperature is as high as the phase inversion temperature of titanium, $\alpha$ phase transforms into $\beta$ phase with obvious coarsening tissue, then becomes acicular $\alpha$ phase during the subsequent cooling process, which makes the plasticity of the base metal TA2 reduced [7]. So it is urgent and hard to acquire a suitable brazing filler for bonding these dissimilar metals TA2 and Q235.

Therefore, the objective of this research is to lower the melting temperature of the Ti-Zr-Cu-Ni brazing filler in order to satisfy the requirement of the welding temperature for brazing TA2 and Q235, and to obtain a brazing filler with good performance appropriate to braze TA2 and Q235. In addition, effects of elements in brazing fillers, performance and microstructure of the fillers and joints will be investigated as well.

## 2. Experimental Work

Simple metals (99.99%) Ti, Zr, Cu, Ni were melted into alloy by high frequency induction heating equipment in argon atmosphere and brazing fillers were prepared by using a single roller rapid solidification apparatus. The experimental parameters can be seen in Reference [8].

Commercially pure titanium TA2 from Baoji Titanium industry CO. and Q235 mild steel in the form of 50 mm × 10 mm × 3 mm were used in hot rolled and annealed condition for brazing with Zr-Ti-Ni-Cu brazing fillers. The chemical composition of commercially pure titanium TA2 is Fe = 0.25, N = 0.01, O = 0.20, Ti balance. And the chemical composition of Q235 mild steel is C = 0.40, Si = 0.28, Mn = 0.52, P = 0.043, S = 0.040, Ni = 0.30, Cr = 0.29, Cu = 0.28, Fe balance.

The phase structure of the brazing filler was tested by D/MAX-1200 X-ray diffractometer. The melting temperatures of brazing fillers were tested by Netzsch DSC 404C differential scanning calorimetry. The brazing of TA2 and Q235 was conducted by vacuum high frequency brazier, the brazing parameters: vacuum degree is 0.1 Pa; welding temperature is 800°C; heating current is 25 A; heating time is 15 s; holding time is 15 s; cooling to room temperature is in furnace. The sample made along the axis of the welded sample was etched with the solution of 3 mL HF + 6 mL HNO$_3$ + 100 mL H$_2$O, the microstructures of brazing fillers were observed by using Olympus GX-71 metallurgical microscope and JSM-6700F type SEM scanning electron microscope, the shear strength of the joints was tested by WE-100 universal testing machine.

## 3. Results and Discussion

### 3.1. Composition of Zr-Ti-Ni-Cu Brazing Fillers

In Ti-Zr-Cu-Ni amorphous brazing fillers, Ni and Cu are stable elements to $\beta$ phase, which can form eutectic with titanium and reduce the melting temperature significantly [6]. Ni can improve the high temperature property and corrosion resistance of joints [9]. Cu can easily form a lot brittle intermetallics with titanium in joint, so the content of Cu should not be too much. Because of the alloying effect of Zr and Ti, Zr becomes one of the main added elements in Ti-based brazing filler. And Zr can form infinite solid solution with Ti, which can improve strength and keep plasticity. When the content of Zr in the alloy is 50%, the melting temperature of titanium alloy shows a minimum. Zr is neutral in titanium alloy, seldom having effect on the $\alpha$-$\beta$ phase inversion temperature, and it also can form eutectic with Ni and Cu [10]. Therefore, Cu, Ni and Zr are added into Ti-based brazing filler to design ZraTibNicCud, and each element has its content: $48 \leq a \leq 60$; $20 < b < 28$; $3 < d < 12$; $19 < c + d < 30$; $0.12 < d/(c + d) \leq 0.5$. In order to obtain the Zr-based amorphous brazing filler with good performance, optimized composition of the filler was designed by means of orthogonal experiment.

The orthogonal experiment $L_9(3^4)$ was arranged to search out the optimum brazing filler. In this experiment, the factors are the content of these four elements. According to the approximate content of each element above, three contents of every element were evenly chosen as the level of every factor. So there are 9 experiments with 9 brazing fillers. Indexes are the melting temperature; the tensile strength; the formability and wettability of every brazing foil. Due to the little difference of the tensile strength; the formability and wettability among all the designed fillers, the most important index is the melting temperature. The factors and levels are showing in **Table 1**.

It is the requirement of this experiment that the melting temperature of the brazing filler should be lower than the phase transition temperature of titanium, 882.5°C, as far as possible. According to this, the most important index, the results from range analysis show that: the dominant factor affecting the melting temperature is the content of Zr, Ti affects less, and Cu affects much less. The compositions of three fillers Zr52Ti22Ni18Cu8, Zr52Ti24Ni16Cu8 and Zr52Ti24Ni13Cu11 are obtained from the largest average combination of every factor's every level. The single effect trends of each element addition on melting temperature are shown in **Figure 1**.

### 3.2. Performance of Zr-Ti-Ni-Cu Brazing Fillers

Element Zr has strong glass-forming ability, so the brazing filler containing Zr, Ti, Ni and Cu can easily present amorphous structure. And this kind of researches has been proved a lot. In this experiment the brazing fillers prepared also have amorphous structure and the good performance of this structure. **Figure 2** shows the X-ray diffraction spectrum patterns of Zr52Ti24Ni16Cu8 brazing filler. In the pictures there is no peak according to crystal phase, but broad diffraction peaks belong to glassy phases only, which indicate the amorphous structure of Zr52Ti24Ni16Cu8 brazing filler. And all the designed fillers almost have the similar X-ray diffraction spectrums.

The nine brazing fillers with thickness of 40 μm - 60 μm, width of 4mm, prepared by a single roller rapid solidification apparatus, have high plasticity, tensile strength and are convenient to use. The tensile strength of the brazing fillers shows in **Table 2**. Because in the periodic table, the elements in the fillers are next to the elements in the base metal and they are easily mutually soluble with each other, the wettability of these 9 brazing fillers on base metals is good.

**Table 1.** Orthogonal factor level table.

| Level | Factor/at% | | | |
|:---:|:---:|:---:|:---:|:---:|
| | Zr | Ti | Cu | Ni |
| 1 | 49 | 22 | 5 | Balance |
| 2 | 52 | 24 | 8 | Balance |
| 3 | 55 | 26 | 11 | balance |

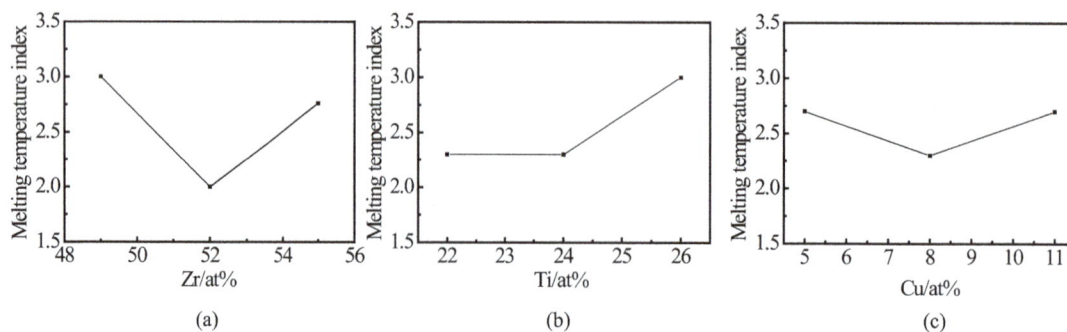

**Figure 1.** Single effect of element addition on melting temperature: (a) Zr; (b) Ti; (c) Cu.

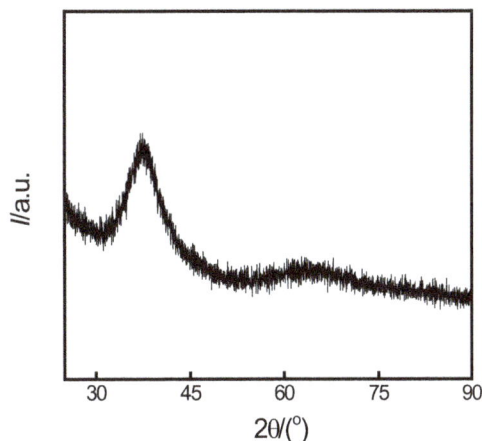

**Figure 2.** XRD spectrum of Zr52Ti24Ni16Cu8 amorphous alloy.

**Table 2.** Strength of extension of the Zr-based fillers.

| Brazing alloy | Tensile strength/(MPa) |
|---|---|
| Zr52Ti22Ni18Cu8 | 318.89 |
| Zr52Ti24Ni16Cu8 | 316.84 |
| Zr52Ti24Ni13Cu11 | 307.96 |

**Table 3** shows the three Zr-based fillers' melting temperature ranges, it can be seen that the melting temperatures of these three fillers are around 600°C, so the welding temperatures are absolutely under 882.5°C, the $\alpha{\to}\beta$ phase inversion temperature of TA2, which meets the requirement of this experiment ensuring that the base metal TA2 can keep its fine microstructure and good properties after brazing. The DSC curves of three fillers are showing in **Figure 3**.

## 3.3. Microstructure of TA2/Zr-Ti-Ni-Cu/Q235 Brazing Joints

The microstructure of overlap brazing joints is observed. **Figure 4(a)** shows the whole morphology of the TA2/Zr52Ti22Ni18Cu8/Q235 joint. There appears three zones from top to bottom: base metal TA2 zone, brazing seam zone and base metal Q235 zone. The base metal TA2 near the brazing seam presents sawteeth shape for it partly converts into lath-like structure of $\beta$ phase. However, the base metal TA2 away from the brazing seam remains the original structure of $\alpha$ phase. Most of the microstructure in the brazing seam zone is dendrite. The light color region in the brazing seam near the base metal Ti is the transition region between the seam and the base metal Ti. There is not an obvious transition region between the seam and the base metal Q235 but a dark color boundary line and the microstructure beside the boundary line is also coarse.

**Figure 4(b)** shows the microstructure of the overlap joint TA2/Zr52Ti24Ni16Cu8/Q235. For the brazing fillers Zr52Ti24Ni16Cu8 and Zr52Ti22Ni18Cu8 have the almost same compositions, the microstructure of the two joints with the two fillers are very similar. The brazing seam zone of the overlap joint TA2/Zr52Ti24Ni16Cu8/Q235 mainly consists of coarse dendrites. A light color transition region emerges between the seam and the base metal Q235, whose microstructure presents sawteeth shape. There is not an obvious transition region between the seam and the base metal Q235 but a dark color boundary line.

It can be seen in **Figure 4(c)** showing the microstructure of the overlap joint TA2/Zr52Ti22Ni13Cu11/Q235, that boundaries between the three zones are clear. And between the seam and the base metal TA2, there is also a transition region that consists of light color upper layer and dark color lower layer. There is a narrow transition region between the seam and the base metal Q235, which is not like the joints with the other two brazing fillers, and some white phases emerge on the boundary between the transition region and the seam.

In the seam of the TA2/Zr52Ti24Ni13Cu11/Q235 joint, white region and black region constitute the substrate, on which white dotted phases with different size do not distribute uniformly and small phases gather to form cluster. The center zone of the seam is showing **Figure 5**. Through spectrum quantitative analysis of point A, it

**Table 3.** Melting range of Zr-Ti-Ni-Cu fillers.

| Brazing alloy | Ts/°C | Tl/°C |
|---|---|---|
| Zr52Ti22Ni18Cu8 | 538 | 698 |
| Zr52Ti24Ni16Cu8 | 640 | 741 |
| Zr52Ti24Ni13Cu11 | 613 | 740 |

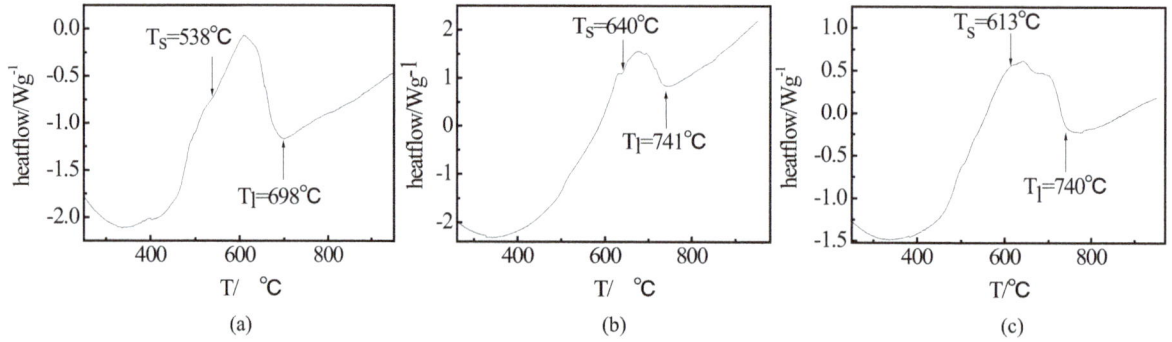

**Figure 3.** DSC curves of brazing ribbons: (a) Zr52Ti22Ni18Cu8; (b) Zr52Ti24Ni16Cu8; (c) Zr52Ti24Ni13Cu11.

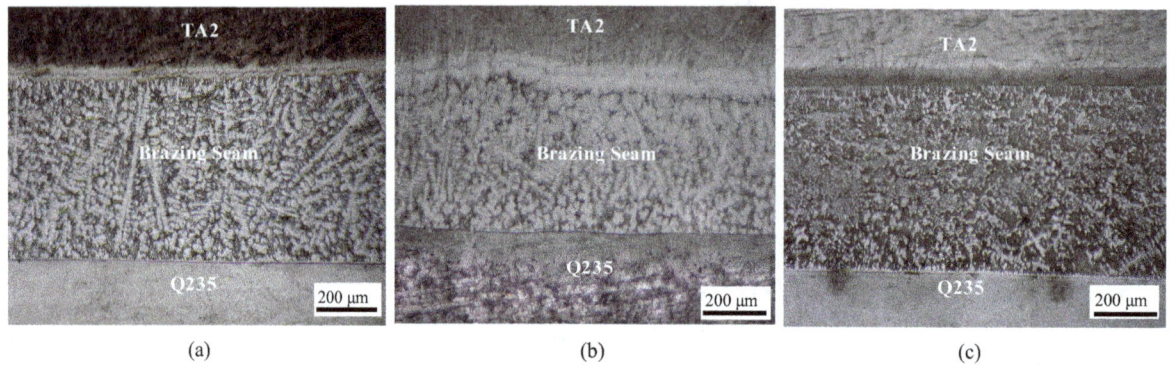

**Figure 4.** The microstructure of TA2/Q235 joints: (a) TA2/Zr52Ti22Ni18Cu8/Q235; (b) TA2/Zr52Ti24Ni16Cu8/Q235 and (c) TA2/Zr52Ti24Ni13Cu11/Q235

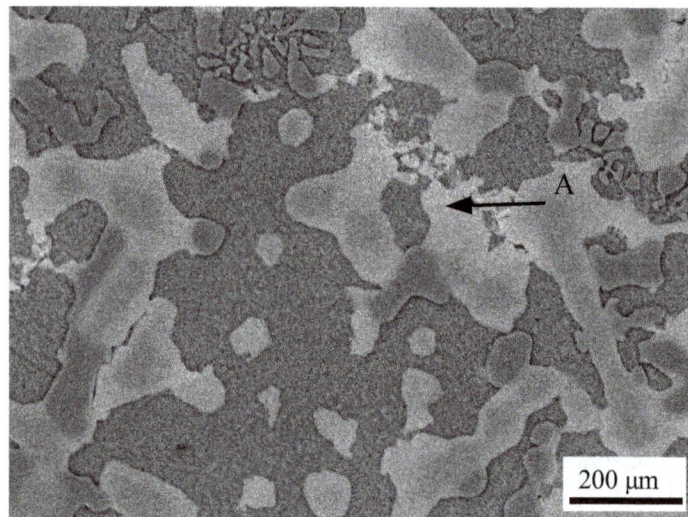

**Figure 5.** Microstructure of TA2/Zr52Ti24Ni13Cu11/Q235.

is identified that the white dotted phase is consist of elements C, Ti, Fe and less Zr, Cu, Ni, and the content of Ti is 40.6%, Fe 20.3%, C 28.02%. So the phase contains TiFe2 and TiC compounds.

By analyzing and comparing the microstructure of TA2/Zr52Ti22Ni18Cu8/Q235, TA2/Zr52Ti24Ni16Cu8/Q235 and TA2/Zr52Ti24Ni13Cu11/Q235 joints, it is found that most microstructures of TA2/Zr52Ti22Ni18Cu8/Q235 and TA2/Zr52Ti24Ni16Cu8/Q235 are coarse dendrites, and no transition region emerges between the seam and the base metal Q235; in TA2/Zr52Ti24Ni13Cu11/Q235 joint, there are white brittle intermetallic compounds TiFe2 and wild phase TiC with different size distributing in the seam, and there is an obvious light color transition region between the seam and the base metal Q235. By comprehensive comparison, the microstructure of TA2/Zr52Ti24Ni13Cu11/Q235 joint is better.

## 3.4. Mechanical Property of TA2/Zr-Ti-Ni-Cu/Q235 Brazing Joints

The shear strength of the brazing joints was tested, and the highest strength is 139 MPa. From the appearance of facture, it can be seen that the fractures are mainly located in the center of brazing seam.

## 4. Conclusions

1) In this paper, Ti-Zr-Cu-Ni amorphous filler was redesigned and optimized by using orthogonal experiment to obtained three Zr-Ti-Ni-Cu amorphous fillers with low melting temperature. The foils were prepared by using a single roller rapid solidification apparatus and high induction frequency brazing of TA2 and Q235 was conducted.

2) The brazing foils with amorphous structure have high tensile strength and low melting temperature, under 882.5°C, which meets the requirement of the welding temperature.

3) When the brazing parameters are welding temperature T = 800°C, heating current I = 25 A, heating time t = 15 s, holding time t = 15 s, the shear strength of the TA2/Zr52Ti24Ni13Cu11/Q235 joint is 139 MPa. The fractures are mainly located in the seam and white brittle intermetallic TiFe, TiFe2 and reinforced phase TiC spread in the center of the seam.

## Acknowledgements

This work was supported by the scientific research project of Shaanxi province science and technology department, the service local special projects of Shaanxi province education department, the integrated innovation plan of Xi'an technology bureau and the western material innovation fund.

## References

[1]   Onzawa, T., Suzumura, A. and Ko, M. (2011) Structure and Mechanical Properties of CP Ti and Ti-6Al-4V Alloy Joints Brazed with Ti-Based Amorphous Filler Metals. *Journal of the Japan Welding Society*, **5**, 205-211.

[2]   Qi, Y., Zhang, Y.H. and Quan, B.Y. (2003) Development and Application of Braze Welding and Ti-Based Braze Material. *Metallic Functional Materials*, **10**, 31-37.

[3]   Huang, Y.J., *et al.* (2008) Formation, Thermal Stability and Mechanical Properties of $Ti_{42.5}Zr_{7.5}Cu_{40}Ni_5Sn_5$ Bulk Metallic Glass. *Science in China Series G: Physics, Mechanics and Astronomy*, **51**, 372-378. http://dx.doi.org/10.1007/s11433-008-0049-y

[4]   Shapiro, A.E. and Flom, Y.A. (2012) Brazing of Titanium at Temperature below 800°C: Review and Prospective Applications. *Welding Journal*, **50**, 1-22.

[5]   Chang, H. and Luo, G.Z. (1995) Development of Ti Alloy Used Brazing Filler Metals. *Rare Metal Materials and Engineering*, **24**, 15-20.

[6]   Zhang, Q.P. and Zhang Y.S. (2005) Technology and the Developmental Situation of the Titanium Alloy. *Aerodynamic Missile Journal*, **7**, 56-64.

[7]   Takemoto, T. (1988) Intermetallic Compounds Formed during Brazing of Titanium with Aluminum Filler Metals. *Journal of Material Science*, **6**, 1301-1308. http://dx.doi.org/10.1007/BF01154593

[8]   Xu, J.F. and Wei, B.B. (2004) Liquid Phase Flow and Microstructure Formation during Rapid Solidification. *Acta Physica Sinica*, **53**, 160-166.

[9]   Elrefaey, A. and Tillmann, W. (2007) Interface Characteristics and Mechanical Properties of the Vacuum-Brazed Joint of Titanium-Steel Having a Silver-Based Brazing Alloy. *Metallurgical and Materials Transactions*, **38**, 2956-2961.

http://dx.doi.org/10.1007/s11661-007-9357-5

[10] Zhai, Q.Y., Xu, J.F. and Cui, J. (2013) A Kind of Amorphous Brazing Fillers Applied to Brazing Series TA Titanium Alloy and Stainless Steel. The Chinese Patent No. ZL2013104891392.

# 10

# Prediction and Verification of Resistance Spot Welding Results of Ultra-High Strength Steels through FE Simulations

**Oscar Andersson[1], Arne Melander[1,2]**

[1]Department of Production Engineering, KTH Royal Institute of Technology, Stockholm, Sweden
[2]Swerea KIMAB, Kista, Sweden
Email: oanderss@kth.se

## Abstract

Resistance spot welding (RSW) is the most common welding method in automotive engineering due to its low cost and high ability of automation. However, physical weldability testing is costly, time consuming and dependent of supplies of material and equipment. Finite Element (FE) simulations have been utilized to understand, verify and optimize manufacturing processes more efficiently. The present work aims to verify the capability of FE models for the RSW process by comparing simulation results to physical experiments for materials used in automotive production, with yield strengths from approximately 280 MPa to more than 1500 MPa. Previous research has mainly focused on lower strength materials. The physical weld results were assessed using destructive testing and an analysis of expulsion limits was also carried out. Extensive new determination of material data was carried out. The material data analysis was based on physical testing of material specimens, material simulation and comparison to data from literature. The study showed good agreement between simulations and physical testing. The mean absolute error of weld nugget size was 0.68 mm and the mean absolute error of expulsion limit was 1.10 kA.

## Keywords

Resistance Spot Welding, FE Simulations, High Strength Steel, Material Modeling, Weld Size

## 1. Introduction

Resistance spot welding (RSW) is the primary joining method in automotive industry due to its low cost and high ability for automation. The main measures for evaluation of the RSW process are weld size, which is indicative of both tensile strength and process robustness, together with the occurrence of expulsion, which risks weld defects and damage to surrounding equipment.

The automotive industry is facing increasing demands for lightweight structures with maintained safety. An efficient solution to meet such demands is the introduction of high strength (HSS) and ultra high strength (UHSS, sometimes also referred to as advanced high strength steels, AHSS) in body-in-white components. While higher strength materials effectively enables reduction of car weight, the material behavior of HSS and UHSS, due to their metallurgy and alloying elements, also imposes challenges for joining in manufacturing [1].

Numerical models are utilized to reduce costs for product verification and optimization in a multitude of fields including material processing. The first numerical analyses of the RSW process were published in the early 1960s. Both a one-dimensional model by Archer [2] and an axi-symmetrical model by Greenwood [3] were presented and calculated the temperature change in the weld zone. However, these early models were not capable of predicting weld sizes. Nied [4] developed an axi-symmetrical FE model which used elastic material behavior and Nishiguchi and Matsuyama [5] implemented elasto-plastic material behavior. Both designated welding simulation software and general simulations software have been used in published work. Sysweld [6], jwrian [7] and sorpas [8] are FEM software designated for welding purposes and incorporates thermo-electro-mechanical modeling and modeling of the sheet contact conditions.

As the models achieved increased level of detail and accuracy, research also focused on comparison and verification of physical experiments. Several studies have been made to verify the prediction of the FE simulations on weld sizes. Simulations were done by Tsai *et al.* [9] for mild steel, by Gupta and De [10] for HSLA steel, by Long and Khanna [11] for AISI 1010 steel and by Moshayedi and Sattari-Far [12] for 304L austenitic steel with good agreement to experiments. Nodeh *et al.* [13] verified result of mild steel with a 3D model with good agreement. Shen *et al.* [14] verified results of 3-sheet joints of combinations of DP600 and mild steel and reported that the model somewhat underestimated weld sizes in all interfaces. In addition, Afshari *et al.* [15] made comparative studies of 6061 aluminum alloy welds with good agreement between simulations and experiments.

Verification of simulations of UHSS is not as extensively investigated. Dancette *et al.* [16] showed good agreement between simulations and experiments for DP980 steel and Radakovic *et al.* [17] showed results of DP780 in a study which focused on failure modes of spot welds. However, no simulations of press-hardened or fully martensitic steels have been found in the literature.

The present paper includes verifications of FE simulations by comprehensive physical experiments. The experiments provide data on weld sizes and the occurrence of expulsion. The FE model used a new material database to describe the behavior of UHSS steels during welding. An investigation of the potential of a process planning tool based on FE simulations has been carried out.

## 2. Experimental Procedure

An extensive experimental procedure was carried out to find weld results for the analysed materials. Six materials were welded, ranging from advanced high strength steels as press hardened boron alloyed steel to lower strength steels. The materials' mechanical properties and alloying content are presented in **Table 1** and **Table 2**, respectively.

The welding was performed as 2-sheet overlap welds on sheets of dimensions 40 mm by 120 mm, with two welds on each sheet pair. The geometry and fixturing of the sheets are illustrated in **Figure 1** and **Figure 2**.

**Table 1.** Materials used in experiments.

| Sheet material | Annotation | Coating | Thickness [mm] | RP0.2 [MPa] | Rm [MPa] | A80 [%] |
|---|---|---|---|---|---|---|
| Boron steel | BO1500 | AlSi | 1.50 | 1100 | 1650 | 5 |
| Rephosphorized steel | RP260 | Uncoated | 1.50 | 283 | 436 | 34 |
| DP600 | DP600 | Z75 | 1.25 | 425 | 647 | 21 |
| DP800 | DP800 | Z75 | 1.50 | 519 | 834 | 18 |
| DP1000 | DP1000 | Z75 | 1.50 | 800 | 1091 | 8 |
| Martensitic steel | MS1400 | Z75 | 1.30 | 1120 | 1329 | 7 |

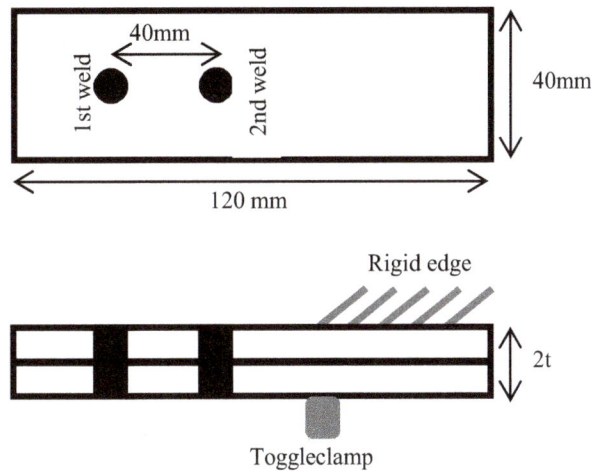

**Figure 1.** Sheet dimension and fixturing.

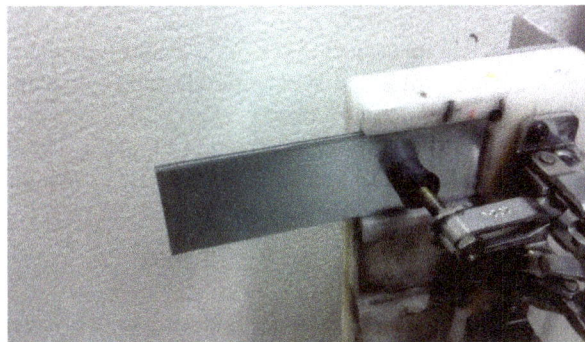

**Figure 2.** Welding fixture before welding.

**Table 2.** Alloying elements of materials.

| Sheet material | C | Si | Mn | P | S | N | Cr | Ni | Cu | Al | Nb | Ti | B |
|---|---|---|---|---|---|---|---|---|---|---|---|---|---|
| BO1500 | 0.004 | 0.200 | 0.540 | 0.072 | 0.006 | 0.000 | 0.020 | 0.040 | 0.010 | 0.030 | 0.000 | 0.080 | 0.000 |
| RP260 | 0.226 | 0.260 | 1.170 | 0.009 | 0.005 | 0.000 | 0.220 | 0.038 | 0.014 | 0.048 | 0.000 | 0.013 | 0.036 |
| DP600 | 0.100 | 0.21 | 1.620 | 0.008 | 0.002 | 0.006 | 0.470 | 0.040 | 0.010 | 0.050 | 0.000 | 0.000 | 0.000 |
| DP800 | 0.150 | 0.200 | 1.720 | 0.012 | 0.003 | 0.005 | 0.420 | 0.040 | 0.010 | 0.040 | 0.020 | 0.000 | 0.000 |
| DP1000 | 0.150 | 0.500 | 1.510 | 0.010 | 0.002 | 0.003 | 0.040 | 0.050 | 0.010 | 0.030 | 0.020 | 0.000 | 0.000 |
| MS1400 | 0.130 | 0.230 | 1.510 | 0.009 | 0.003 | 0.005 | 0.030 | 0.050 | 0.010 | 0.050 | 0.020 | 0.000 | 0.000 |

The welding process can be divided into two time periods, weld time and hold time. The weld time refers to the period when a current is passed between the electrodes and the hold time refers to the period after the weld time when the electrode force are still applied to the sheets but no current is passed between them.

The weld parameters of the experiments, weld time, hold time and electrode force and weld currents, are presented in **Table 3**. The minimum welding current was defined as the current which generated a weld diameter of approximately $5\sqrt{t}$. The maximum welding current was defined as the current which generated two or three expulsion at the second weld location. An increment of 0.3 kA was used within the welding current range. Between each current step, the electrode tips were dressed. Each weld was repeated three times.

Welding was performed with a MFDC PSI6000 inverter and a PSG3050 transformer from Bosch Rexroth. An ISO 8521 B-type electrode of CuCrZr alloy was used.

**Table 3.** Process parameters of experimental welding.

| Material | Weld time [ms] | Electrode force [kN] | Hold time [ms] | Minimum weld current [kA] | Maximum weld current [kA] |
|----------|----------------|----------------------|----------------|---------------------------|---------------------------|
| BO1500   | 330            | 3.6                  | 150            | 5.0                       | 10.1                      |
| RP260    | 400            | 4.5                  | 150            | 6.0                       | 9.9                       |
| DP600    | 340            | 4.1                  | 150            | 6.6                       | 10.2                      |
| DP800    | 330            | 3.8                  | 150            | 7.2                       | 9.9                       |
| DP1000   | 400            | 4.0                  | 150            | 6.6                       | 9.0                       |
| MS1400   | 350            | 4.1                  | 170            | 6.7                       | 10.0                      |

The welded coupons were destructively tested using coach peel tests to measure the weld nugget by a digital calliper, as seen in **Figure 3**. The minimum and maximum diameters were measured to allow for elliptic nuggets and the arithmetic mean of the measurements was used for analysis. Expulsion was visually observed during experiments.

## 3. Finite Element Model

A coupled thermo-electro-metallographical-mechanical axi-symmetrical FE model was developed and computed using FE code SYSWELD. The model is built of axi-symmetric 2D solid elements, with degrees of freedom in displacement (radial and vertical), temperature and voltage. The metal sheets are composed of gradually increasing fineness of elements towards the symmetry line, as seen in **Figure 4**. The sheet elements are 0.1 mm in height for all sheet materials. The thermo-electrical and mechanical models, including governing equations, material models and boundary conditions are treated separately below.

### 3.1. Thermo-Electrical Model

Thermally, the RSW process can be described by the general electro-kinetic equation, as stated in (1), which adds a Joule heating term to the commonly used general heat equation.

$$\rho c \frac{\partial T}{\partial t} - \nabla \left( \lambda \nabla T \right) - \frac{I^2}{\kappa} - Q = 0 \tag{1}$$

where temperature is designated $T$ and time $t$. The first term includes the enthalpy of the system, $c$ is the specific heat and $\rho$ is the density. The second term includes the conductive heat component, where $\lambda$ is the thermal conductivity. The third term includes the electro-thermal Joule heat, where $I$ is the electrical current and $\kappa$ is the electrical conductivity. The fourth term includes the internal heat generation, which is negligible due to pure Joule heating in the RSW process.

For the simulation material data input, the software Thermo Calc was used to determine the specific heat for all materials and microstructures based on thermodynamic equilibrium [18]. Thermo Calc calculated the specific heat from the alloying contents in the materials. Due to small variations between specific heat, a common model was used for all materials, shown in **Figure 5**. The density and electric conductivity were provided through the FE software and are shown in **Figure 5**.

The thermal conductivity data was generated using the material simulation software IDS [19]. The software uses thermal conductivity of the pure alloying elements, which are weighted against the thermal conductivity of pure iron. The thermal conductivity is shown in **Figure 6**.

The electrical boundary conditions are a description of the applied welding current and the physical boundaries of the sheet. No current is flowing at the water channel, at the symmetry line or at the metal sheet surface facing air, as indicated in **Figure 4**. The electric potential is applied between the electrodes, creating the Joule heating.

The metal surface-to-media heat exchanges are defined at the electrode-water interface, at the metal-air sheet interface and at the symmetry line, respectively, as lumped conductance according to (3) to (2) below.

$$-k \frac{\partial T}{\partial n} = 0 \tag{2}$$

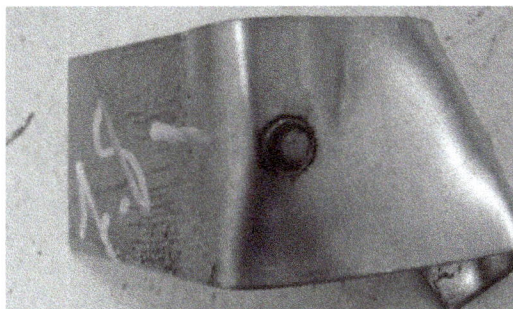

**Figure 3.** Coach-peel tested coupon.

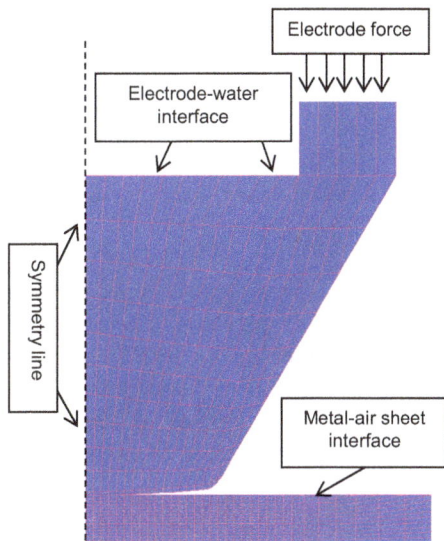

**Figure 4.** Axi-symmetrical mesh of upper electrode and sheet.

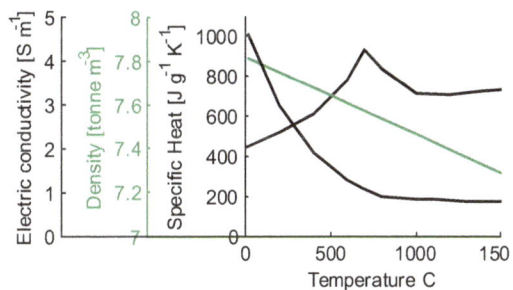

**Figure 5.** Electrical and thermal material properties.

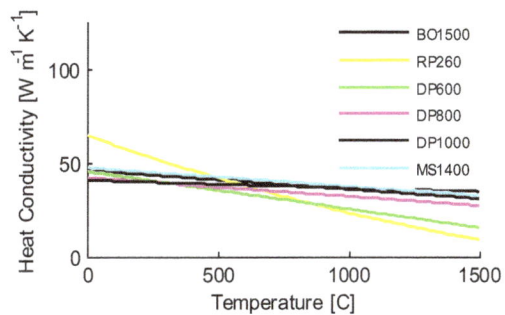

**Figure 6.** Heat conductivity.

$$-k\frac{\partial T}{\partial n} = k'\left(T_w - T\right) \tag{3}$$

$$-k\frac{\partial T}{\partial n} = k'\left(T_\infty - T\right) \tag{4}$$

where $n$ is the surface normal coordinate, $k'$ [W/(mK)] is a controlling conduction factor and $T_w$ is the water temperature (20˚C) and $T_\infty$ is the ambient temperature (20˚C).

## 3.2. Mechanical Model

The solution of the thermo-electric model gives a temperature field which causes mechanical strains due to thermal expansion. The data for simulations is based on dilatometer experiments where the temperature of a test specimen was controlled and the volume changes were measured. The temperature was changed in order to gain all three microstructures; material as delivered, austenite and martensite. Results showed that the standard deviation between materials of the same microstructure is small. Thus, the thermal expansion coefficient is constant at all temperatures at 0.125e−4 $K^{-1}$ for the material as delivered, 0.209e−4 $K^{-1}$ for the austenite and 0.105e−4 $K^{-1}$ for the martensite.

The strain components of the mechanical model can be decomposed as in (5).

$$\varepsilon = \varepsilon_{el} + \varepsilon_{pl} + \varepsilon_{me} \tag{5}$$

where $\varepsilon_{el}$ is the elastic strain component, $\varepsilon_{pl}$ is the plastic strain component and $\varepsilon_{me}$ are strains caused by martensitic transformations due to rapid cooling. The elastic strains are modeled by a temperature dependent Hooke's law through Young's modulus and Poisson's ratio, which were generated from tensile tests at elevated temperatures. The Young's modulus was obtained using least-squares linear regression of the bottom part of stress-strain curves obtained from tensile test curves. As expected, variations of Young's modulus between materials were low and the same elastic model was used for all materials. A linear regression over temperature was generated as seen in **Figure 7**.

A conventional method of defining a yield stress is by locating a specific divergence value between the two curves, *i.e.* a specific plastic strain magnitude. In the present work, the yield stress is defined as the stress where the plastic strain is 0.2%. As for the Young's modulus, the yield stress was modeled using linear regression for each material, see **Figure 8**.

Regarding flow stress, optimized Hollomon curve stress-strain relations were optimized using MATLAB's [20] optimization toolbox to match the results from the tensile tests for each material and temperature. Examples of resulting stress-strain curves for UHSS are shown in **Figure 9**.

The metallurgical transformation strains are defined according to Johnson-Mehl-Avrami's transformation kinetics model. The martensitic transformation temperatures are defined according to **Table 4** and found through dilatometer tests.

The electrode force is applied at the top nodes of the top electrode and a mechanical boundary condition is imposed on the bottom nodes of the bottom electrode by locking all displacement components. Furthermore, the radial displacements are locked along the symmetry line.

**Table 4.** Martensitic transformation temperature.

| Material | Martensitic transformation [˚C] |
|---|---|
| BO1500 | 380 |
| RP260 | 600 |
| DP600 | 412 |
| DP800 | 443 |
| DP1000 | 412 |
| MS1400 | 408 |

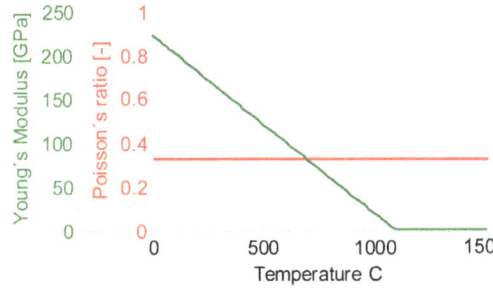

**Figure 7.** Elastic material properties.

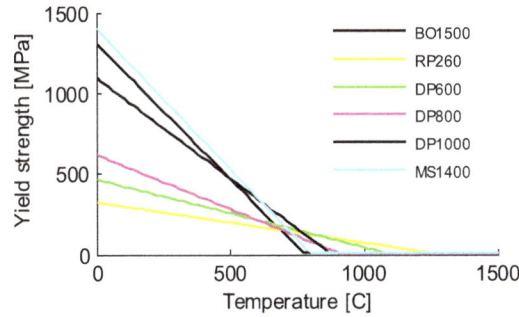

**Figure 8.** Yield strength.

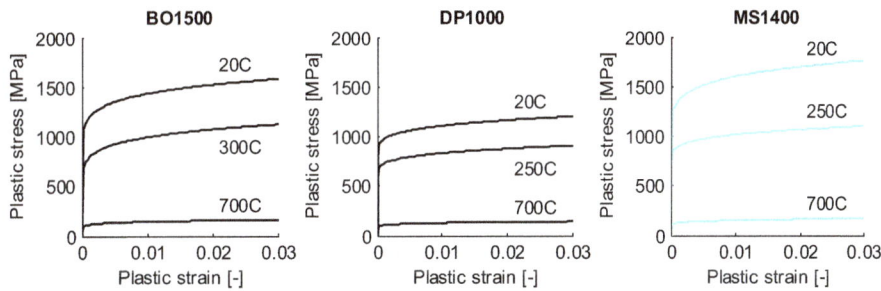

**Figure 9.** Plastic behaviour of UHSS materials at elevated temperatures.

After the complete welding cycle is finished, the electrodes are removed from the sheets and a new set of boundary conditions are applied to the model. Firstly, the radiation component of the sheet surface heat transfer is neglected over the entire sheet. Secondly, the electrical contact condition is removed. Thirdly, both electrodes are set to a vertical displacement of 3 mm away from the sheets.

In post-processing, the weld nugget is defined as the material which has reached a peak temperature above the solidus temperature, defined as in **Table 5** according to Thermo Calc simulations.

It is also of interest to predict the expulsion limit through the FE model. Shen *et al.* [21] introduced the factor $\eta$, the ratio between the weld nugget radius ($R_n$) and contact radius between the sheets ($R_c$) in order to predict expulsion. A requirement for expulsion was formulated as shown in (6).

$$\eta = \frac{R_n}{R_c} > 1 \tag{6}$$

In the FE post-processing, expulsion was reached if the molten zone was in contact with the air gap between the sheets. The present paper evaluated the capability of the $\eta$-factor method on a broad sheet material range.

## 4. Results

The experimental welding results were recorded between a low current which gave a nugget diameter of approximately $5\sqrt{t}$ and a welding current where expulsion occurred repeatedly at both weld locations on the

**Table 5.** Solidus temperatures

| Material | Solidus temperature [°C] |
|----------|--------------------------|
| BO1500 | 1473 |
| RP260 | 1520 |
| DP600 | 1492 |
| DP800 | 1482 |
| DP1000 | 1482 |
| MS1400 | 1486 |

coupon with steps of 0.3 kA between. Due to the high weldability of the stack-ups, an extensive amount of data was collected for verification.

Due to the different alloying elements and microstructures of the materials, the coach-peel tests resulted in different failure modes. The boron steel coach-peel tests resulted in interfacial failures whereas the other steels gave pullout plugs.

Expulsion was identified and recorded during the experimental welding. The expulsion occurrence can be seen as discrete probability distributions for both the first and second weld location. The expected value of the expulsion limit, $I_{\text{lim}}$, can be calculated as in (7).

$$I_{\text{lim}} = \sum I P_{\text{exp}} (I) \tag{7}$$

where $P_{\text{exp}}$ denotes the proportion of expulsion welds at each weld current. When two of three welds at the second location showed expulsions, the weld lobe was ended.

## Simulation Results and Discussion

Each experimental test was compared with a FE simulation using the simulation method described above. The recorded results of the simulation were the weld nugget diameter and the possible occurrence of expulsion in the simulation model.

During welding of the second weld, a proportion of the current will pass through the first weld due to the high conductance at the already bonded material. Thus, the current passing through the faying surface at the second weld location will be reduced compared to the nominal current, resulting in lower heat generation by Joule heating. Consequently, the second weld on the coupon showed a somewhat lower weld nugget size than the first weld in the experiments.

In general, the difference in size between the first and second weld on the coupons with identical current was on average 0.38 mm. This difference is depends on the distance between the weld of 40 mm. The simulation does not take shunting into account and is thus more comparable to the first weld. However, the second weld is of higher interest for process planning since shunting currents occur in almost all industrial welding. Thus, it is of interest to see the accuracy of simulations compared to experiments for both welds.

The results show a good agreement between weld sizes from simulation and physical welding. The agreement is illustrated in **Figure 10** and quantified by mean absolute errors in **Table 6**.

The mean absolute error between the weld size diameter of the FE model and the experiments is 0.53 mm for all welds. The first weld is more accurately modeled with a mean absolute error of 0.42 mm compared to 0.64 mm for the second weld. The better agreement of the first weld is expected due to the similar boundary conditions with regard to shunting currents. The results indicate that shunting currents are significant to model in order to improve accuracy of simulations.

It is of interest to compare the mean absolute error of the FE model to the standard deviation of weld size results in physical testing, since variations in experimental test results occur. Previous research has shown the standard deviation of weld sizes in experiments to be approximately 0.3 mm [22]. Thus, the error of the FE model is 76% larger than the variation of the experiment results themselves. However, for specific materials, the mean absolute error is as low as 0.22 mm (boron steel), 0.38 mm (martensitic steel) and 0.34 mm (rephosphorized steel), which is comparable to the variations in experimental outcome.

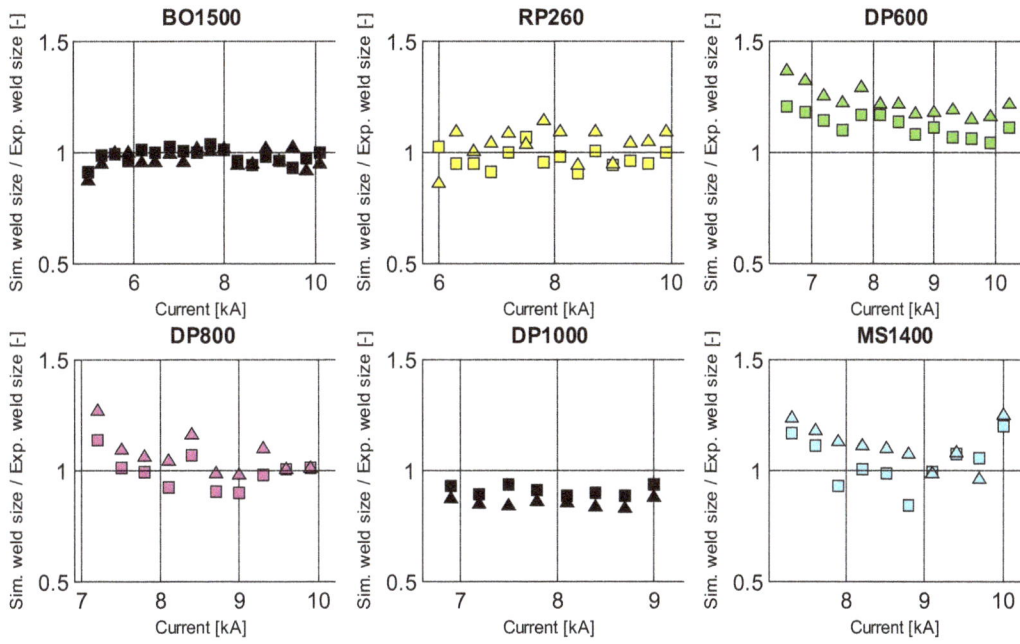

**Figure 10.** Resultsfromsimulationsandphysicaltesting of weld nugget size. Squaresdenotethe 1st physical weld andtrianglesdenotethe 2nd physical weld.

**Table 6.** Mean absolute error between experimental and simulation results of weld sizes.

| Material | First weld on coupon [mm] | Second weld on coupon [mm] | Both welds on coupon [mm] |
|---|---|---|---|
| BO1500 | 0.19 | 0.25 | 0.22 |
| RP260 | 0.27 | 0.41 | 0.34 |
| DP600 | 0.71 | 1.24 | 0.97 |
| DP800 | 0.54 | 0.66 | 0.60 |
| DP1000 | 0.72 | 1.22 | 0.97 |
| MS1400 | 0.34 | 0.41 | 0.38 |
| All materials | 0.42 | 0.64 | 0.53 |

The proportion of expulsions in the experiments gave an expected value of the expulsion limit. Also, the simulations gave an expulsion limit based on the weld zones contact with the air gap, as described above.

Again, the shunting effects during the second welding results in lower weld current and will also affect the expulsion limit. The results showed that the average difference of expected expulsion limit between the first and second weld location is 0.33 kA. Again, this relation is proportional to the shunting distance between the welds of 40 mm.

The agreement between expulsion limits in experiments and simulations is illustrated in **Figure 11**. The results show that the expulsion limit is generally underestimated in simulations and that the mean absolute error between expulsion limit between the simulations and the physical testing is 1.10 kA. The mean absolute error of the first weld location is 1.04 kA and 1.16 kA for the second weld location. Thus, it can be concluded that the expulsion limit better capture the expulsion of the first weld location, which is expected since no shunting effects are modeled in the simulations.

As with weld sizes, expulsion limits has variations in physical testing. By considering the expulsion occurrence as a discrete probability distribution, as with the expected value, the standard deviation of expulsion limit can be measured. The average standard deviation of expulsion limits is 0.34 kA and 0.08 kA for the first and second weld location, respectively. The standard deviation of all welds is 0.21 kA.

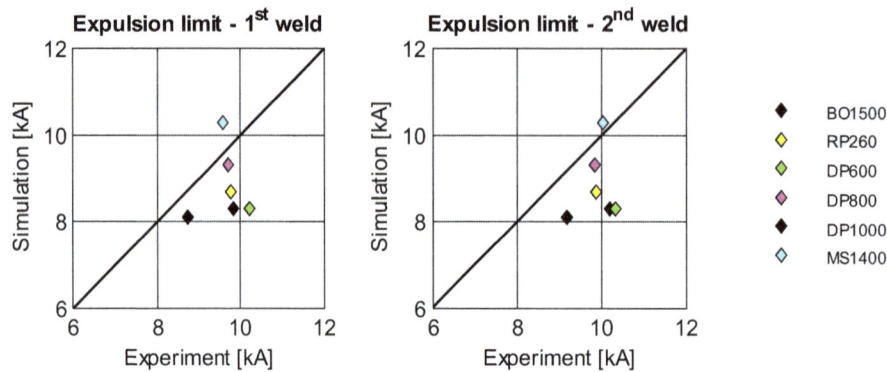

**Figure 11.** Results from simulation and physical testing of expulsion limit.

The mean absolute error of the expulsion limit in simulation compared to experiments is higher than the standard deviation of the expulsion limit in physical testing. For a more detailed study of expulsion limits, smaller current steps than 0.3 kA should be used for analysis, which would give a higher resolution of results.

## 5. Conclusions

The major points of the present paper can be summarized as follows.
- Resistance spot weld experiments of six steel types, including UHSS, have been performed in a semi-industrial environment using weld equipment used in automotive production and assessed using coach-peel tests of weld sizes and expulsion occurrence.
- Material data for FE simulations of the RSW process have been generated from experiment tests of material specimens, material simulations and literature sources.
- An electro-thermo-mechanical FE model has been developed based physical phenomena in resistance spot welding.
- The simulations showed an agreement with the experimental results with a mean absolute error of weld sizes of 0.68 mm with all welds taken into account Comparably, variations of weld sizes in apparently identical physical testing is 0.30 mm.
- The simulations showed an agreement with the experimental results with a mean absolute error of the expulsion limit of 1.10 kA when all welds were taken into account. Comparably, variations of expulsion limit in physical testing are 0.21 kA.

In conclusion, in the present paper weld sizes and occurrence of expulsion have been quantified for a wide range of steel types of different alloying content and heat treatment. The study shows the potential and limitations of the generated temperature-dependent material model and the FE model. The results can be used for implementation of FE simulations in process planning of RSW in modern manufacturing. The results show that the accuracy between physical experiments and simulations is comparable to the variations of weld output in industrial welding. Thus, the prediction accuracy of simulations is useful for engineers for process planning as the possible error is within magnitude of the process variations.

For further analysis of the inaccuracy of the simulations, the experimental methods should be extended with more detailed measurement of temperature history and stress-states in order to identify possible sources of errors in the FE model and material data. Although the accuracy of the simulations varies among materials, it should be noted that the common material model is fairly accurate. For further broadness of the results, it is of interest to extend the material model to other low-weight materials such as aluminum.

Due to the limitations of the FE software, the present study excluded effects of shunting currents in the simulations. For further accuracy of simulations for process planning, effects of shunt current should be included in the model to emulate industrial welding more accurately. Furthermore, a more advanced model of the steel, particularly its behavior in the molten state and the pressure state within the weld would likely increase the accuracy of expulsion prediction and increase the understanding of the causes of expulsion. Finally, as highlighted above, in industrial manufacturing the RSW inherits variations in both weld sizes and expulsion. Such variations are not treated in the simulations, which use nominal input data. To increase the capability of the simulations, a sensitivity analysis of the FE model and material data should be performed.

## Acknowledgements

This work was supported by the Swedish Governmental Agency for Innovation Systems (VINNOVA) [grant number 2009-01585] through the project Spot Light. The authors wish to thank the members of the steering committee of Spot Light for many contributions to the present work. The material testing and simulation was conducted by Johannes Gårdstam, Sven Haglund and Greta Lindwall at Swerea KIMAB.

## References

[1]   Weber, G., Thommes, H., Gaul, H., Hahn, O. and Rethmeier, M. (2010) Resistance Spot Welding and Weldbonding of Advanced High Strength Steels. *Materialwissenschaft und Werkstofftechnik*, **41**, 931-939. http://dx.doi.org/10.1002/mawe.201000687

[2]   Archer, G. (1960) Calculations for Temperature Response in Spot Welds. *Welding Journal*, **39**, 327-s-330-s.

[3]   Greenwood, J. (1961) Tempera in Spot Welding. *British Welding Journal*, **8**, 316-322.

[4]   Nied, H. (1984) The Finite Element Modelling of the Resistance Spot Welding Process. *Welding Journal*, **63**, 123s-132s.

[5]   Nishiguchi, K. and Matsuyama, K. (1987) Influence of Current Wave Form on Nugget Formation Phenomena When Spot Welding Thin Steel Sheet. *Welding in the World*, **25**, 222-244.

[6]   Feulvarch, E., Rogeon, P., Carr, P., Robin, V., Sibilia, G. and Bergheau, J. (2006) Resistance Spot Welding Process: Experimental and Numerical Modeling of the Weld Growth Mechanisms with Consideration of Contact Conditions. *Numerical Heat Transfer Part A*: Applications, **49**, 345-367. http://dx.doi.org/10.1080/10407780500359760

[7]   Murakawa, H., Zhang, J., Fujii, K., Wang, J. and Ryudo, M. (2000) FEM Simulation of Spot Welding Process. *Transactions of JWRI*, **29**, 73-80.

[8]   Zhang, W. (2003) Design and Implementation of Software for Resistance Welding Process Simulations. SAE Technical Papers 2003-01-0978. http://dx.doi.org/10.4271/2003-01-0978

[9]   Tsai, C.L., Dai, W.L., Dickinson, D.W. and Papritan, J.C. (1989) Analysis and Development of a Real-Time Control Methodology in Resistance Spot-Welding. *Welding Journal*, **70**, 339s-351s.

[10]  Gupta, O. and De, A. (1998) An Improved Numerical Modeling for Resistance Spot Welding Process and Its Experimental Verification. *Journal of Manufacturing Science and Engineering, Transactions of the ASME*, **120**, 246-251. http://dx.doi.org/10.1115/1.2830120

[11]  Long, X. and Khanna, S.K. (2003) Numerical Simulation of Residual Stresses in a Spot Welded Joint. *Journal of Engineering Materials and Technology*, **125**, 222-226. http://dx.doi.org/10.1115/1.1543968

[12]  Moshayedi, H. and Sattari-Far, I. (2012) Numerical and Experimental Study of Nugget Size Growth in Resistance Spot Welding of Austenitic Stainless Steels. *Journal of Materials Processing Technology*, **212**, 347-354. http://dx.doi.org/10.1016/j.jmatprotec.2011.09.004

[13]  Nodeh, I.R., Serajzadeh, S. and Kokabi, A.H. (2008) Simulation of Welding Residual Stresses in Resistance Spot Welding, FE Modeling and X-Ray Verification. *Journal of Materials Processing Technology*, **205**, 60-69. http://dx.doi.org/10.1016/j.jmatprotec.2007.11.104

[14]  Shen, J., Zhang, Y., Lai, X. and Wang, P.C. (2011) Modeling of Resistance Spot Welding of Multiple Stacks of Steel Sheets. *Materials and Design*, **32**, 550-560. http://dx.doi.org/10.1016/j.matdes.2010.08.023

[15]  Afshari, D., Sedighi, M., Karimi, M.R. and Barsoum, Z. (2013) On Residual Stresses in Resistance Spot-Welded Aluminum Alloy $6_061$-T6: Experimental and Numerical Analysis. *Journal of Materials Engineering and Performance*, **22**, 3612-3619. http://dx.doi.org/10.1007/s11665-013-0657-1

[16]  Dancette, S., Massardier-Jourdan, V., Fabrègue, D., Merlin, J., Dupuy, T. and Bouzekri, M. (2011) HAZ Microstructures and Local Mechanical Properties of High Strength Steels Resistance Spot Welds. *ISIJ International*, **51**, 99-107. http://dx.doi.org/10.2355/isijinternational.51.99

[17]  Radakovic, D.J. and Tumuluru, M. (2008) Predicting Resistance Spot Weld Failure Modes in Shear Tension Tests of Advanced High-Strength Automotive Steels. *Welding Journal* (*Miami, Florida*), **87**, 96-105.

[18]  (2008) Thermo-Calc for Windows Version 5 User's Guide.

[19]  Miettinen, J., Louhenkilpi, S., Kytönen, H. and Laine, J. (2010) IDS: Thermodynamic-Kinetic-Empirical Tool for Modeling of Solidification, Microstructure and Material Properties. *Mathematics and Computers in Simulation*, **80**, 1536-1550. http://dx.doi.org/10.1016/j.matcom.2009.11.002

[20]  (2007) MATLAB Documentation.

[21]  Shen, J., Zhang, Y. and Lai, X. (2010) Influence of Initial Gap on Weld Expulsion in Resistance Spot Welding of Dual

Phase Steel. *Science and Technology of Welding and Joining*, **15**, 386-392.
http://dx.doi.org/10.1179/136217110X12693513264213

[22]  Andersson, O. and Melander, A. (2011) Statistical Analysis of Variations in Resistance Spot Weld Nugget Sizes. IIW
      Annual Assembly, Chennai.

# 11

# A Study on Microhardness, Microstructure and Wear Properties of Plasma Transferred Arc Hardfaced Structural Steel with Titanium Carbide

**Balamurugan Sivaramakrishnan, Murugan Nadarajan**

Department of Mechanical Engineering, Coimbatore Institute of Technology, Coimbatore, India
Email: balu74_cit@yahoo.co.in, murugan@cit.edu.in

## Abstract

The Plasma Transferred Arc (PTA) hardfacing allows for homogeneous refined microstructure, low distortion and dilution resulting on enhanced surface properties when compared to hardfacing by conventional welding processes. This paper deals with PTA surfacing of a structural steel with a consumable containing TiC. TiC is a very hard refractory material finding increasing usage for wear resistance application. The composition and amount of heat input evidently affect the microstructure and properties of the hardfacing. The microstructure and microhardness of PTA hardfaced structural steel with TiC were investigated at different heat input conditions across the various zones. The influence of hardfacing parameters on the resulting microstructure, microhardness and wear resistance performance was evaluated. Wear resistance of the hardfaced surface was increased significantly.

## Keywords

PTA Hardfacing, Microhardness, Microstructure, Wear Resistance

## 1. Introduction

Hardfacing is a technique used to enhance surface properties of a metallic component as a specially designed alloy is surface welded in order to achieve specific wear properties [1]-[4]. Surface properties and quality of hardfacing depend upon the selected alloy and deposition process. Among the process employed for hardfacing,

Plasma Transferred Arc (PTA) allows for homogeneous mixture and refined microstructure, low distortion and dilution, resulting on enhanced surface properties [5]-[7]. A significant advantage of PTA hardfacing over traditional hardfacing welding processes is that the consumable material used in powder form [8].

The quality of the hardfacing primarily depends upon the hardness of the interface which is primarily governed by the heat input. It is known that one of the most important parameters controlling the wear behavior of the material is hardness [9]. Greater the hardness, greater will be the abrasion resistant of the materials. The strong interaction with the substrate during hardfacing requires analysis of each alloy system to optimize its properties and weldability. These hardfacings are used for several applications such as high temperature turbine for aerospace application, cutting tools, wear resistance surfaces on large agricultural, textile equipment petrochemical and pharmaceutical industries [10]-[13]. As TiC particles are useful for enhancing wear resistance, they have been widely used to improve surface characteristics.

TiC can be deposited by plasma transferred arc welding process and it has a strong interaction with the substrate, which determines the final microstructure and properties of hardfacing [14] [15]. The mechanical properties of hardfacing are directly related to its microstructure. Due to the difference in range of melting temperature between the substrate and hardfacing metal the dilution of the hardfaced layer could be significantly affected. When dilution is altered, it will influence the resulting microhardness and microstructure of the hardfacing.

The heat-affected zone (HAZ) is the area of base material, which has had its microstructure and properties altered by hardfacing operations. The heat from the hardfacing process and subsequent re-cooling causes this change from the weld interface to the termination of the sensitizing temperature in the base metal. The extent and magnitude of property change of base material depend primarily on the base material and the amount and concentration of heat input by the hardfacing process [16].

The thermal diffusivity of the base material plays a large role—if the diffusivity is high, the material cooling rate is high and the HAZ is relatively small. Alternatively, a low diffusivity leads to slower cooling and a larger HAZ. The amount of heat inputted by the hardfacing process plays an important role. To calculate the heat input for arc welding procedures, the following formula is used:

$$Q = \left( \frac{V * I * 60}{S * 1000} \right) * \text{effciency}$$

where $Q$ = heat input (kJ/mm), $V$ = voltage (V), $I$ = current (A), and $S$ = welding speed (mm/min). The efficiency depends on the welding process used. For PTA hardfacing efficiency is taken as 0.6 [17].

In this investigation, TiC was deposited on structural steel plates at low, medium and high heat input conditions by the PTA process and the hardfacings were characterized with the help of microhardness, microstructural and wears resistance analysis.

## 2. Experimental Procedure

The substrate material selected for the PTA hardfacing was IS: 2062 structural steel. The chemical composition of IS: 2062 structural steel and hardfacing alloy is presented in **Table 1**.

Using PTA hardfacing system, Titanium Carbide (TiC) was deposited onto the structural steel plate of size 150 mm × 100 mm × 25 mm. This was done by changing the welding parameters to achieve different heat input conditions: low, medium and high. Single hardfacing bead was laid on each plate. Samples were prepared from each hardfaced plate by cutting them at their centre perpendicular to hardfacing direction. Standard metallurgical procedures were employed to prepare the samples from PTA hardfacing deposited at different heat input condi-

**Table 1.** Chemical composition of base metal and hardfacing alloy.

| Sl. No | Material used | Elements, weight% | | | | | | | |
|--------|---------------|------|------|------|-------|-------|------|------|------|
|        |               | C    | Si   | Mn   | S     | P     | Mg   | Ti   | Fe   |
| 1      | IS:2062 (base metal) | 0.18 | 0.18 | 0.98 | 0.016 | 0.016 | -    | -    | bal  |
| 2      | Titanium carbide (TiC) (PTA powder) | 0.04 | 0.03 | 0.03 | -     | -     | 0.09 | 99.0 | 0.12 |

tions as shown in **Table 2**. Hardfaced plate and typical cross section are shown in **Figure 1** and **Figure 2** respectively.

The transverse sections of all the hardfaced samples were metallographicaly polished and etched. After etching and washing with distilled water, the specimens were subjected to microhardness, microstructure and wear resistance survey.

**Microhardness:** A Mitutoya (MVK-HI) microhardness tester was used to conduct the microhardness analysis

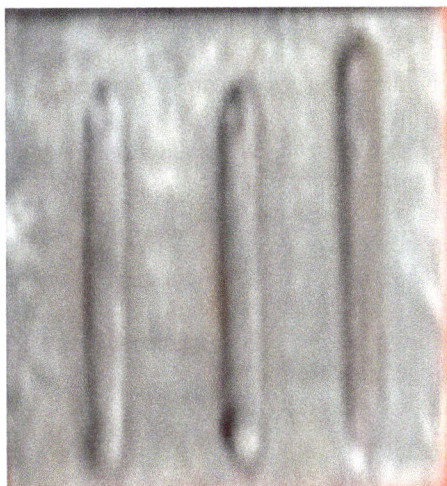

**Figure 1.** Photograph of hardfaced plate.

**Figure 2.** Typical cross section of hardfaced plate.

**Table 2.** PTA hardfacing experimental conditions.

| Sl. No | Parameters | | | | | % Dilution | Heat input, kJ/mm | |
|--------|-----|-----|----|----|-----|------------|-------|-----------------------|
|        | I   | S   | F  | H  | T   |            |       |                       |
| 1      | 160 | 140 | 16 | 10 | 290 | 36.11      | 15.14 | Low Heat Input (LHI)    |
| 2      | 205 | 130 | 14 | 9  | 260 | 34.49      | 18.85 | Medium Heat Input (MHI) |
| 3      | 190 | 120 | 16 | 10 | 290 | 29.81      | 20.98 | High Heat Input (HHI)   |

I = welding current (amps); S = welding speed (mm/min); F = powder feed rate (gm/min); H = oscillation width (mm); T = pre heat temperature (°C).

on the hardfaced specimens starting from the base metal up to the weld metal along the centre line. During microhardness tests, vickers indenter with 100 gm load was applied to make the indentations in all specimens. The microhardness vales obtained from the survey were plotted against the distance across the interface of the weld cross section in graphical form for a thorough analysis.

**Microstructure:** The prepared samples were etched with an etchant consists of 2 - 3 gram sodium molybdate, 5 ml hydrochloric acid (35% concentration) and 1 - 2 gram ammonia bifluoride in 100 ml distilled water for revealing the different zones of the weldments such as hardfaced metal, fusion zone, etc.

**Sliding Wear Test:** The pin-on-disc wear testing apparatus is used for conducting wear tests, under varying sliding speed and applied pressure against steel disc of hardness 500 HV. The pin samples of 25 mm length and 3 mm × 3 mm were prepared from hardfaced plates and are shown in **Figure 3**. The surface of the pin sample and the steel disc were ground using emery paper (grit size 240) prior to each test. Each sample was subjected to sliding wear for an applied test load (3 N, 5 N and 7N) at a constant sliding velocity 2 m/s. Weight loss of the specimen was measured every 200 m of sliding distance travelled by the pin. The test was integrated with WINDUCOM software and the wear loss was recorded. The specimen cleaned with acetone and weighted using an electronic weighing machine with an accuracy of 0.001 g prior to and after each test to determine the weight loss. The difference in weight gives the wear loss of the specimen. The wear rate was calculated from the weight loss measurement and expressed in terms of volume loss per unit sliding distance as given below.

Wear volume, $m^3$ = weight loss/density.

Wear rate, $m^3/m$= wear volume/sliding distance.

## 3. Result and Discussion

### 3.1. Microhardness

It is evident from **Figures 4-6** that the hardness of weldmetal near FBZ is lower than that of the weld metal away from the FBZ indicating the effect of dilution. In **Figure 4**, the hardness adjacent to the interface is found

**Figure 3.** Pin samples for wear tests.

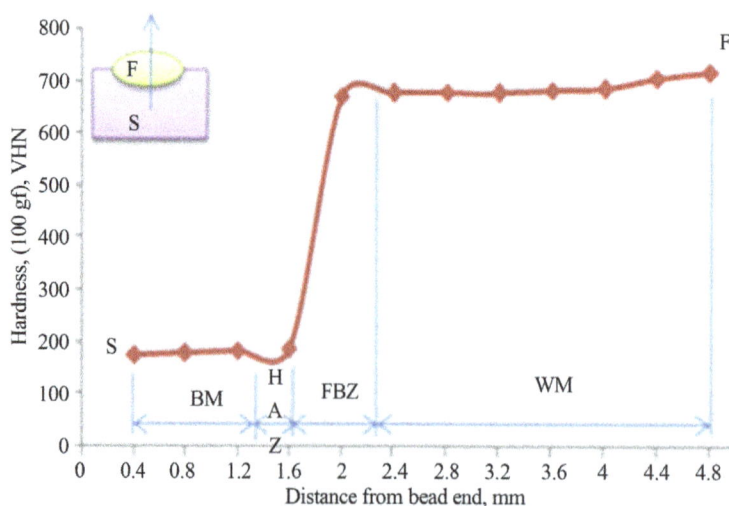

**Figure 4.** Microhardness distribution along various zones of PTA hardfacing for LHI (15.14 kJ/mm).

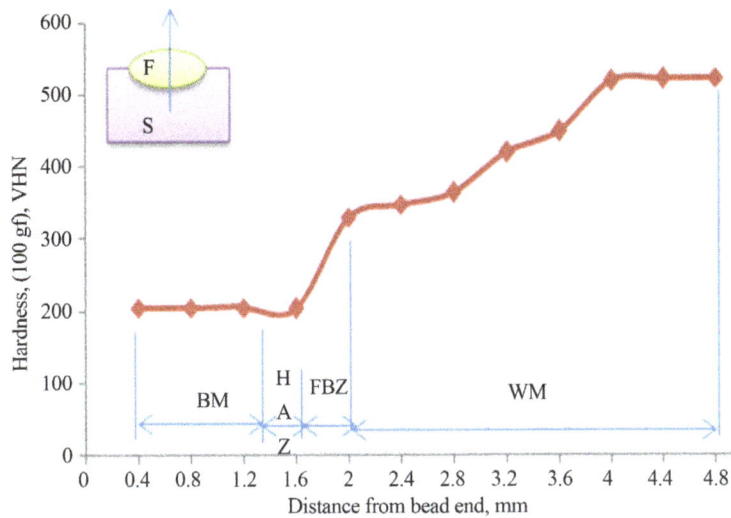

**Figure 5.** Microhardness distribution along various zones of PTA hardfacing for MHI (18.84 kJ/mm).

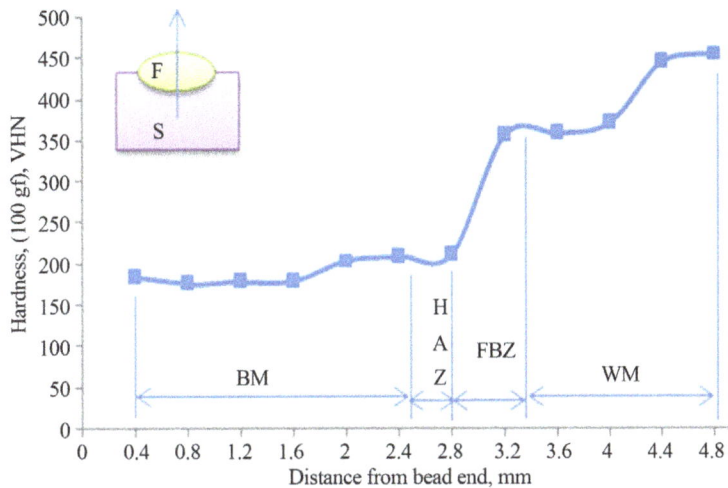

**Figure 6.** Microhardness distribution along various zones of PTA hardfacing for HHI (20.98 kJ/mm).

to be ~650 VHN which is much lower than that of hardness measured away from the interface, *i.e.* ~700 VHN. It is due to the effect of dilution observed near the interface. It is observed from **Figure 5** and **Figure 6** that the microhardness value decreases when heat input is increased. It is attributed to change in dilution caused by the respective magnitude of heat input. Therefore from **Figures 4-6** it is found that an increase in heat input reduces the hardness of PTA hardfacing.

## 3.2. Microstructure

The results obtained from the microstructural observation are presented below:

The photomicrographs taken at different magnification for all the samples welded at low, medium and high heat input conditions are presented.

It is evident from **Figure 7** that the ferrite-pearlite microstructure of the HAZ. The colored microstructure reveals the matrix of ferrite (yellowish, brownish and white) ferrite and dark banded pearlite phases. Typical microstructural changes that occur when IS 2062 structural steel is heated to the vicinity of the eutectoid reaction ~727°C and cooled to ambient temperature. The various regions (*i.e.* HHI, MHI and LHI) where conversion to an austenite matrix (on heating) occurred followed by a retransformation to a ferrite matrix on cooling. It is

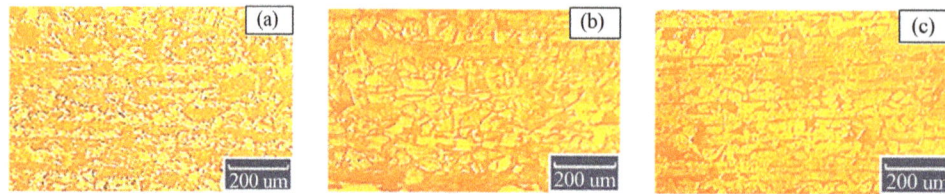

**Figure 7.** Photomicrograph showing the microstructure of the HAZ: (a) LHI (b) MHI and (c) HHI.

found that heat input increases grain coarsening occurs.

It is evident from **Figures 8-10** that microstructures of hardfaced plates at different heat input like HHI, MHI and LHI conditions reveal coarser precipitates of carbides and TiC. When the heat input increases, a slight tendency for the elements (C, Mn, Si) to decrease in the composition. It can be seen that the most TiC particles had a faceted morphology and were uniformly dispersed in the steels, but a slight agglomeration was observed in the steel with the increase of carbon content in the matrix steels the volume fraction of TiC particles increased while the size of them decreased.

The reinforcing particles are idiomorphic crystals of polygonal almost uniformly dispersed in the ferrous matrix, which consists of a fine ferrite microstructure. The form of the reinforcing particles is clearly different from the globular form of the powder grains used for the preparation of the hardfacing; their respective size distribution is also quite different. These remarks confirm that during the PTA melting process the TiC powder was dissolved completely in the melt coming from the substrate. Therefore, the microstructure of the hardfacing resulted from a solidification process and is characteristic of an "*in situ* composite."

The reinforcing particles' volume fraction was found that it could be somewhat higher on the external surface of the coatings, because the TiC particles have a lower density than the ferrous melt and tend to rise to the surface. This phenomenon is, however, limited due to the high solidification rate of the process. The hardness of the transition zone was found to be below 400 VHN due to low-carbon levels and lower dilution achieved in hardfacing. Hardfacing solidified initially with planar or cellular structure and then gradually changed to cellular-dendritic structure depending upon the heat input condition and the dilution involved. Color metallography revealed three modes of solidification of stainless steel hardfacing and observed modes of solidification were in good agreement.

**Figures 11-13** show that the presence of interface between structural steel substrate and hardfaced. This band formation is due to the attainment of high temperature during hardfacing. It is clear from the microstructures that with increase in heat input the width of the fusion boundary zones also increases. The microstructure consisted of TiC dendrites embedded in an eutectic matrix consisting of small polygonal TiC crystals and ferrite. The micro-structure was interpreted in terms of the ternary Fe-Ti-C phase diagram, which allowed anticipating that, during the solidification process, TiC separated first from the melt as primary dendritic crystals and then, at a lower temperature range, polygonal TiC crystals of smaller size formed within a ferrous matrix along a eutectic valley of the ternary diagram.

Being alloyed with titanium, in addition to the small, homogeneously distributed carbides contains a titanium additive, giving the steel structure reinforcement with very fine and extremely hard titanium carbides (TiC) that have hardness. These carbides give the steel a significantly higher wear resistance. It has been designed to make use of several metallurgic aspects: Hardness increase in use; the strengthening of the steel structure with titanium carbides.

The process requires only sufficient application of heat to obtain a sound bond with the parent metal. Admixtures of the two metals are minimized. Using too much heat during hardfaced welding will dilute the hardfaced alloys with the base metal, thereby reducing effective wear resistant properties of the material. Excessive heats will vaporize the alloys and will cause oxidation. Overly heated weld puddles can further result in such fluid condition of the alloys that they create a thinner deposit than may be intended. Insufficient welding heat, on the other hand, will cause weak bonds with the base metal and can possibly result in spalling.

### 3.3. Wear Properties

It may be important to explain the difference in wear behavior of hardfacing produced under different heat input conditions. The average wear and wear rate are show in **Table 3**.

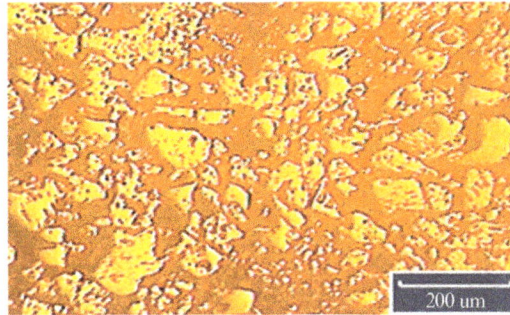

**Figure 8.** Photomicrograph showing the structure of the hardfaced at LHI condition.

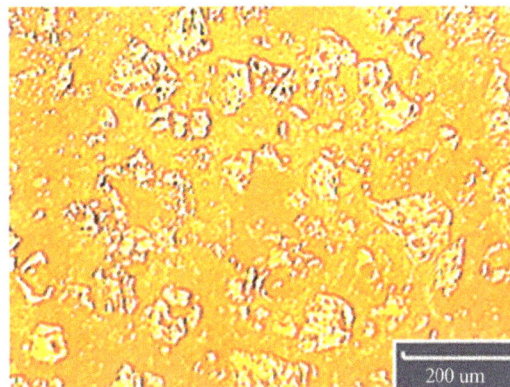

**Figure 9.** Photomicrograph showing the structure of the hardfaced MHI condition.

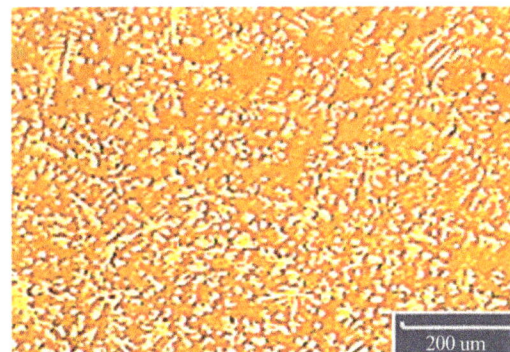

**Figure 10.** Photomicrograph showing the structure of the hardfaced at HHI condition.

**Table 3.** Wear properties of hardfacing.

| Sl. No | Normal load (N) | Sliding velocity (m/s) | Wear properties | | | | | |
|--------|-----------------|------------------------|-----|-----|-----|-----|-----|-----|
| | | | Wear ($\mu$m) | | | Wear rate, $10^8$ ($m^3$/m) | | |
| | | | LHI | MHI | HHI | LHI | MHI | HHI |
| | 3 | 2 | 0.63 | 0.69 | 0.81 | 1.07 | 1.08 | 1.11 |
| 1 | 5 | 3 | 1.55 | 1.56 | 1.60 | 1.18 | 1.31 | 1.41 |
| | 7 | 4 | 2.11 | 2.23 | 2.30 | 1.32 | 1.45 | 1.58 |

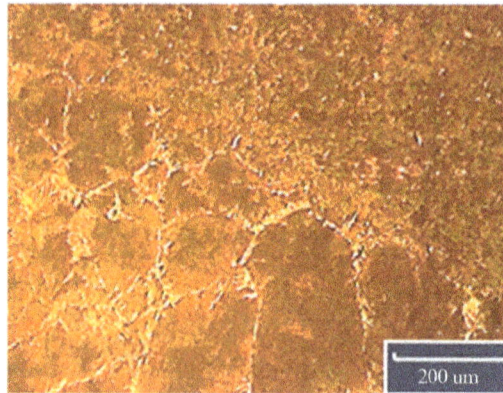

**Figure 11.** Photomicrograph showing the structure of the interface surface at LHI condition.

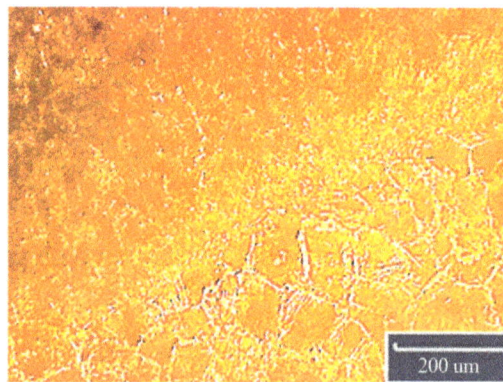

**Figure 12.** Photomicrograph showing the structure of the interface surface at MHI condition.

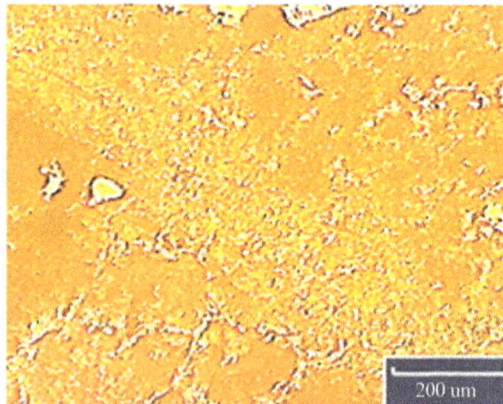

**Figure 13.** Photomicrograph showing the structure of the interface surface at HHI condition.

It is found that for a low heat input hardfacing, at 3 N and 5 N loads up to an initial sliding distance of 400 m, there is no significance in wear. Increase in normal load generally increases the wear rate under identical sliding conditions. At a higher load 7 N wear increases from the beginning itself and the amount of wear is much greater than that occurred for low loads. The weight loss for as hardfaced MHI and HHI is more than that of LHI. The increased weight loss obtained in the case of high heat input hardfaced layer resulting from high heat input provided during the process on the abrasive wear resistance.

## 4. Conclusions

- Dilution of TiC deposited by PTA hardfacing at different heat input conditions has an significant effect on the microstructure and hardness of the deposit.
- Hardness varies across the deposit/substrate interface of low heat input where as it gradually varies for medium and high heat input.
- Maximum hardness of fusion zone is larger for low heat input than that of the medium and high heat input.
- The volume fraction of precipitates of carbides and TiC is higher in the case of low heat input.
- Dendritic formations are seen in the high heat input hardfaced layer.
- Primary carbides are refined gradually with the increase in titanium content. The morphology changes from a bulk form to a refined one.
- Sliding abrasive wear increases with increase in normal load under identical sliding conditions. The hardfaced layer deposited at low heat input condition has lower wear rate than that of medium and high heat input conditions.

## References

[1]    Su, Y.L. and Chen, K.Y. (1997) Effect of Alloy Additions on Wear Resistance of Nickel in Pulsed GMAW. *Welding Journal*, **76**, 143s-150s.

[2]    Zhang, Q.H. and Yin, B.Y. (2006) Microstructure and Wear Properties of TiC-VC Reinforced Iron Based Hardfacing Layers. Institute of Materials, Minerals and Mining Published by Maney on Behalf of the Institute.

[3]    Zollinger, O.O., Beckham, J.E. and Monroe, C. (1998) What to Know before Selecting Hardfacing Electrodes: Special Emphasis: Developments in Welding Electrodes. *Welding Journal*, **77**, 39-43.

[4]    Jha, A.K., Prasad, B.K., Dasgupta, R. and Modi, O.P. (1999) Influence of Material Characteristics on the Abrasive Wear Response of Some Hardfacing Alloys. *Journal of Materials Engineering and Performance*, **8**, 190-196. http://dx.doi.org/10.1361/105994999770347034

[5]    Davis, J.R., Davis and Associates (1994) Hardfacing, Weld Cladding and Dissimilar Metal Joining. *ASM Handbook—Welding, Brazing and Soldering*, Vol. 6, 10th Edition, ASM Metals Park, OH, 699-828.

[6]    Lugscheider, E., Morkramer, U. and Ait-Mekideche, A. (1991) Advances in PTA Surfacing. *Proceeding of the 4th National Thermal Spray Conference*, Pittsburgh.

[7]    D'Oliveira, A.S.C.M. (2005) Pulsed Current Plasma Arc Welding. *Journal of Material Processing Technology*, **171**, 167-174. http://dx.doi.org/10.1016/j.jmatprotec.2005.02.269

[8]    Wang, X.B. (1998) Metal Powder Thermal Behavior during the PTA Surfacing Process. *Surface and Coatings Technology*, 156-161.

[9]    Liu, C.S., Huang, J.H., Zhao, Y. and Liu, M. (2000) *Transactions of Nonferrous Metals Society of China*, **10**, 405-407.

[10]   Gonzalez, C. and Llorca, J. (2001) Micromechanical Modeling of Deformation and Failure in Ti-6Al-4V/SiC Composites. *Acta Materialia*, **49**, 3505-3519. http://dx.doi.org/10.1016/S1359-6454(01)00246-4

[11]   Fu, Y.-C., Shi, N.-L., Zhang, D.-Z. and Yang, R. (2004) Preparation of SiC/Ti Composites by Powder Cloth Technique. *The Chinese Journal of Nonferrous Metals*, **14**, 465-470.

[12]   Ding, H.-M., Liu, X.-F., Lin, Y. and Zhao, G.-Q. (2007) The Influence of Forming Processes on the Distribution and Morphologies of TiC in Al-Ti-C Master Alloys. *Scripta Materialia*, **57**, 575-578. http://dx.doi.org/10.1016/j.scriptamat.2007.06.028

[13]   Hill, D., Banerjee, R., Huber, D., Tiley, J. and Fraser, H.L. (2005) Formation of Equiaxed Alpha in TiB Reinforced Ti Alloy Composites. *Scripta Materialia*, **52**, 387-392. http://dx.doi.org/10.1016/j.scriptamat.2004.10.019

[14]   Falat, L., *et al.* (2005) Mechanical Properties of Fe-Al-M-C (M = Ti, V, Nb, Ta) Alloys with Strengthening Carbides and Laves Phase. *Intermetallics*, **13**, 1256-1262. http://dx.doi.org/10.1016/j.intermet.2004.05.010

[15]   Dóllar, A. and Dymek, S. (2003) Microstructure and High Temperature Mechanical Properties of Mechanically Alloyed Nb₃Al Based Materials. *Intermetallics*, **11**, 341-349. http://dx.doi.org/10.1016/S0966-9795(03)00002-5

[16]   Weman, K. (2003) Welding Processes Handbook. CRC Press LLC, New York.

[17]   Puntharani, K. and Murugan, N. (2010) Finite Element Simulation for Prediction of Bead Geometry and Resudual Stesses in Stellite Hardfacing Gate Valve by PTAW Process. *International Journal of Advanced Manufacturing Technology*.

# 12

# Impact Strength Analysis of Plate Panels with Welding-Induced Residual Stress and Deformation

Yehia Abdel-Nasser[1], Ninshu Ma[2], Sherif Rashed[2], Hidekazu Murakawa[2]

[1]Naval Architecture and Marine Engineering, Faculty of Engineering, Alexandria University, Alexandria, Egypt
[2]Joining and Welding Research Institute, Osaka University, Osaka, Japan
Email: yehia-nasser@hotmail.com, ma3liang@yahoo.co.jp

## Abstract

The paper describes the simulation of impact loads applied on plate panels with welding-induced residual stresses and deformation (WSD). Numerical simulations using FEM are carried out to study the influence of welding-induced residual stresses and deformation on the impact strength of plate panels. Welding is simulated using a three dimensional thermal mechanical coupled finite element method. The welding stress and deformation are taken as the initial imperfections in the impact strength analysis and their influence on the behavior of plate panels subjected to impact loadings. The impact loadings from the three directions, the lateral direction and two in-plane directions of the plate panels are studied. Results show a certain reduction in the impact strength due to the existence of welding stress and deformation in the plate panels. It is found that the reduction of impact force is strongly influenced by the welding deformation and the impact directions in the plate panels. This reduction is more significant when the impact force is in the lateral direction.

## Keywords

Lateral Impact, In-Plane Impact, Welding Deformation, Plate Panels, Impact Force, FEM

## 1. Introduction

The structural design of ships concerning grounding and collision requires an accurate prediction of the damage of plate panels under impact loading. Several experimental works on laterally loaded panels have been conducted in order to derive analytical expressions [1]. Collision is considered as a time-depended nonlinear dynamic phenomenon. The majority of researchers have focused on deriving the resultant damage of the ship collisions via analytical, experimental, and finite element methods. Hagiwara *et al.* [2] proposed a method for predicting low-energy ship collision damage based on experiments and determined the initiation of plate fracture. Cho and Lee [3] developed a simplified method for the prediction of the extent of damage of stiffened plates due to lateral collisions. Ehlers *et al.* [4] performed numerical simulations of the collision response of ship side structures and investigated sensitivity of the various failure criteria. Villavicencio and Soares [5] studied numerically the deflection and failure of small panels subjected to lateral impact using different stiffening systems and impact locations. However, literatures on the investigation of the effect of welding induced initial welding stresses and deformation (WSD) on the impact strength of plate panels are limited. Generally, the evaluation of impact strength of plate panels is based on the shape of initial welding imperfections and material properties with the consideration of the plastic strain. Ma, *et al.* [6] and Takada *et al.* [7] proved that the material failure must be considered in the impact simulation in order to accurately evaluate the impact strength. Also, the effects of dimension error, plastic strain and residual stresses due to welding have to be studied. In this paper, the influence of welding induced residual stresses and deflections is taken into considerations for plate panels subjected to impact loadings. Two different directions of impacts such as lateral impact and in-plane impact are modeled to investigate the behavior of plate panels. Finite element analysis is a useful tool to predict the impact strength of plate panels. The paper summarizes the results from numerical simulations of plate panels subjected to in-plane and lateral impact loadings. The influence of the initial welding imperfection on the force-displacement response is reviewed. Results show a reduction in the impact strength due to the occurrence of initial welding imperfections in the plate panels. The reduction of impact force is found to be dependent on the mode of initial deflection and direction of loading in the plate panels. However, the nonlinear dynamic analysis should be compared with experimental scaled tests to verify the results.

## 2. Computation Methods and Procedures

This research work is aimed to investigate the influence of welding-induced residual stresses and deformation (WSD) on the impact strength of plate panels. Since experimental investigations on partial and/or full scale welded plate panels are difficult, FEM is often used to simulate the impact behaviors of plate panels subjected to a striking object such as a ball or a rigid frame. The previous study [8] showed that simulation results on the deformation and impact forces of hybrid plate panels agreed very well with experimental ones. Therefore, in this study, the simulation methods, simulation models and their results are focused.

The computation methods and procedures used in analyzing impact strength of plate panels with welding-induced residual stresses and deformation are summarized in **Figure 1** as follows:

- Three dimensional welding heat conduction analysis on a butt welded plate panel is carried out using a research version of the in-house FEM solver JWRIAN developed by authors [9].
- A coupled thermal-elastic plastic analysis is performed to estimate the welding residual stresses and deformation using the in-house software JWRIAN.

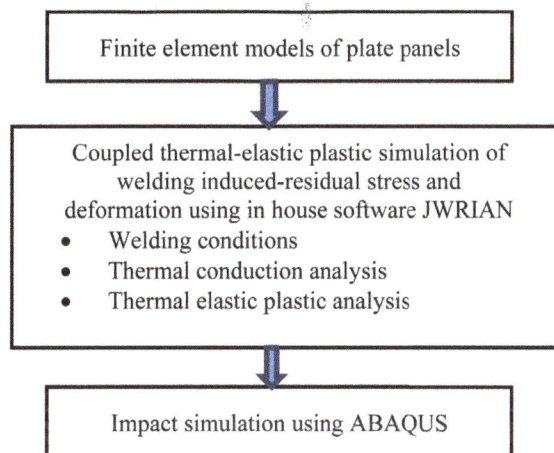

**Figure 1.** Computational methods and procedures.

- Implicit dynamic analyses using ABAQUS Ver. 6-13 [10] are carried out to simulate the deformation behaviors of the plate panel under impact loadings in different directions.
- Material properties of plate panels and its characteristics are as shown in **Table 1** [11]. Strain rate effect was not considered in this analysis.

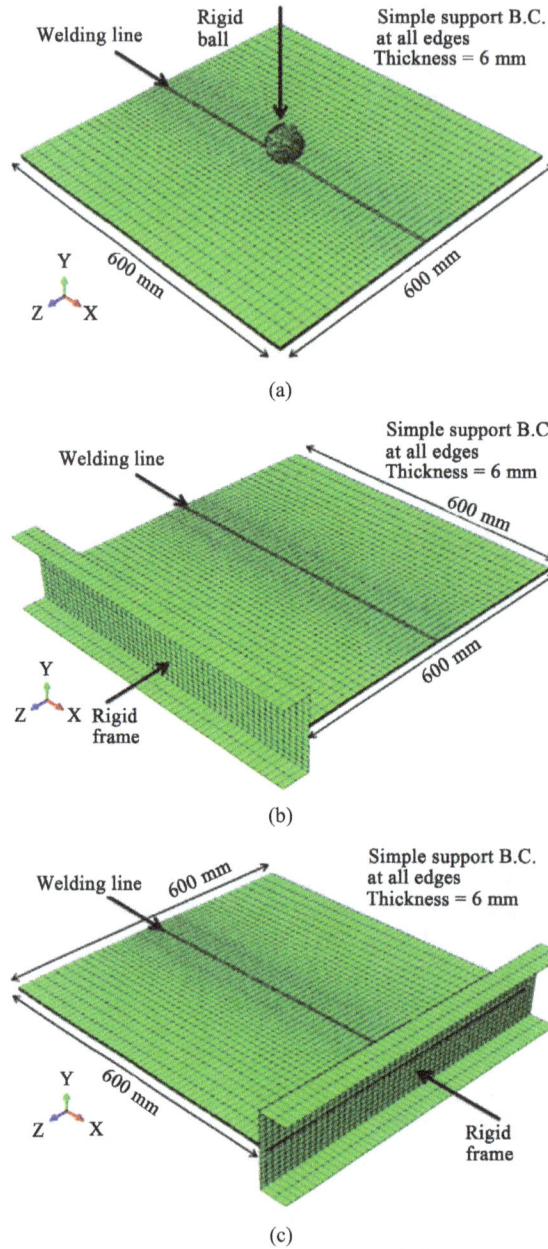

(a)

(b)

(c)

**Figure 2.** (a) Plate panel impacted by a rigid ball in direction Y; (b) Plate panel impacted by a rigid frame in-plan direction-Z; (c) Plate panel impacted by a rigid frame in-plan direction-X.

**Table 1.** Mechanical properties of the material.

| Mechanical properties of the material | | | | |
|---|---|---|---|---|
| Young's Modulus | Poisson Ratio | Yield Stress | Tensile Stress | Failure Strain |
| 210 Gpa | 0.3 | 320 Mpa | 480 Mpa | 35% |

## 3. Plate Panel and FEM Models

A square plate panel 600 × 600 × 6 mm is modeled using eight node hexa elements as shown in **Figure 2(a)**. A fine mesh of size 2 × 2 mm is adopted at the welding line with four elements through the plate thickness. A gradually coarsened mesh towards the plate edges is adopted to save computational time. The plate panel is assumed simply supported at all edges. The impactor (ball or frame) is modeled as a rigid body. An artificial mass is concentrated at the center of the impactor. An initial impact velocity was assigned to the impactor in the motion direction and other degrees of freedom were constrained. Different directions of impact loadings are investigated, namely lateral and in-plane loadings as shown in **Figures 2(a)-(c)**, respectively.

## 4. Welding Residual Stresses and Deformation

The effect of initial imperfections, welding residual stresses and deformation (WSD), are taken into account in the plate panel when subjected to impact loadings. A volumetric heat source is applied through the welding line. The problem is treated as a de-coupled thermal-mechanical analysis. However, temperature dependent mechanical properties are used. First, a thermal analysis was performed to predict the temperature history of the plate panel. Subsequently, thermal loads induced by transient welding temperature fields were applied to the plate panel and residual stresses were predicted using a nonlinear thermal elastic plastic FEM [12]. Results of analyses such as temperature, welding residual stresses and deformation are shown in **Figure 3**.

## 5. Impact Analyses

### 5.1. Lateral Impact in Y-Direction

By considering the welding residual stresses and deformation (WSD) shown in **Figure 3** as initial welding im-

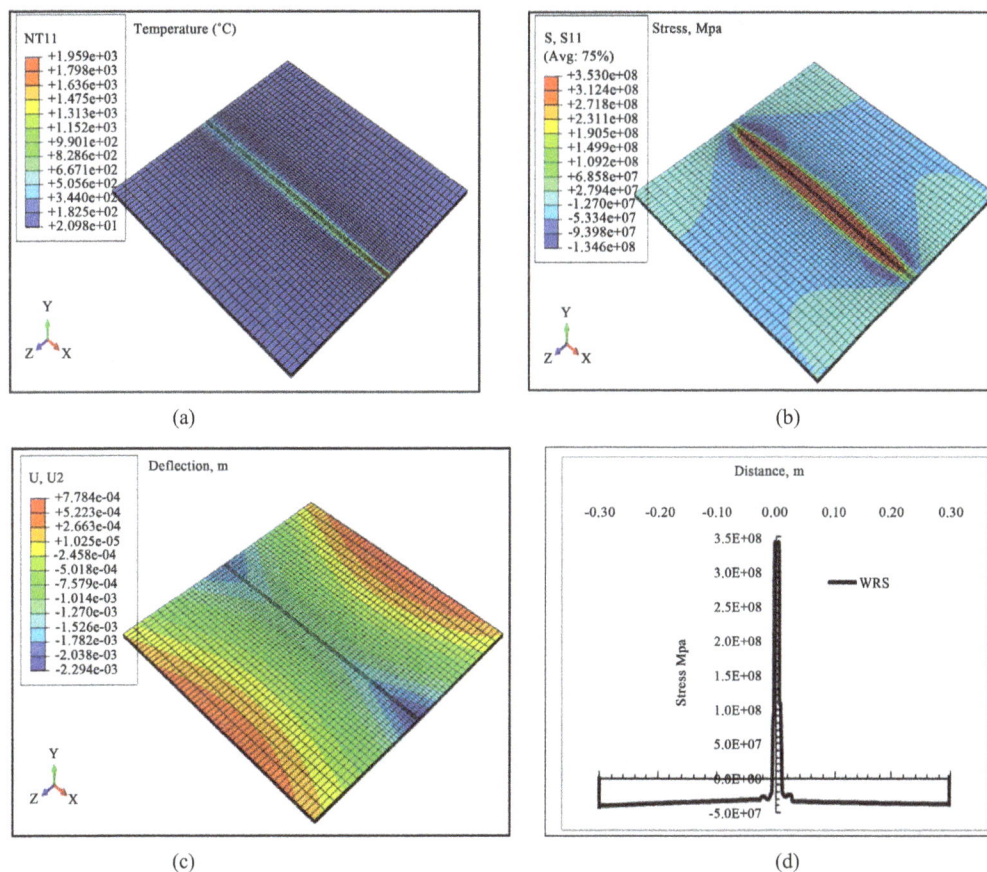

**Figure 3.** Welding residual stresses and deformation computed by thermal elastic plastic FEM. (a) Temperature distribution; (b) Distribution of residual stress-X; (c) Deflection mode; (d) Residual stress-X distribution in Z-direction.

perfections, a lateral impact loading in Y-direction is applied as shown in **Figure 2(a)**. A virtual mass of 200 kg and an initial velocity of 6 m/sec are concentrated at the center of the impactor (rigid ball). Here, a plate panel without WSD is referred to as "flat plate" while the plate panel with the considered initial imperfections is referred to as "plate with WSD". **Figure 4** and **Figure 5** show the deformation due to impact loading of the flat plate and the plate with WSD, respectively. It is observed that the two plates have the same deflection mode. However, the maximum deflection at the center of the plate with WSD reached 57 mm which is larger than the maximum deflection of 49 mm in the case of the flat plate. After the velocity of the impact ball becomes equal to zero, the ball rebounds and the deflection of the plate panel decrease. **Figure 6** shows the stress distribution during the impact loading. **Figure 7** shows relationships between impact force and displacement of the ball. The plate with WSD attained lower impact force than that of the flat plate. This is because that the stiffness of the plate panel with WSD is reduced by initial welding imperfections from the beginning of loading. In this type of impact, the initial welding imperfections reduced the impact strength of the plate panels significantly. After yielding, plasticity spreads in the central part of the plate panels which increases the reduction of the impact force for the both plates.

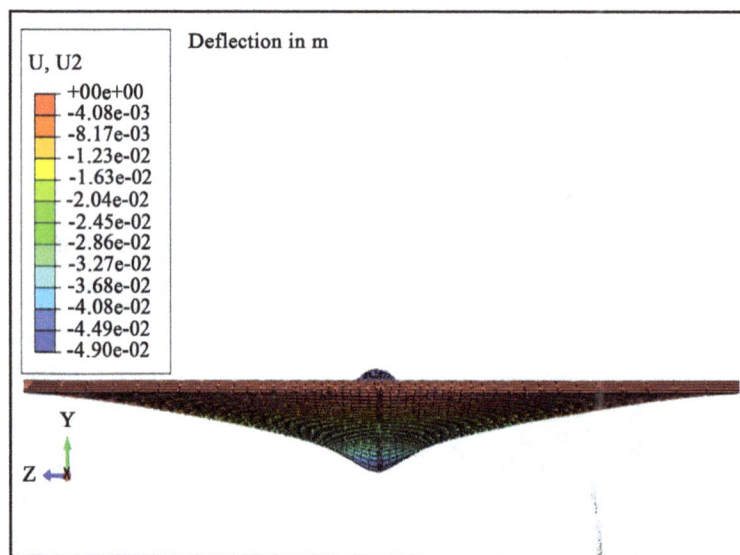

**Figure 4.** Deflection mode for flat plate.

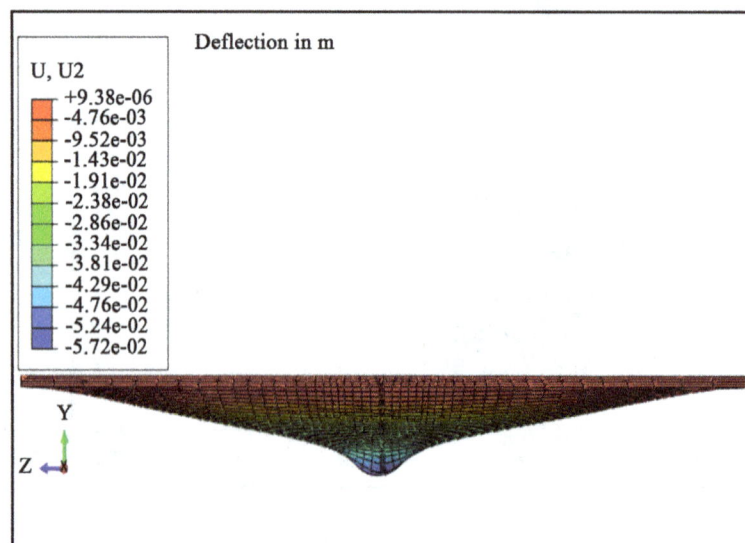

**Figure 5.** Deflection mode for plate with WSD.

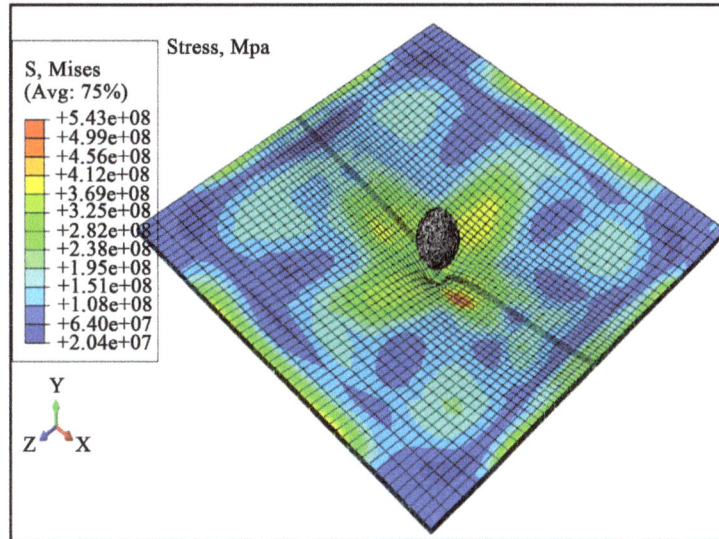

**Figure 6.** Stress and deformation of plate with WSD.

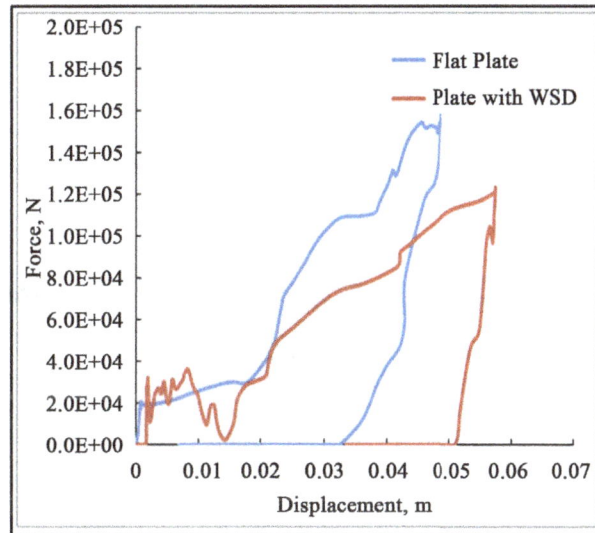

**Figure 7.** Force-displacement relationships.

## 5.2. In-Plane Impact in Z-Direction

In this section, in-plane impact loading in Z-direction as shown in **Figure 2(b)** is investigated. This case is typically similar to the case of the plate of web frame and subjected from outside to impact loading. A virtual mass of 10,000 kg and an initial velocity of 6 m/sec are concentrated at the reference point of the impactor (Rigid frame). For plate panels with WSD, the maximum deformation as shown in **Figure 8** is observed at the center of the plate panel (at the weld line) with a value equal to 50 mm. While the maximum deflection of the flat plate as reached about 35 mm and it is far from the center of the plate as shown in **Figure 9**. Here, the deflection mode of the flat plate is somewhat different from that of the plate with WSD. The deflection mode of the flat plate looks like a buckling mode. Mises stress and deformation of the plate with WSD produced by impact loading are shown in **Figure 10**. **Figure 11** shows relationships between the impact force and displacement of the rigid frame. During the elastic state, force and displacement relationships are similar. Deviations are observed after the spread of plasticity. The two plates have almost the same maximum impact force. However, the plate with WSD showed fast decrease of the impact force after the spreads of plasticity. On the other hand, the flat plate has sustained additional load after yielding and then the impact force decreased. It can be understood that the

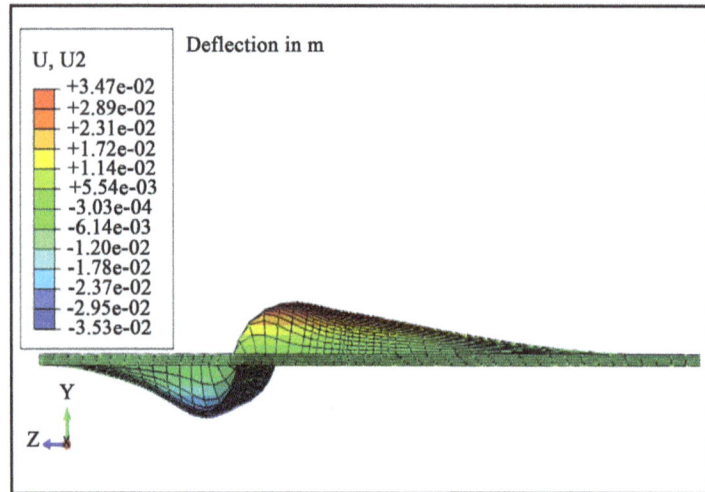

**Figure 8.** Deflection mode for flat plate.

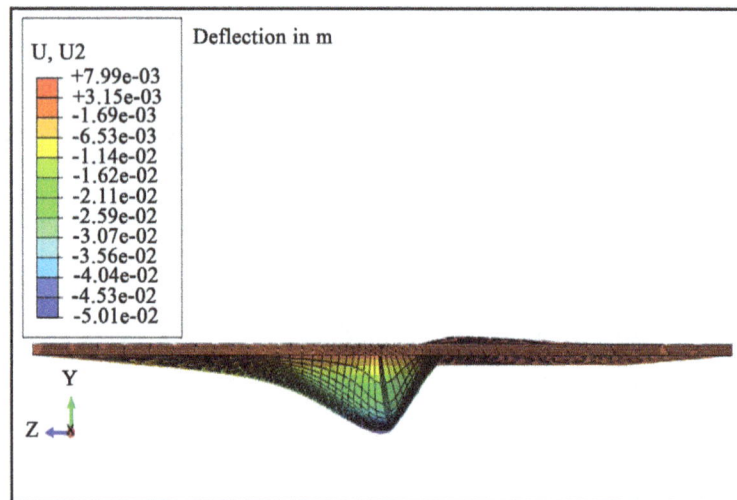

**Figure 9.** Deflection mode for plate with WSD.

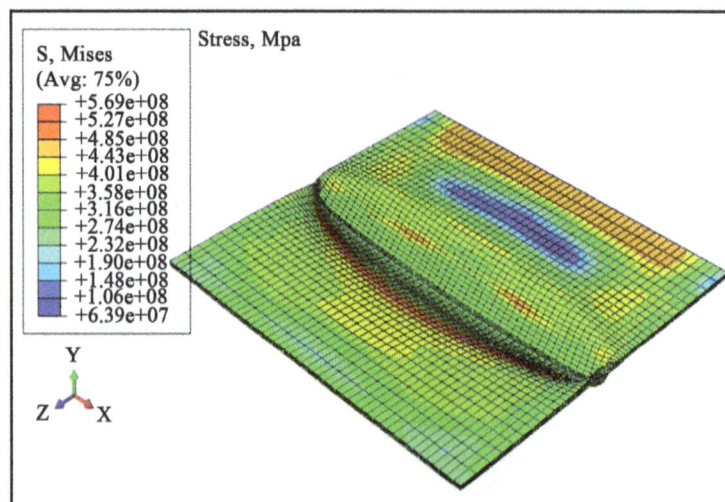

**Figure 10.** Stress and deformation of plate with WSD.

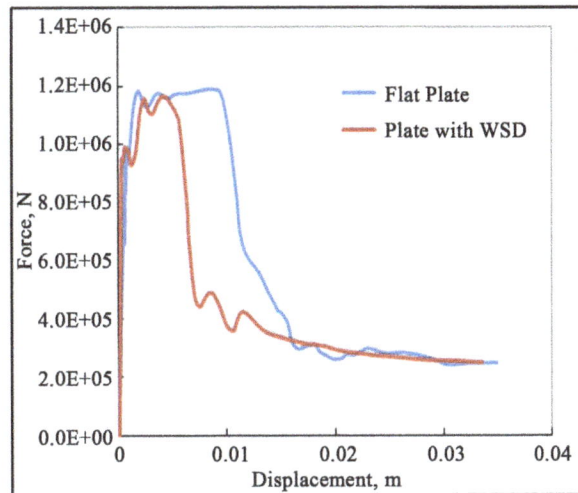

**Figure 11.** Force-displacement relationships.

two plates have almost the same initial stiffness. The influence of the initial welding imperfections on impact deformation near the center of the plate can be clearly observed. The deflection mode of the plate with WSD follows the mode of initial deflection, while the flat plate has a strong effect on a post buckling mode.

## 5.3. In-Plane Impact in X-Direction

In this section, in-plane impact loading in X-direction as shown in **Figure 2(c)** is investigated. Previous initial conditions due to welding of the plate panel are applied. Also, the same values of virtual weight and initial velocity are concentrated at the center of the rigid frame. For both the plate panel with WSD and the flat plate, the maximum deflections are observed far from the center of the plate panel. Although the plate panel has same initial conditions and same acting impact load, values of the maximum deflections in this case of impact are less than those in the case of impact in Z-direction. Also, the deformation modes of the plate WSD and the flat plate are similar as shown in **Figure 12** and **Figure 13**. Both plates deflect in the post buckling mode. **Figure 14** shows the stress and deformation of plate panel with WSD during impact loading. **Figure 15** shows relationships between impact force and displacement. Initially, the two plates have almost the same stiffness. Also, the two plates have almost the same maximum impact force. When plastic deformation has occurred, the load carrying capacity of the plate with WSD starts to decrease causing the impact force to rapidly decrease. However, the flat plate sustains a little more load after yielding. In this type of impact loading, the influence of the initial welding imperfections is less significant on the behavior of the plate panels than in the case of impact in Z-direction.

## 6. Conclusions

The paper refers to the simulation impact of plate panels with welding stress and deformation (WSD). Two different directions of impact loadings on plate panels such as lateral impact and in-plane impacts are investigated. The following related aspects are considered to be essential for such a research work:

- Whatever the direction of impact loading, the occurrence of welding stress and deformation has a significant role in the reduction of the impact strength of the plate panels. The reduction of impact force is found to be dependent on the mode of initial deflection and direction of impact loading in the plate panels. The reduction is more significant when the impact force is in the lateral direction.
- Regarding to lateral impact, the stiffness of the plate panel with WSD is reduced from the beginning of impact loading. After yielding of the plate panel, the impact force is drastically decreased and its value reaches to zero force.
- Regarding to in-plane impact in Z-direction, the deflection mode for the plate panel with WSD follows the mode of initial deflection, while the flat plate behaves post buckling mode. For in-plane impact in X-direction, the influence of the initial welding imperfections is less significant on the impact strength and behavior of the plate panels.

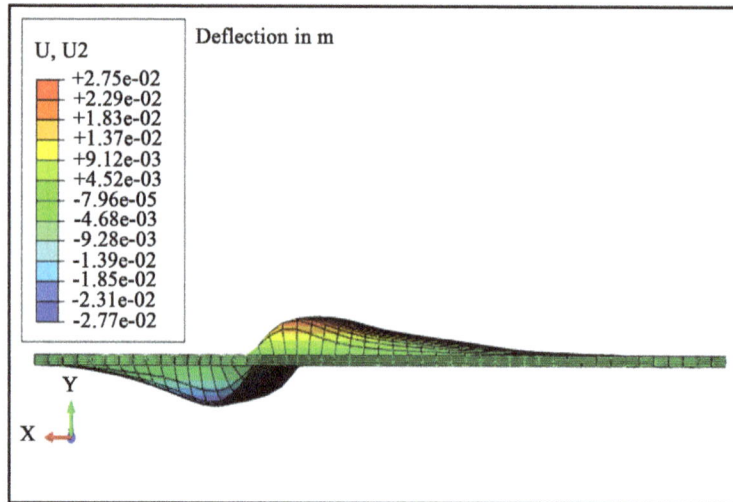

**Figure 12.** Deflection mode for flat plate.

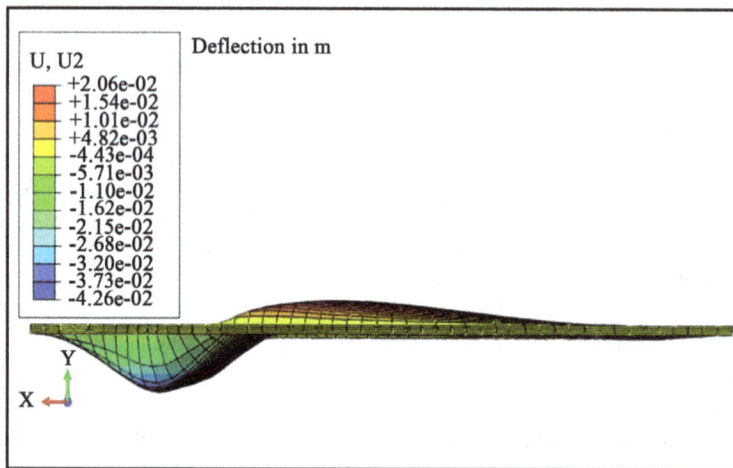

**Figure 13.** Deflection mode for plate with WSD.

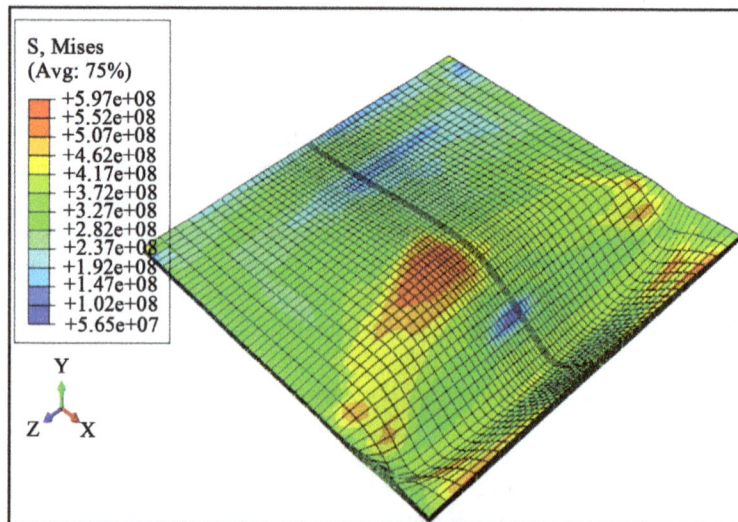

**Figure 14.** Stress and deformation of plate with WSD.

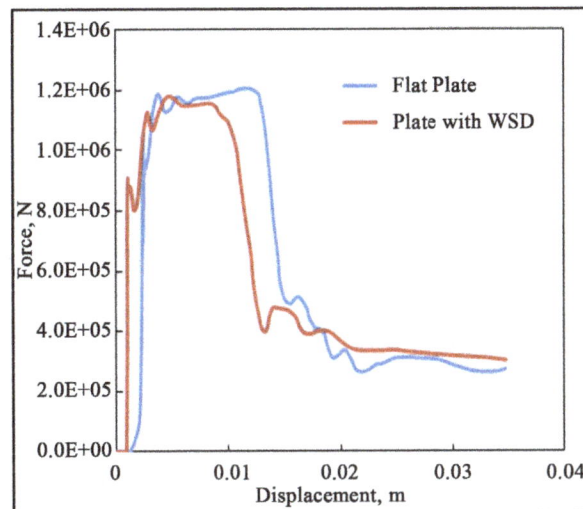

**Figure 15.** Force-displacement relationships.

- As the value of initial deflection is increased in the plate panels, as the reduction in the impact force is increased.

## Acknowledgements

Authors would like to express their acknowledgments to Mr. Hui Huang, JWRI, Osaka University for his support in simulation works.

## References

[1] Brown, A.J. (2002) Modeling Structural Damage in Ship Collisions. SSC -1400 Report.

[2] Hagiwara, K., Takanabe, H. and Kawano, H. (1983) A Proposed Method of Predicting Ship Collision Damage. *International Journal of Impact Engineering*, **1**, 257-279. http://dx.doi.org/10.1016/0734-743X(83)90022-2

[3] Cho, S.R. and Lee, H.S. (2009) Experimental and Analytical Investigations on the Response of Stiffened Plates Subjected lo Lateral Collisions. *Marine Structures*, **22**, 84-95. http://dx.doi.org/10.1016/j.marstruc.2008.06.003

[4] Ehlers, S., Broekhuijsen, J., Alsos, H.S., Biehl, F. and Tabri, K. (2008) Simulating the Collision Response of Ship Side Structures: A Failure Criteria Benchmark Study. *International Shipbuilding Progress*, **55**, 127-144.

[5] Villavicencio, R., Liu, B. and Guedes Soares, C. (2012) Response of Stiffeners with Attached Plate Subjected to Lateral Impact. Maritime Engineering and Technology, Taylor & Francis Group, London.

[6] Ma, N., Takada, K. and Satoh, K. (2014) Measurement of Local Strain Path and Identification of Ductile Damage Limit by Simple Tensile Test. *Engineering Procedia*, **81**, 1402-1407. http://dx.doi.org/10.1016/j.proeng.2014.10.164

[7] Takada, K., Sato, K. and Ma, N. (2015) Fracture Prediction for Automotive Bodies Using a Ductile Fracture Criterion and a Strain-Dependent Anisotropy Model, Fracture Prediction for Automotive Bodies Using a Ductile Fracture Criterion and a Strain-Dependent Anisotropy Model. *SAE International*, **8**, 803-813. http://dx.doi.org/10.4271/2015-01-0567

[8] Abdel-Nasser, Y., Ma, N., Murakawa, H. and El-Malah, I. (2015) Impact Analysis of Aluminum-Fiber Composite Lamina, *Quarterly Journal of Japan Welding Society*, **33**, 166s-170s. http://dx.doi.org/10.2207/qjjws.33.166s

[9] Ueda, Y., Murakawa, H. and Ma, N. (2012) Welding Deformation and Residual Stress Prevention. 1st Edition, Elsevier, Butterworth-Heinemann.

[10] ABAQUS/Imlicit Software, Version 6.13, Osaka, Japan.

[11] Ninshu, M.A., Chiyonobu, M. and Hisamori, T. (2008) Analysis of Welded and Stamped Members under Impact Loading. *Proceedings of International Symposium on Structure under Earthquake, Impact and Blast Loading*, 2008, 62-65.

[12] Murakawa, H., Ma, N. and Huang, H. (2015) Iterative Substructure Method Employing Concept of Inherent Strain for Large-Scale Welding Problems. *Welding in the World*, **59**, 53-63.

# 13

# Research on Interlayer Alloys for Transient Liquid Phase Diffusion Bonding of Single Crystal Nickel Base Superalloy DD6

Qiuya Zhai, Jinfeng Xu, Tianyu Lu, Yan Xu

School of Materials Science and Engineering, Xi'an University of Technology, Xi'an, China
Email: qiuyazhai@xaut.edu.cn

## Abstract

Transient Liquid Phase Diffusion bonding (TLP bonding) is an effective method to achieve excellent joint of DD6, which is a new generation single crystal superalloy to manufacture aero-engine turbine blades. In this paper, the interlayer alloys for DD6 TLP bonding were designed. The alloy foils with thickness 40 μm ~ 60 μm, width 4 mm were prepared by using a single roller rapid solidification apparatus and the TLP bonding of DD6 was conducted. Then the joint microstructure and alloying elements diffusion behaviors were analyzed. The results indicate that microstructures of interlayer alloys prepared are fine and homogeneous, the melting point range of alloys from 1070°C to 1074°C and their melting temperature interval is merely 20°C, when the chemical composition of alloys are 1.5 ~ 2.0Cr, 3.2 ~ 4.0W, 3.7 ~ 4.5Co, 2.2 ~ 3.0Al, 0.7 ~ 1.0Mo, 3.2B, remain Ni (wt%). When the welding parameters are bonding temperature 1200°C, holding time 8.0 hour and welding pressure 0.3 MPa, the compacted joints obtained and the microstructure of TLP bonding seams were similar to base metal. The bonding joint is composed of weld center zone, isothermal solidification zone and diffusion-affected zone. Within joint, the elements diffusion is sufficient and borides in the diffusion zone are fewer.

## Keywords

Interlayer Alloy, DD6 Single Crystal Superalloy, TLP Bonding, Microstructure of Joint

## 1. Introduction

Ni-based single crystal super alloys are key materials to manufacture aero-engine turbine blades owing to their fine thermostability, heat resistance and high-temperature structural stability. Considering thermo dissipation and energy saving, aero-engine turbine blades are always designed in complex hollow structures, and it's inevitably involves welding problems [1] [2] with during the manufacturing process of that hollow structures. Since the melting weld seam is apt to suffer from hot cracking and the brazed joining intensity is always insufficient, Transient Liquid Phase Diffusion (TLP) Bonding is an appropriate method to carry out the welding of Ni-based single crystal super alloys [3]. Except the process factors, the interlayer alloy's chemical components as well as its applied forms are closely related to the welding quality and joint microstructures of TLP bonding DD6 alloy [4] [5]. So according to the chemical composition of DD6 alloy and the requirement of property matching, in this paper, the composition of the interlayer alloy is designed and the alloy foils is prepared, which is used for DD6 TLP bonding, thus the excellent quality bonding joint of DD6 single crystal alloys would be obtained.

## 2. Design of Interlayer Alloys

In order to realize the TLP bonding of DD6 single crystal alloys, there are several sound principles to follow. First, assuring that the joint strength matching to the base metals, the main alloying elements of interlayer alloys must be as much as possible approximate to the parent metal [6]. Secondly, insuring that the welding process go without a hitch, the lower melting point elements and easy to diffusion elements would be added [7].

### 2.1. Selection of Lower Melting Point Elements

The grains of superalloy DD6 to begin coarsening at 1320°C, therefore, TLP bonding should be conducted lower than that temperature. References [8] [9] show that the appropriate TLP bonding temperature of single crystal alloy DD6 is perforce controlled in less than 1200°C.

Due to the fact that the TLP bonding temperature should be higher than that of the interlayer alloy's melting point 100°C ~ 150°C, the applicable melting temperature of the interlayer alloy should be 1050°C ~ 1100°C.

On the basis of the chemical composition of base metal, the interlayer alloys must contain some certain elements in order to bring about eutectic action with matrix metal Ni, so the melting point of interlayer alloy lowered. B, Si, Nb and Mn *et al.* are the alternative alloying elements to lower the melting point of interlayer alloys used to TLP bonding DD6 alloy. Among them, B has the most significant effect to drop interlayer alloy's melting point, due to B atom has the smaller radius with easy to diffuse conveniently and is able to react with Ni form eutectic with lower melting temperature. Whereby B is selected as main additive element to interlayer alloys to lower their melting points and the additive amount is set nearby to the Ni-B eutectic point. Ni-B binary phases diagram is shown in **Figure 1**.

### 2.2. Selection of the Diffuse Element

The essence of transient liquid phase diffusion bonding is elements' diffusion. During the TLP bonding process, with the temperature rising, the alloys at interface melt and the elements to lower the melting point diffuse toward the base metal rapidly, causing the melting point of base metal at the interface lower and the base metal melt, so the liquid phase area become wider. When the density of the elements to lower the melting point at the liquid/solid interface lower to the liquidus temperature, with the elements to lower the melting point diffusing further, the melting point of the liquid phase area at the liquid/solid interface rise, the liquid phase area begin to solidify from the interface to the center of welding seam. Hence, elements' diffusion must be made as precondition for successful TLP bonding and the interlayer alloys' elements to lower the melting point must possess the

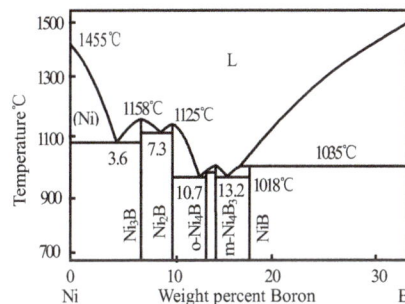

**Figure 1.** Ni-B binary phase diagram [10].

character of diffusing rapidly. Generally speaking, diffuse elements are B and Si. B atom has small radius and high speed of diffusion. Besides lowering the melting point, B is the main diffuse element.

## 2.3. Selection of Main Alloying Elements

In order to obtain the joint whose structure and performance can be matched with the base metal, the kind of the elements which the interlayer alloys contain should be similar or close to that of the base metal's. The interlayer alloys for DD6 TLP bonding contain W, Cr, Co, Mo, Al as well as other alloy elements. Additionally, the interaction between the alloying elements and the elements to lower the melting point and diffuse should also be considered when determine the content of the alloying elements. If the main alloying elements can react with the elements to lower the melting point and diffuse into brittle compounds with high melting point, the joint mechanical properties will reduce dramatically so that the content of this element is unfavorable and overmuch. Contrary to that, the elements can be added appropriately. Whether the alloying elements can generate borides or not and the stability of the borides is the key to determining the content of alloying elements.

The criteria for judging the stability of the borides are:

1) According to the phase diagram, the lower the eutectic temperature of the alloys' elements and B is, the easier B melt and avoid generating the borides with high melting point. Therefore, the higher the eutectic temperature of the alloys' elements is, the less the content of the alloys' elements is.

From **Figure 2**, the Cr-B binary phases, it is found that the lowest temperature when the eutectic action of Cr-B take place is 1630°C and the reaction temperature is high, then the melting point of the generated $Cr_2B$ is very high. Accordingly, the eutectic temperature of Co-B is 1100°C ~ 1350°C and the melting temperature of the generated borides is lower, shown in **Figure 3**.

2) According to the differential value of elements electronegativity, the lower the differential value of elements electronegativity of the alloys' elements and B is, their bond is weaker, the easier the generated borides resolve. However, the higher the differential value of elements electronegativity of the alloys' elements and B is, the more difficult the generated borides resolve, so the content of the alloys' elements should be controlled in lower range. The differential value of elements electronegativity of the alloys' elements and B is shown in **Table 1**. It can be seen from **Table 1** that the differential value of elements electronegativity of the alloys' elements and B from high to low can be ranged as Al > Cr > W > Co > Mo.

Referring to the chemical composition of the DD6 superalloy and considering the eutectic temperature and the

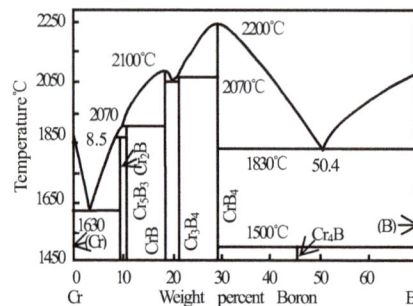

**Figure 2.** Cr-B binary phase diagram [10].

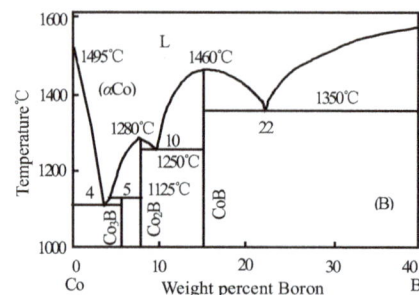

**Figure 3.** Co-B binary phase diagram [10].

**Table 1.** The electronegativity of elements.

| Alloys' elements X | The electronegativity of X | The electronegativity of B | The differential value of elements electronegativity of X and B |
|:---:|:---:|:---:|:---:|
| W | 2.36 | 2.04 | 0.32 |
| Al | 1.61 | 2.04 | 0.43 |
| Cr | 1.66 | 2.04 | 0.38 |
| Co | 1.88 | 2.04 | 0.16 |
| Mo | 2.16 | 2.04 | 0.12 |

differential value of elements electronegativity's influence on the stability of the borides, the composition range of the interlayer alloy is designed as: $1.0 \leq Cr \leq 2.0$, $2.5 \leq W \leq 4.0$, $3.0 \leq Co \leq 4.5$, $1.5 \leq Al \leq 3.0$, $0.5 \leq Mo \leq 1.0$, $3.0 \leq B \leq 3.6$ and remain Ni (wt%). For easier research, the interlayer alloys with the composition above were divided into two groups for testing, shown in **Table 2**.

## 3. Preparation of the Interlayer Alloys

### 3.1. Preparation of the Interlayer Alloy Foils

The interlayer alloys' materials for testing are simple metals (99.9%): Ni, Al, W, Cr, Mo, Co and B. First, purify the alloys with $B_2O_3$. Then melt the interlayer alloys purified and B needed with high frequency induction heating equipment. As B cannot be dissolved completely when melt the interlayer alloys and B, the alloys should be weighted before melting and melt repeatedly, until the ratio of B and the alloy meets the requirement. Finally, The alloy foils with thickness 40 μm ~ 60 μm, width 4 mm were prepared by using a single roller rapid solidification apparatus. The testing parameters are shown in Reference [11].

### 3.2. Analysis of the Interlayer Alloys Foils' Microstructure

The homogeneity of interlayer alloys' composition has great effect on the welding result. **Figure 4** shows the microstructure of the interlayer alloys foils along thickness, from which it can be seen that the composition of the interlayer alloy foils prepared is homogeneous and the microstructure of it is fine meeting the requirement for interlayer alloys for TLP welding.

### 3.3. DTA Analysis for Interlayer Alloy Foils

**Figure 5** shows the DTA curve of the interlayer alloys, it can be seen that the crystallization of interlayer alloys starts at 454°C and finish at 466°C showing a crystallization peak. With the temperature rising continuously, alloy foils begin to melt when it is up to 1074°C and finish at 1094°C. Their melting temperature interval is about 20°C. The melting temperature of the interlayer with small melting temperature interval ranges from 1050°C to 1100°C meeting the operating requirement for DD6 TLP welding.

## 4. DD6 TLP Bonding

### 4.1. Welding Method and Process

#### 4.1.1. Material for Testing

The base metal for testing is as-cast DD6 with dimension of $\Phi 16 \times 7$ mm and its chemical composition is: 4.34Cr, 9.01Co, 1.98Mo, 8.44W, 6.14Ta, 5.4Al, 3.02Re, 0.98Nb, 0.12Hf, remain Ni (wt%).

#### 4.1.2. Weldment Assemble

Before welding, remove the oxide film on the surface of the prepared base metal mechanically and clear the weldment in the acetone under ultrasonic vibration, then assemble them according to **Figure 6**. Particular way is: first, put several interlayer alloy foils with width about 4 mm between the end face for welding of the two weldments parallelly and horizontally. Then, fix the weldments in heat-resistant ceramic moulds, which are put in vacuum furnace. Finally, press weights on the assembled weldments with welding pressures 0.3 MPa on.

**Table 2.** The chemical composition of interlayer alloys.

| No. | The chemical composition of interlayer alloys (wt%) | | | | | | |
|-----|---------|----------|---------|---------|---------|-----|---------|
|     | Cr      | W        | Co      | Al      | Mo      | B   | Ni      |
| 1   | 1.0 ~ 1.5 | 2.5 ~ 3.2 | 3.0 ~ 3.7 | 1.5 ~ 2.2 | 0.5 ~ 0.7 | 3.2 | balance |
| 2   | 1.5 ~ 2.0 | 3.2 ~ 4.0 | 3.7 ~ 4.5 | 2.2 ~ 3.0 | 0.7 ~ 1.0 | 3.2 | balance |

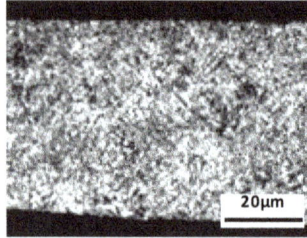

**Figure 4.** Microstructure of interlayer alloy.

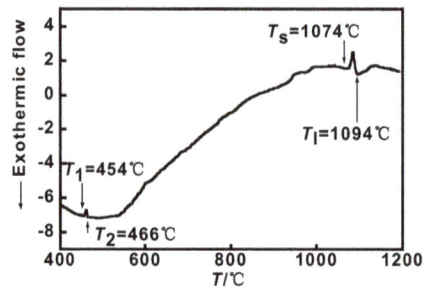

**Figure 5.** The DTA curve of interlayer alloy.

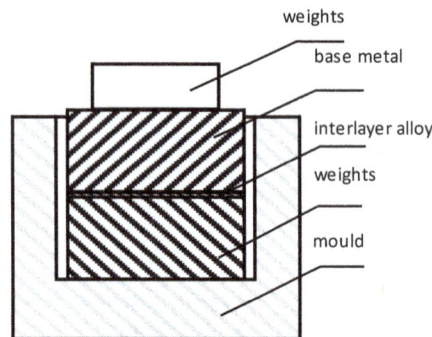

**Figure 6.** Welding assembly diagram.

### 4.1.3. Welding Process

TLP bonding parameters are: welding temperature 1200°C, holding time 8 h, welding pressures 0.3 MPa, vacuum degree $5 \times 10^{-2}$ Pa. Heating specifications are: heating to 1000°C at the heating rate of 10°C/min, holding time 30 min; welding temperature 1200°C, holding time 8h, cooling down to room temperature within the furnace.

### 4.2. Analysis of Joints' Microstructure

By standard metallographic techniques, the metallographic specimens of the welding joints' cross-section were prepared in the welding bars' longitudinal direction. The surface of joints should be etched with reagent (Compositions: 5 ml $HNO_3$ + 15 ml HCL) for about 3 s. DTA analysis the interlayer alloys with CRY-2P Differential Thermal Analyzer. Study the microstructure of the joints with JSM-6700F scanning electron microscope (SEM) and EDAX.

### 4.2.1. DD6 TLP Bonding Joints' Structure

**Figure 7(a)** shows the structure of a typical joint made with DIF2 interlayer alloys for DD6 TLP bonding under the condition of welding temperature 1200°C, holding time 8h and pressure 0.3 MPa. The welding joint is composed of three zones of weld center zone, isothermal solidification zone and diffusion-affected zone whose amplifications were shown in **Figures 7(b)-7(d)** respectively. Among the three, the weld center zone is composed of fine grey $\gamma$ phase, black $\gamma'$ phase (consistent with the isothermal solidification zone) and the larger black $\gamma'$ phase. The diffusion-affected zone contains white acicular W-Mo-Boride enriched MyB phase. Each phase' EDS testing point is shown in **Figure 7(a)** and analysis results shown in **Table 3**. The larger black $\gamma'$ phase in weld center zone precipitates in the cooling process of residual liquid phase because the isothermal solidification time is not enough. The boride phase in the diffusion-affected zone is formed because B diffuses into the base metal in the interlayer. The brittle phase borides are emerged when the content of B is higher than the solubility limit of B in the base metal.

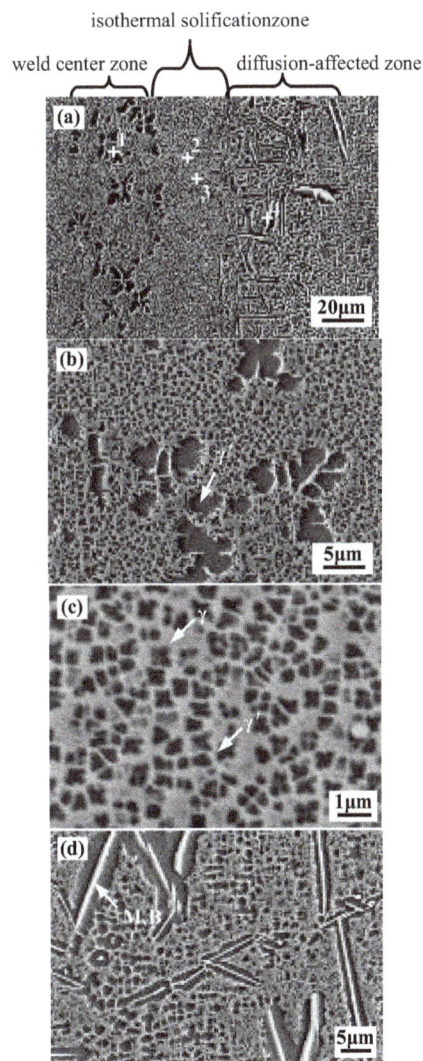

**Figure 7.** Microstructures of DD6 alloy TLP welding joint. (a) The structure of joint; (b) Amplification of weld center zone; (c) Amplification of isothermal solification zone; (d) Amplification of diffusion-affected zone.

### 4.2.2. Influence of Alloying Elements on the Joints' Structure

The structure of welding joints is mainly related with the diffusion of B. Under the condition of the fixed parameters, the diffusion of B is mainly controlled by the contents of the alloying elements in interlayer alloys.

The contents of the alloying elements have certain effect on the structure of the diffusion-affected zone in welding joint which is shown in **Figure 8**. It can be seen that the acicular borides in diffusion-affected zone increases with the increasing of the alloying elements in interlayer alloys which demonstrates that the more the alloying elements in interlayer alloys are, the greater they prevents the solute atomic diffusion and the slower the solute atomic diffusion is. For the same holding time, the more the residual borides in the diffusion-affected zone of the welding joints, the more the brittle borides are generated, which is more harmful to the welding joints.

Therefore, when the welding parameters are welding temperature 1200°C, holding time 8 h and welding pressure 0.3 MPa, the brittle phase borides are fewer in the diffusion zone of the bonding joint made with the interlayer alloys for DD6 TLP bonding with lower contents of alloying elements.

## 5. Conclusions

1) The microstructures of interlayer alloys prepared with the composition range: $1 \leq Cr \leq 2$, $2.5 \leq W \leq 4$, $3 \leq Co \leq 4.5$, $1.5 \leq Al \leq 3$, $0.5 \leq Mo \leq 1$, $3 \leq B \leq 3.6$ and remain Ni (wt%) are fine and homogeneous. The melting point of the alloy foils having better welding performance with thickness 40 μm ~ 60 μm ranges from 1070°C to 1074°C and their melting temperature interval is merely 20°C.

2) Under the condition of welding temperature 1200°C, holding time 8h and pressure 0.3 MPa, the structure of a typical joint made with the interlayer alloys for DD6 TLP bonding when the chemical composition of alloys are 1.5 ~ 2Cr, 3.2 ~ 4W, 3.7 ~ 4.5Co, 2.2 ~ 3Al, 0.7 ~ 1 Mo, 3.2B, remain Ni (wt%), is composed of weld center zone, isothermal solidification zone and diffusion-affected zone.

3) Under the condition of welding temperature 1200°C, holding time 8 h and pressure 0.3 MPa, the brittle

**Table 3.** Chemical compositions of phases by EDS analysis (wt%).

| Measuring location | Chemical compositions (wt%) | | | | | | | Phase |
|---|---|---|---|---|---|---|---|---|
| | Ni | Co | W | Ta | Al | Cr | Mo | |
| 1 | 66.5 | 5.7 | 5.5 | 13.0 | 7.4 | 1.9 | - | $\gamma'$ |
| 2 | 77.7 | 7.1 | 5.7 | - | 4.1 | 3.8 | 1.6 | $\gamma'$ |
| 3 | 75.0 | 6.8 | 8.0 | - | 6.8 | 3.4 | - | $\gamma$ |
| 4 | 26.8 | 4.9 | 39.6 | 13.9 | - | 2.1 | 12.7 | $M_yB$ |

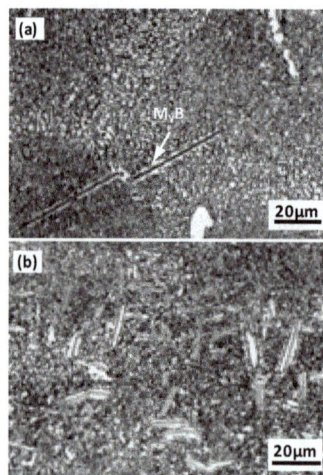

**Figure 8.** Microstructures of the diffusion zone of DD6 TLP bonding joint at 1200°C for 8 h with various interlayer alloys.

phase borides are fewer in the diffusion zone of the bonding joint made with the interlayer alloys for DD6 TLP bonding when the chemical composition of alloys are $1.5 \sim 2Cr$, $3.2 \sim 4W$, $3.7 \sim 4.5Co$, $2.2 \sim 3Al$, $0.7 \sim 1Mo$, $3.2B$, remain Ni (wt%).

## Acknowledgements

This work was supported by the scientific research project of Shaanxi province department, the industrialization cultivation project of Shaanxi province education department, the overall innovation plan of Xi'an technology bureau and the western material innovation fund. The authors would like to express the heartfelt thanks to Professor Wei Bing-Bo to offer the benefit help.

## References

[1]  Wang, G., Zhang, B.G. and Feng, J.C. (2008) Research Progress in Pepair Welding Technology of Ni-Based Superalloy Blands. *Welding & Joining*, **1**, 20-23.

[2]  Chen, R.Z. (1995) Development Status of Single Crystal Superalloys. *Materials Engineering*, **8**, 83-12.

[3]  Li, X.H., Mao, W., Guo, W.L. and Xie, Y.H. (2005) Transient Liquid Phase Diffusion Bonding of a Single Crystal Superalloy DD6. *Transactions of the China Welding Institution*, **26**, 51-54.

[4]  Pouranvari, M. and Ekrami, A. (2008) Microstructure Development during Transient Liquid Phase Bonding of GTD-111 Nickel-Based Superalloy. *Journal of Alloys and Compounds*, 641-647.

[5]  Ojo, O.A., Richards, N.L. and Chaturvedi, M.C. (2004) Isothermal Solidification during Transient Liquid Phase Bonding of Inconel 738 Superalloy. *Science and Technology of Welding and Jioning*, **9**, 532-540.

[6]  Zhai, Q.Y., Tong, D.X., Li, W.W. and Gong, S.T. (2009) Preparation and Welding Performance of Amorphous Interlayer Alloys for TLP Bonding. *Transactions of the China Welding Institution*, **30**, 17-20.

[7]  Wang, F.S., Chen, S.J., Gao, Z., Wen, S.L. and Chen, L. (2009) Applications of Transient Liquid-Phase Bonding in Nickel and Nickel-Base Alloy Welding. *Electric Welding Machine*, **39**, 38-42.

[8]  Liu, S., Olsen, D.L. and Martin, G.P. (1991) Modeling of Brazing Processes That Use Coatings. *Welding Journal*, **70**, 207-215.

[9]  Shi, D.K. (2003) Materials Science. Mechanical Industrial Press, Beijing.

[10]  Dai, Y.N. (2009) Atlas of Binary Alloy Phase. Beijing Science Press.

[11]  Xu, J.F. and Wei, B.B. (2004) Liquid Phase Flow and Microstructure Formation during Rapid Solidification. *Acta Physica Sinica*, **53**, 160-166.

# 14

# Residual Stresses and Micro-Hardness Testing in Evaluating the Heat Affected Zone's Width of Ferritic Ductile Iron Arc Welds

**Georgios K. Triantafyllidis\*, Dimitrios I. Zagliveris, Dionysios L. Kolioulis, Christos S. Tsiompanis, Titos N. Pasparakis, Athanasios P. Gredis, Melina L. Sfantou, Ioannis E. Giouvanakis**

Department of Chemical Engineering, Aristotle University, Thessaloniki, Greece
Email: *gktrian@auth.gr

## Abstract

Shielded Metal Arc Welding (SMAW) in Ductile Irons (DI) is often required by foundries for practical manufacturing reasons. The mechanical properties of the welded structures are strongly dependent on their HAZ's width. A model based on the behaviour of the ferritic matrix of high-Si DIs in order to make an approach in measuring their HAZ's width is developed in this study. A series of thermal treatments on 3.35 and 3.75 wt% Si as-cast DIs and spot SMAWs is applied on these materials. The applied SMAWs are done on non-preheated and preheated samples (150°C - 300°C). For welding we modify the amperage (100 - 140A). The micro-hardness Vickers changes in the ferrite of the as-cast samples and inside the HAZ of the welded ones can be attributed to the existence of residual stresses (RS) in the ferritic matrix and assist in estimating the HAZ's width.

## Keywords

Welds, Heat Affected Zone, Residual Stresses, Micro-Hardness Vickers, Heat Affected Zone's Width

## 1. Introduction—Theoretical Background

Heat treatments impose residual stresses (RS) in metallic materials. Arc welds transfer large amounts of heat on

---

*Corresponding author.

the welded metals imposing, so, RS in addition to crystallographic changes. The heat affected zone (HAZ) is that part of the base metal subjected to such transformations due to the passage of the welding heat front. Crystallographic transformations and RS, as well, define the HAZ's width.

Ductile Irons (DIs) is a broad family of iron based casting alloys [1]. Their welding metallurgy is presented in [2]. Weld microstructures, heat-affected zone, partially melted region, fusion zone, weldability and preweld testing are well analyzed and are taken into consideration in this study.

Their mechanical properties are given in the specification EN 1563. Microstructures (from fully ferritic to fully pearlitic) underline progressive change in their mechanical properties [3]. Higher amounts of silicon have being added (3.35 - 3.75 wt%) in the melts in order to strengthen the ferritic matrix by solid-solution hardening after solidification. The family EN-GJS-500-10 (fully ferritic), so, has been incorporated to the EN 1563. In comparison to the category EN-GJS-500-7 (ferritic-pearlitic), it offers higher elongation (10%) by keeping the $\sigma_{UTS}$ at the same level (500 MPa).

SMAW under proper welding conditions in this type of materials is often required by foundries for practical manufacturing reasons [4].

For an electrode arc weld, before any further qualification test (as the tensile test), two parameters are important and desirable in characterizing its quality:

1) A smooth micro-hardness profile without great changes in hardness values transverse to the seam [5].

2) Small HAZ's width. It depends on the heat-source intensity [6].

Continuous cooling conditions (and not isothermal ones) are prevailing during the passage of the welding heat front through the base metal [7]. They cause a series of microstructural modifications of the BM in the HAZ, and accumulation of RS as well.

## 2. Experimental Procedure

### 2.1. General Information

The target of our study was the quality evaluation of spot SMAWs performed in DIs with metallography, micro-Vickers hardness testing and X-rays diffraction (XRD) of as-cast samples.

So, we operated two experimental cycles with differentiation in some conditions. The detailed description of each one's procedure is following.

"Hitiria Makedonias S.A.", a foundry facility located in Industrial Area of Sindos/Thessaloniki/Greece, produced cylindrical parts (diameter: 30 mm, length: 300 mm) of the DIs (EN-GJS-500-10) used in the study in the as-cast condition.

### 2.2. 1st Experimental Cycle for Si Content 3.35 wt%

Following are detailed all the steps that have taken place during this cycle:

1) Slices of 5 mm in width were cut out under cooling water from the cylinders and these pieces were divided into quadrants forming totally 28 samples, as they are presented (with their hardness values) in **Table 1**. They were subjected to full annealing-(*FA-samples*) and normalizing-type (*N-samples*) heat treatments for 5 minutes at a range of 450°C - 1100°C with 50°C step. These conditions were chosen according our experience. At the full annealing-type treatment the samples were inserted in the furnace at the set point temperature and left to be cooled in the furnace after the end of each thermal cycle. At the normalizing-type the sample were inserted in the furnace at the set point temperature and after 5 minutes they were withdrawn and left to be cooled in the atmosphere in still air.

2) One extra sample was thermally treated for 5 minutes at 1100°C and then water quenched at 25°C (*QS sample*).

3) All heat treatments were performed in a CARBOLITE furnace. No decarburization surface effects were noticed.

4) All these samples were mounted in bakelite by using a PRESI Mecapress 3 mounting machine. They were properly grinded, polished and etched in Nital 4% for metallographic observation.

5) Another extra sample of the as-cast condition was mounted, grinded, polished, etched and metallographically observed (*AC sample*).

6) Cylinders of 25 mm in length were cut out under cooling water from the as-cast cylinders. At the center of

**Table 1.** Micro-hardness Vickers (HV) values of heat treated samples.

| Sample No. | Sample Details | | | |
| :---: | :---: | :---: | :---: | :---: |
| | Heat Treatment Temperature (°C) | Cooling Conditions | Average Hardness (HV) | Standard Deviation |
| 1 | 450 | FA-type | 169 | 2.1 |
| 2 | 450 | N-type | 180 | 3.6 |
| 3 | 500 | FA-type | 188 | 6.4 |
| 4 | 500 | N-type | 189 | 6.4 |
| 5 | 550 | FA-type | 183 | 8.7 |
| 6 | 550 | N-type | 190 | 9.8 |
| 7 | 600 | FA-type | 186 | 13.5 |
| 8 | 600 | N-type | 185 | 4.0 |
| 9 | 650 | FA-type | 196 | 3.8 |
| 10 | 650 | N-type | 184 | 8.5 |
| 11 | 700 | FA-type | 206 | 2.0 |
| 12 | 700 | N-type | 197 | 4.6 |
| 13 | 750 | FA-type | 201 | 3.1 |
| 14 | 750 | N-type | 202 | 7.2 |
| 15 | 800 | FA-type | 196 | 2.0 |
| 16 | 800 | N-type | 198 | 2.0 |
| 17 | 850 | FA-type | 193 | 9.9 |
| 18 | 850 | N-type | 198 | 4.0 |
| 19 | 900 | FA-type | 197 | 8.9 |
| 20 | 900 | N-type | 293 | 10.6 |
| 21 | 950 | FA-type | 187 | 6.0 |
| 22 | 950 | N-type | 384 | 11.0 |
| 23 | 1000 | FA-type | 171 | 5.1 |
| 24 | 1000 | N-type | 422 | 11.9 |
| 25 | 1050 | FA-type | 181 | 12.5 |
| 26 | 1050 | N-type | [300][a] | |
| 27 | 1100 | FA-type | 153 | 9.5 |
| 28 | 1100 | N-type | [293][a] | |
| As-cast (AC) (29) | - | - | 204 | 4.0 |
| QS (30) | 1100 | Quenched | [756][b] | |

[a]Samples 26 & 28 exhibited inhomogeneity in its hardness values, due to melting phenomena. We performed 3 measurements in #26 and 5 in #28 with the following results: (348, 406, 234, 216, 263) and (260, 297, 343), respectively. [b]The QS (30) exhibited mixed ferritic-pearlitic microstructure with HV at several places: (749, 763, 710, 808, 808, 371, 927, 890, 1008, 589, 618, 618, 808, 808, 808, 808). Microstructures intermediate in hardness (from pure carbides to martensite, retained austenite and ferrite).

their flat surface a blind groove with 6.5 mm diameter and about 5 mm depth had being drilled. By using a KEMPPI Minarc 150 Arc Welding machine shielded metal spot welds were performed in order to fill up the grove with an ESAB OK 92.58/NiFe-Cl-A electrode with a diameter of 3 mm. Weldments were performed by using 90 - 140 A current with step 10 A. After welding the surface was grinded, polished and etched in Nital 4% for metallographic observation.

7) The characterization mainly of the ferritic matrix of the samples and the structures transverse to the welding seam was performed by optical metallography by using an OLYMPUS A70 Metallograph and micro-Vickers hardness testing by using an AFFRI Microhardness Tester.

8) XRD experiments using a Seifert 3003 TT apparatus with Fe-K$\alpha$ radiation were carried out on the samples AC, QS, and 16, 20 and 24.

## 2.3. 2st Experimental Cycle for Si Content 3.35 wt% and 3.75 wt%

For not expatiate, we will refer only the points of the procedure that show differentiations in relation to the first cycle:

- An amount of 20 samples with 3.35% wt. Si were subjected to normalizing type heat treatment at a range of 700°C - 1150°C with 50°C step for 1 min (10 samples) and 2 mins (10 samples).
- Another 20 samples with 3.75% wt. Si were subjected to the same heat treatment conditions.
- The 3.35-Si samples were subjected to weldments by using a 110 A current. They were preheated for 1 min at 150°C - 300°C with a step of 50°C.
- For the 3.75-Si samples the amperage ranged between 100 and 140 A with a step of 10 A.
- No XRD measurements were performed in this experimental cycle.

## 3. Results

### 3.1. 1st Experimental Cycle

A number of 28 heat treated samples and their HV hardness values are presented in **Table 1**, above. No decarburization phenomena were observed except for the quenched sample. In all cases 5 measurements were made with quite limited standard deviation.

In **Figure 1**, the radial hardness Vickers profiles for 3.35-Si specimens *welded at various amperages* are depicted. These hardness data were the criterion for evaluating the HAZ's width, which is presented in **Figure 2**.

Microstructures of a welded specimen transverse to the welding seam is presented in **Figure 3** (140 A, 200X).

**Figure 4** & **Figure 5** are presented the results of the XRD measurements in some selected samples.

### 3.2. 2nd Experimental Cycle

**Figure 6** & **Figure 7** show the microstructures in the HAZ's vicinity of welded specimens (3.35- and 3.75-Si, respectively) at the same amperage (110 A) for comparison reasons.

In **Figure 8**, the radial hardness Vickers profiles for 3.35-Si specimens *preheated at various temperatures* are depicted. These hardness data were the criterion for evaluating the HAZ's width, which is presented in **Figure 9**.

In **Figure 10**, the radial hardness Vickers profiles for 3.75-Si specimens *welded at various amperages* are depicted. These hardness data were the criterion for evaluating the HAZ's width, which is presented in **Figure 11**.

## 4. Conclusions

The XRD spectra of the QS (**Figure 5**) revealed a mixed fcc (retained austenite) + bcc microstructure (Peaks: 56°: (111)-bcc//65°: (200)-fcc//85°: (200)-bcc//100°: (220)-fcc//112°: (211)-bcc). All the other X-Rays specimens (**Figure 4**) followed a typical bcc structure (Peaks: 56°: (111)-bcc//85°: (200)-bcc//112°: (211)-bcc). Metallographic study revealed ferritic structures except for the QS that exhibited mixed ferritic and pearlitic structure.

Spot SMAW under the referred conditions produced welds with important characteristics:

- All full-annealing type heat treatments resulted in hardness values of the matrix ferrite of less than 200 HV (thermally unaffected ferrite).
- All normalizing-type heat treatments until 850°C resulted in hardness values of matrix ferrite of less than 200 HV (thermally unaffected ferrite).

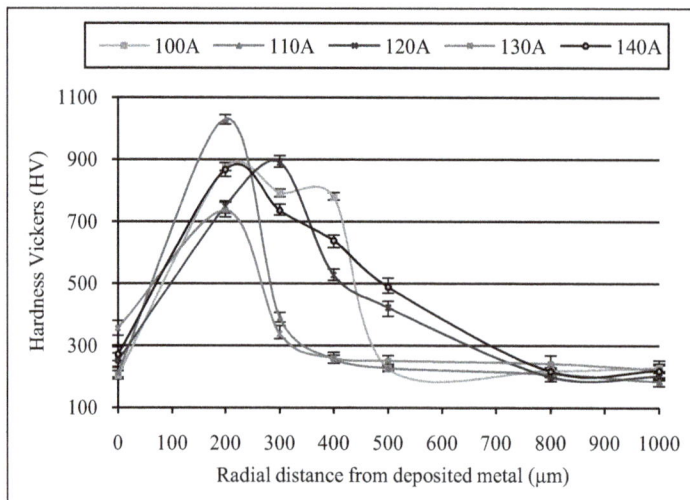

**Figure 1.** Hardness profiles for various welding amperages (3.35-Si).

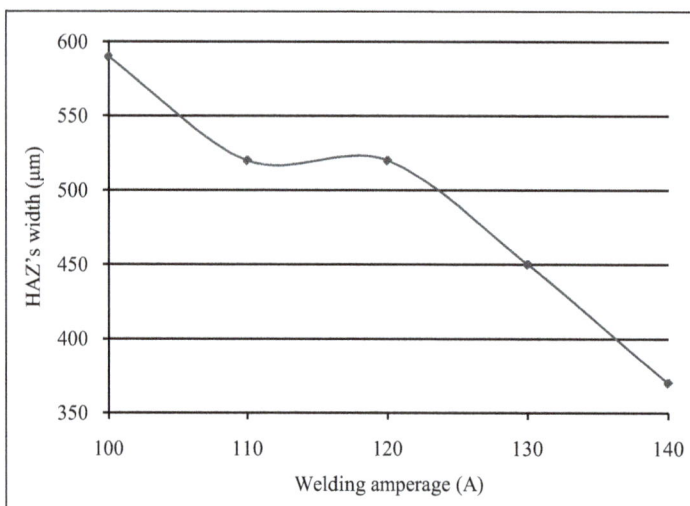

**Figure 2.** Estimated HAZ's width for all welding amperages.

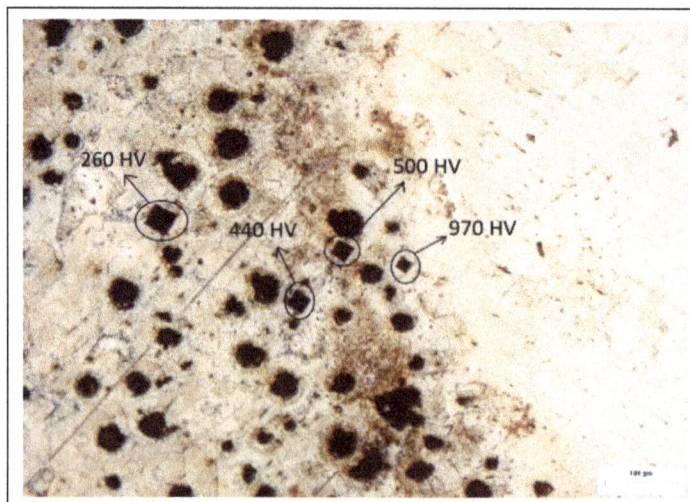

**Figure 3.** Microstructure transverse to the weld seam (140 A, 200 X).

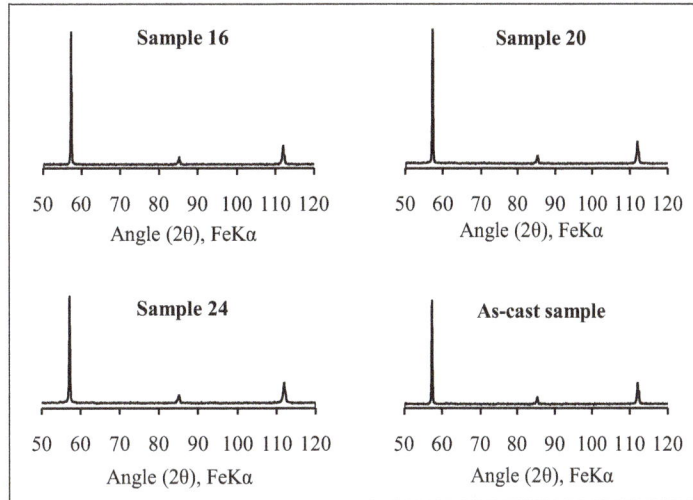

**Figure 4.** The XRD spectra of some of the samples with bcc structure.

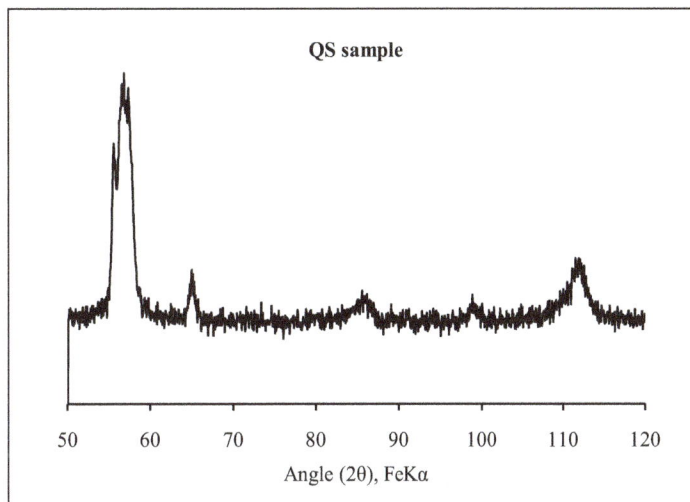

**Figure 5.** The XRD spectrum of the QS-sample. Mixed fcc & bcc.

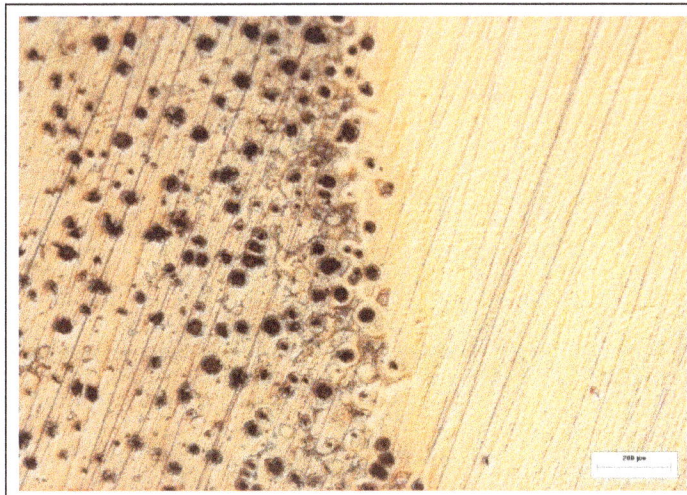

**Figure 6.** Microstructure of welded 3.35-Si sample (200 X).

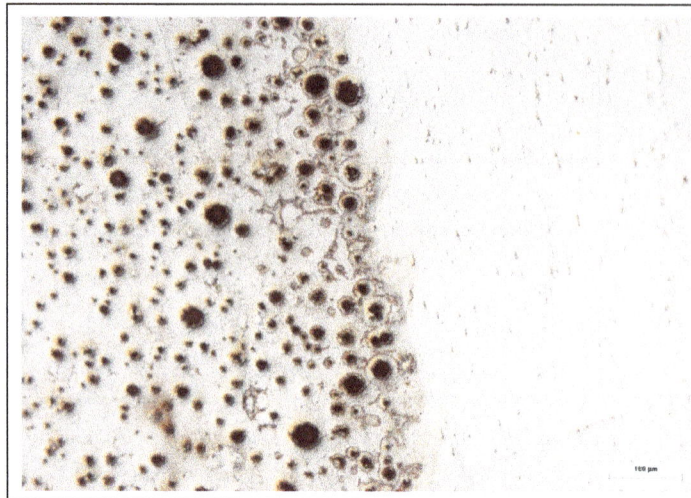

**Figure 7.** Microstructure of welded 3.75-Si sample (200 X).

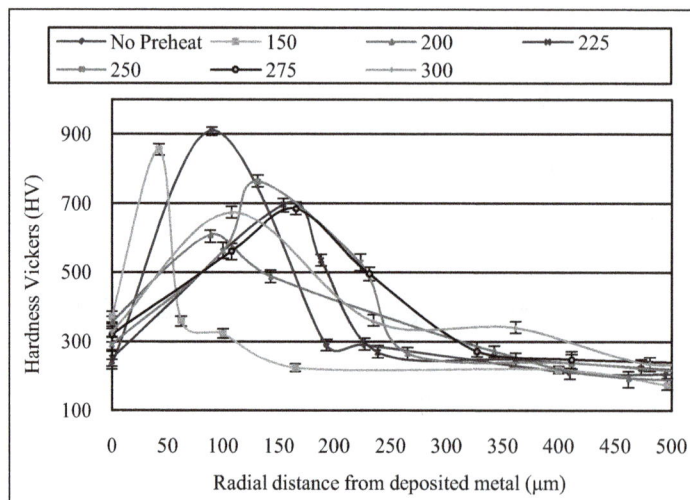

**Figure 8.** Hardness profiles for various preheating temperatures (3.35-Si).

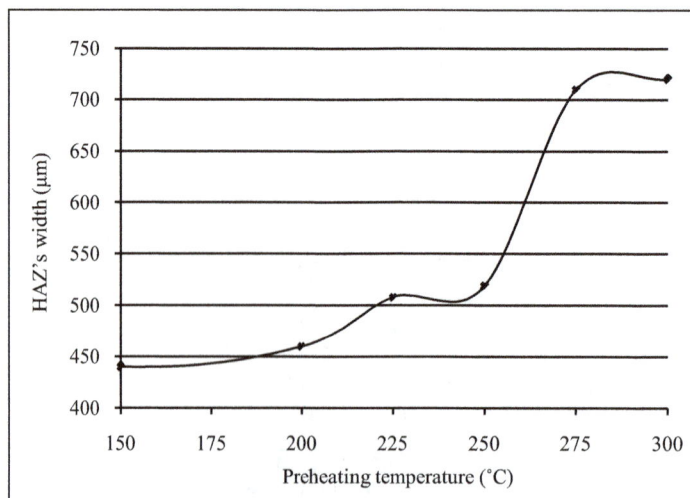

**Figure 9.** Estimated HAZ's width for all preheating temperatures.

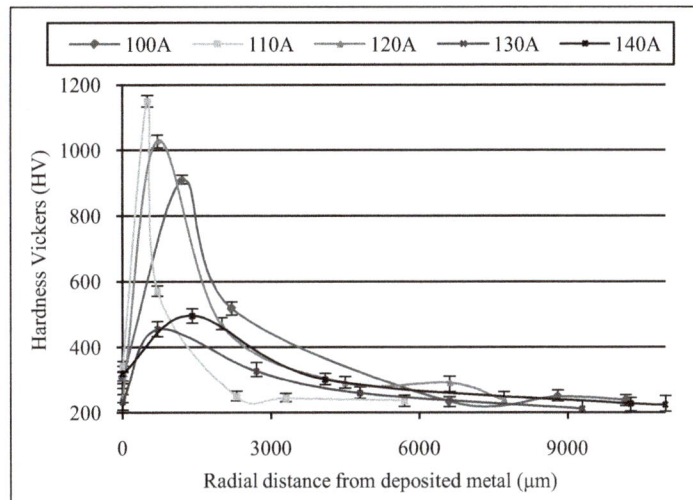

**Figure 10.** Hardness profiles for various welding amperages (3.75-Si).

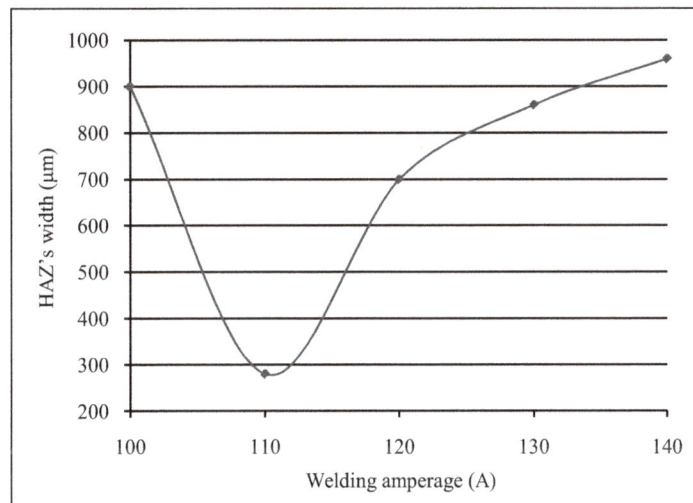

**Figure 11.** Estimated HAZ's width for all welding amperages.

- At higher temperatures normalizing-type treatment resulted in higher hardness values. These values were formed according to the effect of RS.
- After 900°C at normalizing-type heat treatments the matrix ferrite was thermally affected.
- A relatively normal hardness profile transversed to the seam with values fluctuating between 350 HV (DM) and 200 HV (matrix ferrite, BM).
- Ferritic structures with HV values of 420 to 390 underlined thermally affected ferrite (HAZ).
- This normal profile was abnormally disturbed by the presence of carbides at the partially melting zone of the weld (white iron, 900 - 1000 HV). They could not be manipulated without a proper pre- or post-weld heat treatment.
- Inside the HAZ, the hardness changed progressively from the carbidic zone (900 - 1000 HV) to the unaffected ferrite zone (less than 280 - 260 HV).
- HAZ stopped at the point where the HV values of the ferritic matrix dropped beyond the 280/260 threshold due to the presence of RS (unaffected ferrite).
- The HAZ's width was measured between this point and the point where carbides were present, just near the deposited metal area (partially melted zone).
- It was strongly affected by the welding conditions. Higher currents (130, 140 Amps, base value the 90 Amps) for the 3.35-Si samples produced thinner HAZs.

## 5. Discussion

SMAW in general imposes crystallographic changes and RS in the ferritic matrix of the DI. They lead, both, in local increase of the mechanical strength of the material. Any local increase of the hardness, so, could be attributed to these two factors. Crystallographic changes are clearly defined by metallography. RS are practically invisible and could be detected only by indirect means. An abnormal increase, so, of the micro-hardness of the ferritic grains in the HAZ is directly related to RS. All our micro-hardness measurements are compatible to this model.

HAZ starts at the point where carbides are observed with very high HV values (more than 900 HV). This point belongs to the partially melted zone of the weld. From that point a characteristic zone with crystallographic modifications of the BM is observed. It incorporates RS also, but they cannot be clearly distinguished by optical microscopy.

By approaching the BM, beyond the previous referred zone, the hardness of the ferrite expresses higher than normal values. These values could be a strong evidence for the existence of thermally imposed RS.

By applying this model to our SMAW welds we focus our attention to the following:
- Preheating results in absence of carbides for the 3.35-Si samples but in broadening of their HAZ's width.
- The Si content affects the HAZ's width in a contradictory manner. For the 3.75-Si samples the lowest HAZ's width is observed at 110 A. With increasing amperage the HAZ's width broadens.

These notes could be explained by considering a substantial modification of the physical properties of the ferritic matrix due to the addition of Si. But the explanation of the data of **Figure 9** and **Figure 11** could not be directly related only on the percentage of Si. Further research based on the influence of the welding conditions on related materials seems to be necessary to be done.

Normalizing-type heat treatments of the as-cast samples at temperatures referred follow conditions closer to the ones of continuous cooling. They assist in distinguishing the thermally affected ferrite due to RS by micro-hardness measurements and measuring the HAZ's width. All full annealing-type heat treatments left the matrix ferrite unaffected.

The model presents a low-cost procedure and is suitable for the industry (foundries) in order to evaluate welding conditions of this type (selection of welding electrodes etc) before doing more expensive mechanical tests.

These conclusions could assist, also, in the application of (NDT) XRD for characterizing of RS in high-Si ferritic DI grades SMAWs, especially where failures are observed [9].

## Acknowledgements

We express our thanks to:
- Professor Georgios Litsardakis (Aristotle University of Thessaloniki, Faculty of Engineering, Laboratory of Materials for Electronics) for his kind contribution in performing the XRD spectra of our study.
- Mr. Dimitrios Repanis MSc of "Hitiria Makedonias S.A." for the preparation of the as-cast alloys.
- Dr. Panagiotis K. Tsolakisof Materials Industrial Research & Technology Center—MIRTEC S.A. (Volos Industrial Area, Greece) for his kind contribution in performing the chemical analysis of the material with OES.

## References

[1]　Stefanescu, D.M. (1990) Classification and Basic Metallurgy of Cast Iron, Properties and Selection: Irons, Steels, and High-Performance Alloys. Vol. 1, ASM Handbook, ASM International, 3-11.

[2]　Davis, J.R. and Associates (1996) Cast Irons, Secondary Processing of Cast Irons, Welding and Brazing. ASM Specialty Handbook, 217-219.

[3]　Larker, R. (2009) Solution Strengthened Ferritic Ductile Iron ISO 1083/JS/500-10 Provides Superior Consistent Properties in Hydraulic Rotators. Overseas Foundry.

[4]　Repair and Maintenance Welding Handbook. 2nd Edition, ESAB.

[5]　(1995) The Metals Blue Book. Vol. 3, Welding Filler Metals, CASTI Publishing Inc. and American Welding Society.

[6]　Eagar, T.W. (1993) Energy Sources Used for Fusion Welding. ASM Handbook, Vol. 6, Welding, Brazing and Soldering, First Printing.

[7]  Fonda, R.W., Vandermeer, R.A. and Spanos, G. (1998) Continuous Cooling Transformation (CCT) Diagrams for Advanced Navy Welding Consumables. Naval Research Laboratory, United States Navy.

[8]  Krauss, G. (2007) Steels: Processing, Structure and Performance. 3rd Printing, ASM International.

[9]  Pineault, J., Manufacturing, P. and Mich, T. (2015) Measuring Residual Stresses via X-Ray Diffraction Helps Optimize Engine Component Fabrication. *Advanced Materials and Processes, ASM International*, **173**, 20-22.

# 15

# FEM Simulation of Distortion and Residual Stress Generated by High Energy Beam Welding with Considering Phase Transformation

**Y.-C. Kim[1]\*, M. Hirohata[2], K. Inose[3]**

[1]Osaka University, Osaka, Japan
[2]Graduate School of Engineering, Nagoya University, Nagoya, Japan
[3]IHI Corporation, Yokohama, Japan
Email: \*kimyc@ark.zaq.jp

## Abstract

A series of experiments was carried out so as to elucidate the effect of the phase transformation in the cooling process on welding distortion and residual stress generated by laser beam welding (LBW) and laser-arc hybrid welding (HYBW) on the high strength steel (HT780). Then, the experiments were simulated by 3D thermal elastic-plastic analysis with FEM (Finite Element Method) which was performed with using the idealized mechanical properties considering the transformation superplasticity. From the results, the effects of the phase transformation on welding distortion and residual stress generated by LBW and HYBW were elucidated. Furthermore, the generality of the idealization of the mechanical properties was verified.

## Keywords

Welding Distortion, Residual Stress, Laser Beam Welding, Laser-Arc Hybrid Welding, Phase Transformation, FEM

## 1. Introduction

For improving function and reducing weight of steel structures, aggressive use of high strength steel has been tried. In order to achieve this trial, it is essential that the performance of joints of high strength steel is improved.

---

\*Corresponding author.

For this objective, researches of joining with laser beam have been conducted [1].

Although joints with high quality can be manufactured by laser beam welding (LBW), it requires quite severe gap control between welded plates for avoiding generation of under fill. Therefore, application of laser-arc hybrid welding (HYBW) has been investigated for improving manufacture efficiency. By supplying deposit metal by arc welding, the gap control of HYBW is expected to be easier than that of LBW without deposit metal.

The heat input characteristics of LBW are largely different from those of existing arc welding such as heat energy, bead width, penetration depth and so on. Furthermore, those of HYBW with dual heat source differ from those of welding by using only laser or arc.

On the other hand, in welding of high strength steel, the phase transformation in the cooling process largely affects the generation of welding distortion and residual stress [2].

Concerning to this, a series of welding experiments has been conducted so as to elucidate residual stress generated by arc welding on high strength steel (HT780). The experiment has been simulated by 3D thermal elastic-plastic analysis based on FEM (Finite Element Method). In the analysis, the mechanical properties of the high strength steel were idealized by considering the phase transformation. As a result, the validity of simulation by FEM was verified for predicting the welding distortion and residual stress [1] [3] [4].

By the way, from the viewpoints of both the welding process and the phase transformation, it is unknown how distortion and residual stress are generated by LBW and HYBW on high strength steel.

In this paper, a series of welding experiments on high strength steel (HT780) by LBW and HYBW is performed. And then, the experiment is simulated by FEM of which the validity was verified by the previous researches [5] [6]. The effects of the phase transformation on the generation of welding distortion and residual stress of LBW and HYBW are elucidated. Moreover, the generality of the idealized mechanical properties with considering the phase transformation is verified.

## 2. Welding Experiment

### 2.1. Test Specimen

The shape and dimension of the specimen are shown in **Figure 1**.

In order to obtain distortion and residual stress generated by butt welding with high accuracy, it is necessary that a specimen is made so that a linear misalignment due to tack welding does not occur [6]. Therefore, one-pass bead-on-plate welding is selected.

The material is a high strength steel of which the tensile strength is over 780 MPa (HT780). The thickness is 12 mm. **Table 1** shows the chemical compositions and the mechanical properties of the material.

**Table 2** shows the welding conditions of LBW and HYBW. In both processes, a fiber laser is used. In HYBW, $CO_2$ MAG arc welding with 780 MPa class filler wire is also used.

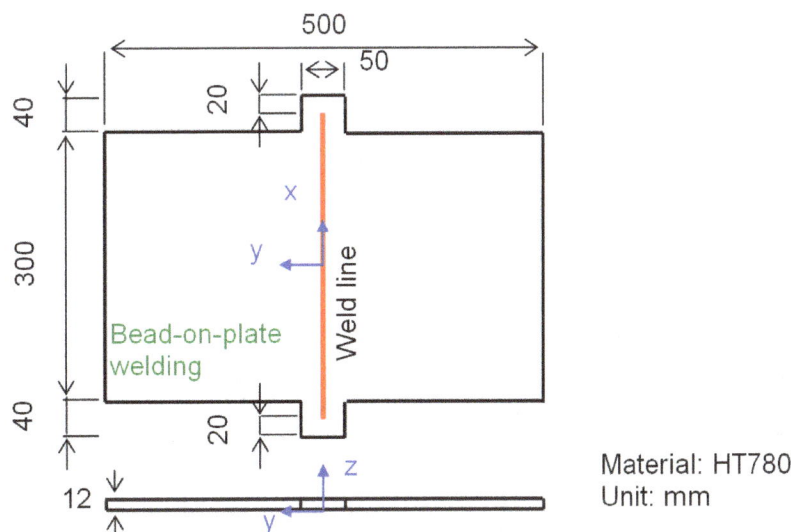

**Figure 1.** Test specimen.

**Table 1.** Chemical compositions and mechanical properties.

| Chemical compositions (mass %) | | | | | Mechanical properties | |
| --- | --- | --- | --- | --- | --- | --- |
| C | Si | Mn | P | S | Yield point (MPa) | Tensile strength (MPa) |
| 0.08 | 0.22 | 0.96 | 0.007 | 0.002 | 813 | 840 |

**Table 2.** Table type styles.

| Process | Laser | | Arc | | | Speed (m/min) |
| --- | --- | --- | --- | --- | --- | --- |
| | Type | Power (kW) | Type | Filler | Energy (kW) | |
| LBW | Fiber laser | 6.0 | - | - | - | 0.6 |
| HYBW | | 6.8 | $CO_2$ MAG | 780 MPa class | 5.4 | 0.8 |

## 2.2. Experimental Procedure

During the welding, the temperature histories are measured by the thermo couples attached to 4 points (y = 15, 30, 50, 80 (mm)) of the upper side of the specimen (z = 12 (mm)) in the center of the welding direction (x = 15 (mm)).

After the welding, the penetration shapes are confirmed by obtaining the macrographs. These are also used for making grids in simulating the experiments by FEM.

The out-of-plane deformation is measured and the residual stress is obtained by a stress relaxation method in both LBW and HYBW specimens.

## 2.3. Results of Experiments

### 2.3.1. Macrographs

**Figure 2** shows the examples of the macrographs of LBW and HYBW.

In LBW, the bead width is around 5 mm and the penetration depth is 9 mm. In HYBW, the bead width at the upper side is wide (around 10 mm) due to the supply of weld metal by arc welding. The penetration depth is almost the same as that by LBW.

### 2.3.2. Welding Distortion

**Figure 3** shows the welding out-of-plane deformation. The angular distortion of LBW is smaller than that of HYBW (**Figure 3(a)**). The longitudinal bending distortions of both LBW and HYBW are extremely small (**Figure 3(b)**).

### 2.3.3. Welding Residual Stress

**Figure 4** shows the welding residual stress.

The residual stress generated by LBW is noted.

The stress component along the weld line; $\sigma_x$ by LBW (**Figure 4(a)**) is tensile but under the yield stress (813 MPa) in the weld metal on the upper side. The magnitude is around 600 MPa. On the lower side, the tensile stress around 80 MPa occurs. On the upper and lower sides, small compressive stress under 20 MPa occurs in the base metal. The stress component perpendicular to the weld line; $\sigma_y$ by LBW (**Figure 4(b)**) is compressive around 90 MPa in the weld metal on both the upper and lower sides. Scarcely any stress occurs in the base metal.

In the same way, the residual stress generated by HYBW is noted. The stress component along the weld line; $\sigma_x$ by HYBW (**Figure 4(c)**) is almost the same as that of LBW in the weld metal on the upper side. That is tensile and the magnitude is around 600 MPa. The region in which the tensile stress occurs is wider than that of LBW. On the lower side, the tensile stress around 300 MPa occurs, which is larger than that of LBW because the total heat input of HYBW is larger than that of LBW. The compressive stress in the base metal on the upper and lower sides is small as well as that of LBW. The stress component perpendicular to the weld line; $\sigma_y$ by HYBW (**Figure 4(d)**) is tensile from 100 to 150 MPa in the weld metal on the upper side. That on the lower side is around 70 MPa. The stress in the base metal is small as well as that of LBW.

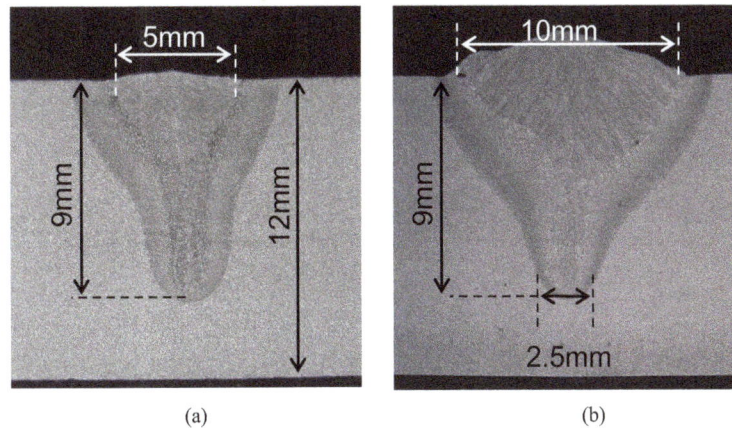

Figure 2. Macrograph. (a) LBW; (b) HYBW.

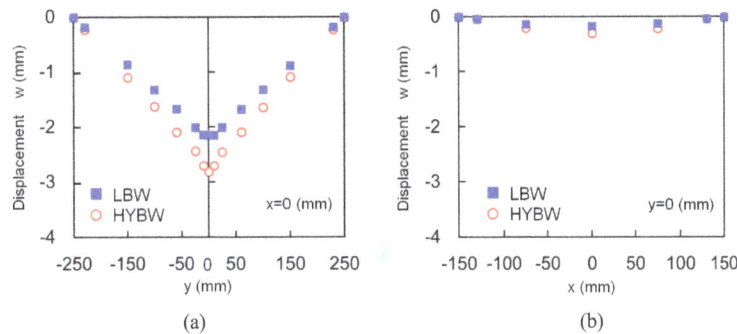

Figure 3. Welding out-of-displacements generated by LBW and HYBW. (a) Angular distortion; (b) Longitudinal bending distortion.

## 3. Idealization of Mechanical Properties

For simulating the welding on high strength steel, idealized mechanical properties considering the phase transformation in the cooling process are used. Here, the idealization of mechanical properties in the phase transformation range in the cooling process proposed by the author is explained. The details of idealization have been shown in the references [3] [4].

The starting and finishing temperatures of phase transformation in the cooling process; $M_s$ and $M_f$, were measured by Formastor test with changing the cooling rates (40, 100, 150 degrees Celsius/s). **Figure 5** shows the result in the case that the cooling rate is 100 degrees Celsius/s. The variations of them were within 20 degrees Celsius in the three cases of cooling rates.

The averages of them were adopted, *i.e.*, $M_s$ was 441 degrees Celsius and $M_f$ was 324 degrees Celsius. The range between them was defined as the phase transformation range.

The average of thermal expansion coefficient; $\alpha$, was $-22 \times 10^{-6}$ per degrees Celsius in the phase transformation range. This was uniformly used in that range.

A series of tensile tests at high temperature was carried out. Specimens were heated up to 1250 degrees Celsius and the temperature was kept in some seconds. After cooling the specimens to the test temperatures, the tensile tests were carried out. Although the cooling rates were changed (40, 100, 150 degrees Celsius/s), it was confirmed that the effect of cooling rate on the mechanical properties was small.

As an example of the results of the tensile tests, temperature dependencies of Young's modulus are shown in **Figure 6**. The above-mentioned transformation expansion occurs in the phase transformation range and it is known that transformation superplasticity also occurs in this range. It means the phenomenon that strength is remarkably lowered and extraordinary ductility is observed when the phase transformation occurs and progresses [2]. Therefore, the yield stress and Young's modulus in the phase transformation range cannot be identified because the strain due to loading and that due to the transformation expansion and the transformation superplasticity cannot be separated.

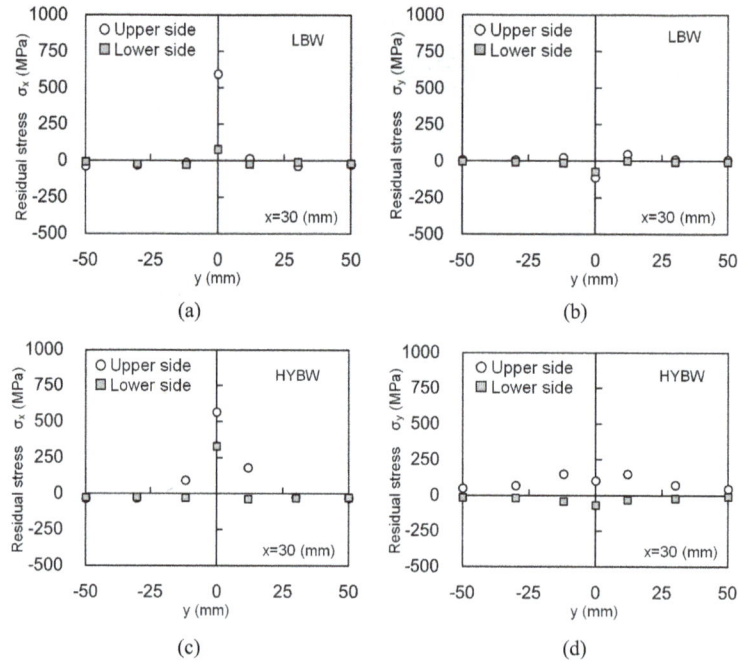

**Figure 4.** Distributions of residual stress generated by LBW and HYBW. (a) Stress component, $\sigma_x$ by LBW; (b) Stress component, $\sigma_y$ by LBW; (c) Stress component, $\sigma_x$ by HYBW; (d) Stress component, $\sigma_y$ by HYBW.

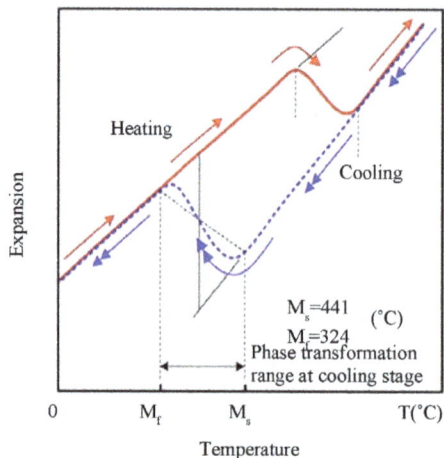

**Figure 5.** Expansion-temperature diagram.

In simulating the experiment by the thermal elastic-plastic analysis, the three types of idealized mechanical properties as shown in **Figure 7** are used.

1) Model A: The phase transformation is not considered. That is, the mechanical properties in the heating process are also used in the cooling process.

2) Model B: Only the transformation expansion is considered by using the negative value of the thermal expansion coefficient; $\alpha$ in the phase transformation range is used. However, the yield stress; $\sigma_y$ and Young's modulus; E are the same in the heating and cooling processes.

3) Model C: Not only the transformation expansion but also the transformation superplasticity are considered. Although the thermal expansion coefficient; $\alpha$ and the yield stress; $\sigma_y$ are the same as those in Model B, Young's modulus; E between $M_s$ and $M_f$, is idealized with considering the extraordinary ductility due to the transformation superplasticity. That is, Young's modulus; E at $M_s$ is lowered around zero and it is linearly recovered up to $M_f$ in the cooling process.

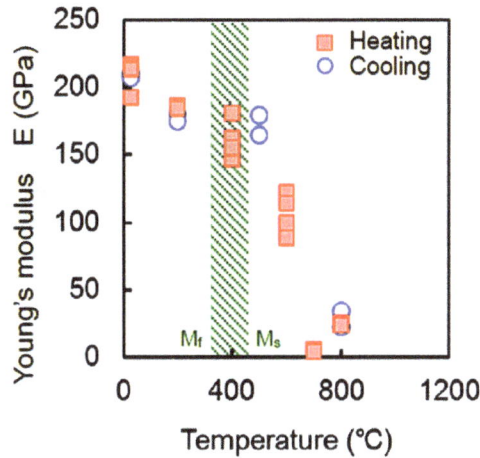

**Figure 6.** Temperature dependencies of Young's modulus.

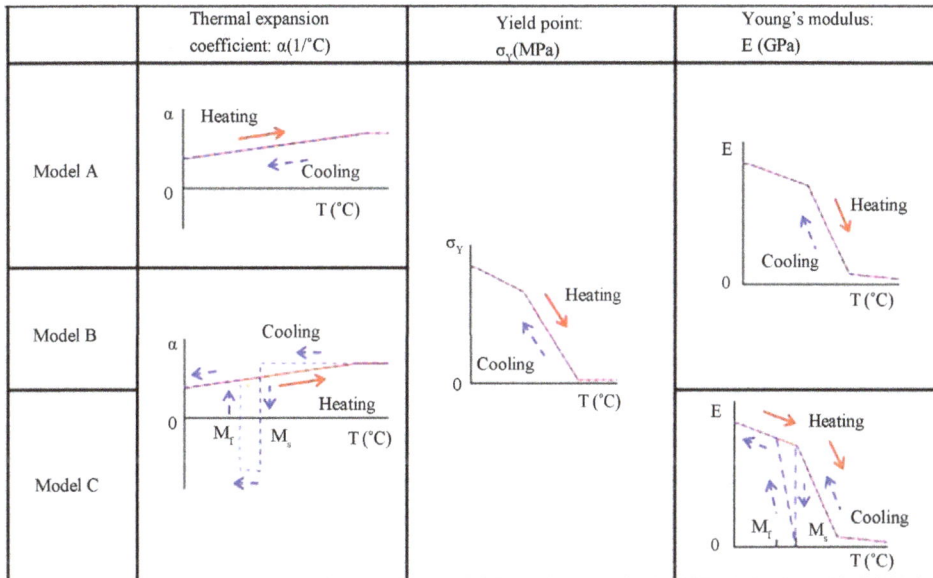

**Figure 7.** Idealization of mechanical properties in phase transformation range at cooling stage.

In the thermal elastic-plastic analysis described below, these mechanical properties are used in the cooling process in the elements of which the temperature exceeds the $A_1$ transformation temperature (around 720 degrees Celsius) in the heating process, that is, the weld metal and heat affected zone (HAZ). In the elements except them, the mechanical properties in the heating process are also used in the cooling process.

## 4. Simulation by 3D Thermal Elastic-Plastic Analysis

### 4.1. Model for Simulation

**Figure 8** shows the models for simulation by 3D thermal elastic-plastic analysis.

The isoparametric solid elements with 8 nodes are used. The half model is adopted. The heat input elements are decided by referring the macrographs [6].

In the case of LBW model, the heat energy by laser is given into the penetration shape. On the other hand, in the case of HYBW model, the penetration shapes by laser and arc are separated referring to the macrograph of LBW (**Figure 2(a)**). In each part, the magnitude of heat energy by laser ($Q_L$) and arc ($Q_A$) are given respectively.

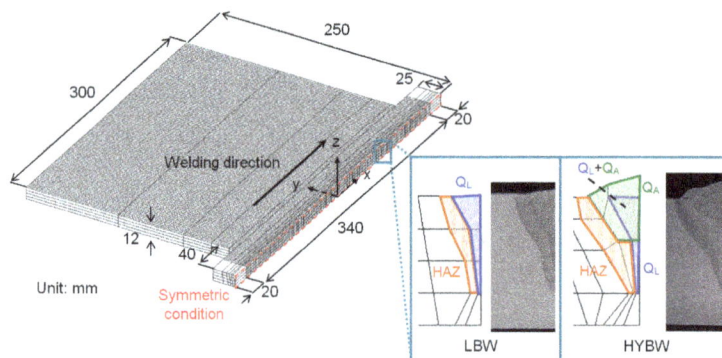

**Figure 8.** Model for thermal elastic-plastic analysis.

## 4.2. Results of Analysis

### 4.2.1. Temperature Histories

The non-steady heat conductivity analyses are carried out on LBW and HYBW models.

**Figure 9** shows the temperature histories. They are measured on the upper side at the cross-section of x = 15 (mm). The distances from the weld line (y) are 15 30, 50 and 80 (mm). The symbols represent the results obtained by the experiment and the lines represent those obtained by the analysis.

The temperature histories obtained by the analysis are successfully agreed with those by the experiments in LBW and HYBW.

It can be confirmed that the maximum temperatures in each measured point of HYBW are higher than those of LBW due to the difference of the total heat input between LBW and HYBW.

### 4.2.2. Welding Distortion

The thermal elastic-plastic analysis is carried out with using the temperature histories obtained by the non-steady heat conductivity analyses.

**Figure 10** shows the welding out-of-plane distortion obtained by the experiment and the analysis.

Here, only the angular distortion is noted because the longitudinal bending distortion is too small to discuss the meaning of the results of the analysis.

In both cases of LBW (**Figure 10(a)**) and HYBW (**Figure 10(b)**), the distortions of Model A (green broken lines), in which both the transformation expansion and the transformation superplasticity are not considered, are extremely larger than the experimental results (circular symbols).

Although those of Model B (red dotted lines), in which only the transformation expansion is considered, are smaller than those of Model A, they are larger than the experimental results.

The distortions of Model C (blue solid lines) agree with the experimental results in which both the transformation expansion and the transformation superplasticity are considered.

The results indicate that the phase transformation in the cooling process largely affects and controls the welding distortion. The effects are caused by not only the transformation expansion but also the transformation superplasticity.

### 4.2.3. Welding Residual Stress

**Figure 11** shows the distribution of residual stress obtained by the experiment and the analysis.

Here, only the stress component along the weld line; $\sigma_x$ on the upper side is noted because the effect of the phase transformation on it is the largest in all stress components on the upper and lower sides.

In both cases of LBW (**Figure 11(a)**) and HYBW (**Figure 11(b)**), the residual stress of Model A (green broken lines), in which both the transformation expansion and the transformation superplasticity are not considered, and Model B (red dotted lines), in which only the transformation expansion is considered, are extremely large tension in the weld metal. That of Model C (blue solid lines), in which both the transformation expansion and the transformation superplasticity are considered, is drastically changed from tensile yield stress to zero in the narrow region of the weld metal. This is because the phase transformation, *i.e.*, the tensile stress generated in the cooling process is released in the phase transformation range [4].

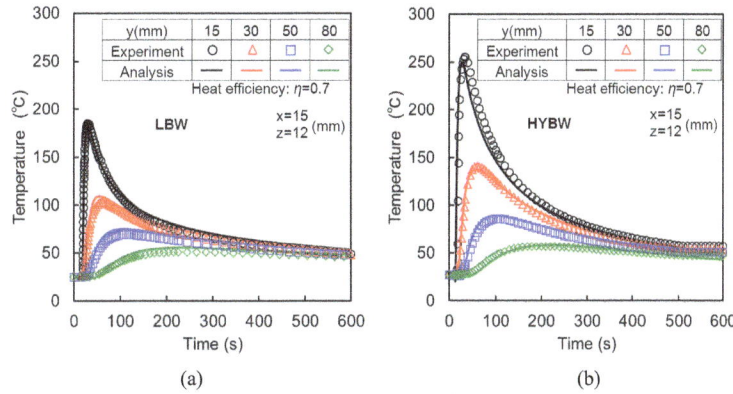

**Figure 9.** Temperature histories obtained by experiment and analysis. (a) Temperature histories of LBW; (b) Temperature histories of HYBW.

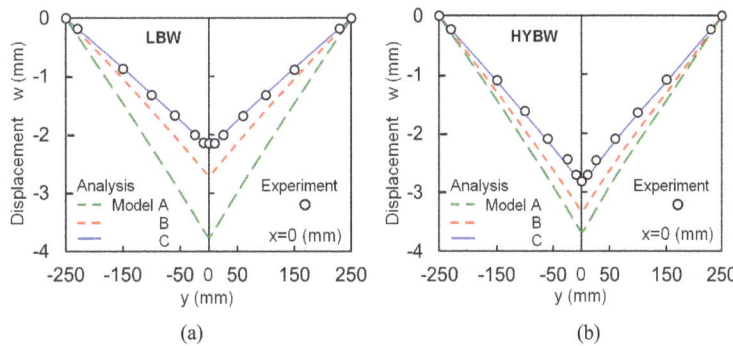

**Figure 10.** Welding out-of-plane displacements obtained by experiment and analysis. (a) Angular distortion of LBW; (b) Angular distortion of HYBW.

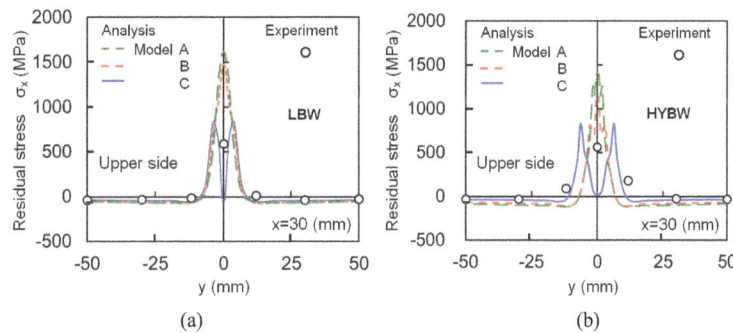

**Figure 11.** Distributions of residual stress obtained by experiments and analysis. (a) Stress component, $\sigma_x$ by LBW; (b) Stress component, $\sigma_x$ by HYBW.

Although the experimental result in the weld metal is only one, the magnitude is smaller than the tensile yield stress. By considering the accuracy of the stress relaxation method, the analysis results of Model C simulate the experimental results.

When noting the results of Model C, the region in which the phase transformation affects the distribution of the residual stress of HYBW is wider than that of LBW. The reason is that the total heat input of HYBW is larger than that of LBW.

In any case, the phase transformation in the cooling process largely affects the generation of the residual stress as well as the welding distortion.

From the view point of both the welding distortion and residual stress, the validity of idealizing the mechanical properties with considering the phase transformation could be verified. The generality could also be verified because the experimental results of LBW and HYBW were simulated by the analyses with using these idealized mechanical properties.

## 5. Conclusions

In order to investigate the effects of the phase transformation on the generation of welding distortion and residual stress of LBW and HYBW, the experiment was carried out.

1) Welding distortion of LBW was smaller than that of HYBW because the total heat input of HYBW with dual heat source was larger than that of LBW.

2) In both cases of LBW and HYBW, the welding residual stress $\sigma_x$ (component along the weld line) generated in the welds was tensile but smaller than the yield strength in the weld metal.

Then, the experiments were simulated by the 3D thermal elastic-plastic analyses based on FEM with using the idealized mechanical properties considering the phase transformation in the cooling process.

3) The experimental results could be simulated with high accuracy by the analyses in both cases of LBW and HYBW. That is, the generality of the idealized mechanical properties with considering the phase transformation could be verified. It meant that FEM simulation was valid for predicting the welding distortion and residual stress generated in high strength steel by LBW and HYBW.

4) It was confirmed that the phase transformation in the cooling process largely affected and controlled welding distortion. The effects were caused by not only the transformation expansion but also the transformation superplasticity.

5) Tensile residual stress generated in the cooling process was released by the phase transformation. The effects could be confirmed in both LBW and HYBW. However, the region in which the phase transformation affected on residual stress of HYBW was wider than that of LBW due to the difference of the magnitude of the heat input.

6) The effects of the phase transformation on the generation of welding distortion and residual stress were the same in both cases of LBW and HYBW even though the characteristics of the heat input were different from each other.

## References

[1]   Inose, K., Lee, J.-Y., Nakanishi, Y. and Kim, Y.-C. (2008) Characteristics of Welding Distortion/Residual Stress Generated by Fillet Welding with Laser Beam and Verification of Generality of Its High Accurate Prediction. *Quarterly Journal of the Japan Welding Society*, **26**, 61-66. (in Japanese)

[2]   John, S.V. (2001) Superplasticity: Mechanisms and Application. *Journal of the Minerals, Metals and Materials Society*, **53**, 22.

[3]   Kim, Y.-C., Hirohata, M. and Hageyama, Y. (2009) Modeling of Phase Transformation in Cooling Process and Verification of Its Validity. *Conference on High Strength Steel for Hydropower Plants*, Takasaki, 20-22 July 2009, 15.1-15.4.

[4]   Kim, Y.-C., Hirohata, M. and Inose, K. (2012) Effects of Phase Transformation on Distortion and Residual Stress Generated by LBW of High Strength Steel. *Welding in the World*, **56**, 64-70.

[5]   Kim, Y.-C., Hirohata, M. and Inose, K. (2011) Modeling of Laser-Arc Hybrid Welding Considered Phase Transformation. *64th Annual Assembly of International Institute of Welding* (*IIW*), Chennai, 17-22 July 2011, XV-1380-11.

[6]   Kim, Y.-C., Lee, J.-Y. and Inose, K. (2010) Determination of Dominant Factors in High Accuracy Prediction of Welding Distortion. *Welding in the World*, **54**, 234-240.

# 16

# A New Characterization Approach of Weld Nugget Growth by Real-Time Input Electrical Impedance

**Yoke-Rung Wong\*, Xin Pang**

School of Mechanical & Aerospace Engineering, Nanyang Technological University, Singapore City, Singapore
Email: \*wong0663@e.ntu.edu.sg

## Abstract

The in-process changes of weld nugget growth during the Resistance Spot Welding were investigated based on the resistance of input electrical impedance. To compute the time varying resistance of input electrical impedance, the welding voltage and current signals are measured simultaneously and then converted into complex-valued signals by using Hilbert transform. Comparing with the dynamic contact resistance as reported in literature, it showed that the time varying resistance of input electrical impedance can be accurately correlated with the physical changes of weld nugget growth. Therefore, it can be used to characterize the in-process changes of weld nugget growth. Several new findings were reported based on the investigation of spot welds under no weld, with and without weld expulsion conditions.

## Keywords

In-Process, Input Electrical Impedance, Dynamic Contact Resistance, Weld Nugget, Hilbert Transform

## 1. Introduction

Resistance Spot Welding (RSW) has been broadly used for the structural joints due to its advantages such as easy-to-use, low cost, clean working environment and short process time. In practice, a weld nugget is formed at the sheet-to-sheet interface (faying interface) of metal sheets due to the heat generation by the resistances to the flow of high electric current within a short period of time. During the welding process, electrode force is applied through the electrodes to bring the metal sheets into contact. On the other hand, water is used to cool down the

---

\*Corresponding author.

electrodes (as shown in **Figure 1**).

The RSW is a coupled electrical-thermal-mechanical process which involves heat generation due to the presence of electric current and resistances according to Joule's Law and the weld nugget growth as a result of metal melting [1] [2]. Therefore, the RSW process is a dynamic and complicated system which requires real time measurement for system characterization. In order to investigate the dynamic behavior of RSW process, the dynamic contact resistance was firstly introduced by Dickinson *et al.* [3].

As shown in **Figure 1**, the dynamic contact resistance measured during the welding process is the summation of resistances, which can be distinguished into 4 groups: 1) R1 and R7, electrode resistances; 2) R3 and R5, bulk resistances; 3) R2 and R6, electrode to sheet interface resistances; 4) R4, contact resistance at sheet-to-sheet interface (faying interface). Dickinson *et al.* divided the contact resistance into 6 stages in order to characterize the weld nugget growth (see **Figure 2**). At Stage a, the metal sheets are in intimate contact as a result of material softening and collapse of surface contaminants at the faying interface. It causes a rapid drop in the resistance. However, the resistance stops to reduce further because the resistivity of metal increases as temperature is increasing at Stage b. Due to the continuous raising of temperature, the resistance reaches its maximum point at Stage d. Once the melting temperature is reached, the metal starts to melt at the faying interface and a weld nugget is formed. At Stage e, the surface indentation at the electrode to sheet interface causes the resistance to reduce gently. In case of weld expulsion, the ejection of molten metal from the weld nugget results into a sudden drop of resistance at Stage f.

To investigate the in-process changes of weld nugget growth, this dynamic contact resistance has to be a time-varying curve. However, Gedeon *et al.* [4] revealed that it is difficult to obtain such dynamic curve due to the presence of inductive noise in the welding system. In order to overcome this problem, some researchers

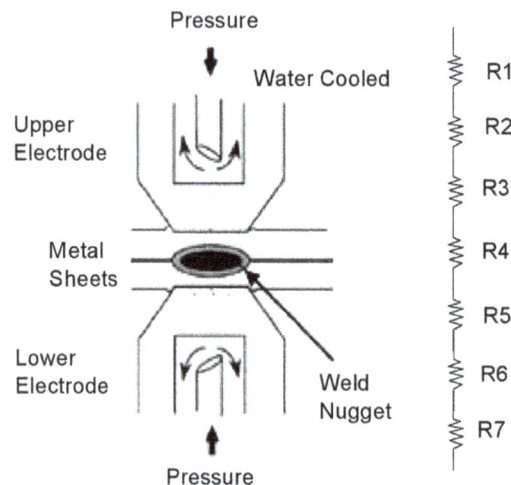

**Figure 1.** Schematic diagram of RSW.

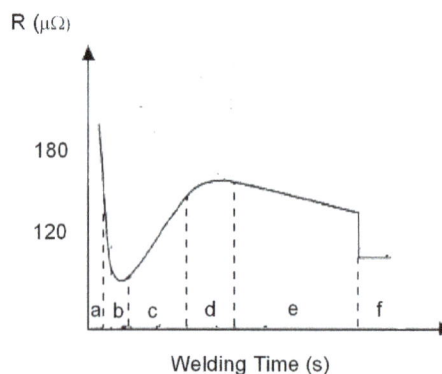

**Figure 2.** Dynamic contact resistance curve.

suggested that the instantaneous voltage is divided by the instantaneous current when the current reaches its peak value and then joins these resistance points from each half cycle to form the resistance curve. In this case, substantial error is unavoidable as a result of linear curve fitting approach.

In 2002, Cho and Rhee [5] developed a technology to obtain the primary dynamic resistance in primary circuit of welding machine. They claimed that this dynamic resistance can reflect the dynamic contact resistance in the secondary circuit of welding machine which is measured in an in-process system environment. However, the primary dynamic resistance is still obtained based on the instantaneous voltage and current measured at the primary circuit of welding machine and therefore the accuracy is poor. On the other hand, Wang and Wei [6] presented a numerical solution to predict the dynamic contact resistance based on the temperature distribution. Garza and Das [7] also proposed a method using Recursive Least Squares (RLS) algorithm to estimate the dynamic contact resistance. Both methods claimed that the predicted dynamic contact resistance is accurate and *in-situ*. Later in 2004, however, both results were reviewed by Tan *et al.* [8] and found that the first peak of resistance is not located exactly at the beginning of welding process when he investigated the relationship between the dynamic contact resistance with the micro-structural change of Ni sheets by small scale RSW. Instead, the first peak happens sometime after the welding process is started. In the paper, they suspected that the different location of first peak in dynamic contact resistance curve by AC current weld of large scale RSW is due to the high current increasing rate and thick surface film of metal sheets.

Recently, the use of input impedance as the monitoring signature for manufacturing processes has been in focus for ultrasonic welding [9], wire bonding [10] and micro drilling [11]. Furthermore, the development of real-time quality monitoring method has been done successfully for RSW based on the input electrical impedance of RSW [12]. They reported that the resistance of input electrical impedance is more accurate than the dynamic contact resistance because it is a time varying resistance curve which is directly measured during the welding process. Through pattern recognition and classification process, several features of the resistance of input electrical impedance were correlated non-destructively with the weld quality by employing an Artificial Neural Network (ANN) model. Since the correlation of the weld quality with the features of resistance of input electrical impedance was only done by ANN model, it was not able to reveal the in-process changes of weld nugget growth. As an extension of previous work reported in [12], a detail study of the relationship between the resistance of input electrical impedance with the dynamic changes of weld nugget growth and the underlying arguments will be investigated and discussed in this paper.

## 2. Input Electrical Impedance of RSW, $Z_{in}(t)$

According to Ling *et al.* [12], the purpose of using Hilbert transform is to obtain the analytic voltage and current signal so that the time varying input electrical impedance can be computed. The input electrical impedance, $Z_{in}(t)$ is calculated by taking the quotient of voltage to current signal in their analytic forms:

$$\hat{V}(t) = V(t) + jh[V(t)] \tag{1}$$

$$\hat{I}(t) = I(t) + jh[I(t)] \tag{2}$$

which the second term of the equations is the imaginary part of the real-valued signal, voltage $V(t)$ and current $I(t)$, obtained by performing Hilbert transform [13]. Basically, the Hilbert transform is defined as

$$\tilde{x}(t) = h[x(t)] = \int_{-\infty}^{\infty} \frac{x(u)}{\pi(t-u)} du \tag{3}$$

where $\tilde{x}(t)$ is the convolution integral of any real-valued signal, $x(t)$ and $(1/\pi t)$. In practice, the product of Hilbert transform can be obtained easily when the convolution integral is done in frequency domain.

$$\tilde{X}(f) = X(f) \begin{cases} e^{-j(\pi/2)} & f > 0 \\ e^{j(\pi/2)} & f < 0 \end{cases} \tag{4}$$

where $\tilde{x}(t)$ is the Fourier transform of $x(t)$ shifted by $(\pi/2)$. In other words, the product of Hilbert transform is the original signal shifted by $+90°$. After that, the $Z_{in}(t)$ is obtained as [14]:

$$Z_{in}(t) = \frac{\hat{V}(t)}{\hat{I}(t)} = Z_r(t) + j\omega Z_x(t) = R(t) + j\left(\omega L(t) - \frac{1}{\omega C(t)}\right) \qquad (5)$$

In the above, the first term of $Z_{in}(t)$ is known as real part or resistance of $Z_{in}(t)$ and the second term is the imaginary part or reactance of $Z_{in}(t)$. As compared to the conventional method for impedance calculation by using Fast Fourier Transform (FFT), the Hilbert transform allows the complex-valued signals, $\hat{V}(t)$ and $\hat{I}(t)$ to be processed in time domain so that the time variations of $Z_{in}(t)$ is retained. Furthermore, the $Z_r(t)$ is indeed the dynamic resistance and inherently isolated from the influence of inductive noise which is reflected by the $Z_x(t)$. Therefore, a unique method for measuring the accurate dynamic contact resistance is developed.

## 3. Experimental Setup

**Figure 3(a)** and **Figure 3(b)** show the experimental setup to measure the $Z_{in}(t)$ of RSW. A 50 Hz single phase AC, 75 KVA rated power pedestal type welding machine was used. For acquiring the welding voltage and current signal, a simple BNC to crocodile clip cable and a toroidal coil type current probe (AmpFlex, Flexible AC Current Probes, 10000-24-2-0.1) were used. The clip cable was attached to the electrodes and the current probe was hooked on the copper wire at the secondary circuit of the welding machine. These signals were then acquired simultaneously by the DAQ system (Nicolet Scope, 12 bits, 8 channels) at 10 kHz of sampling rate. Analog filters were used before the signal sampling process and their cutoff frequency was set at 100 Hz to avoid signal aliasing. Bi-polar (windows) triggering mode was used and the sampling period was set as same length as the welding time in order to prevent signal leakage at the first welding cycle.

Carbon steel (AISI 1023) with 1.0 mm thickness metal sheets were used. As shown in **Figure 4(a)**, the di-

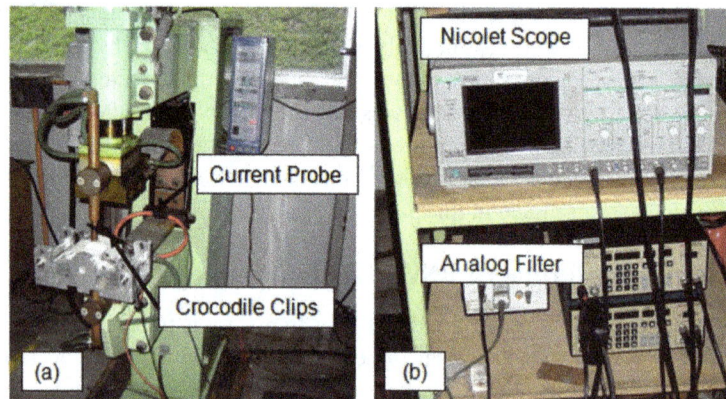

**Figure 3.** Experimental setup of (a) signals measurement; (b) DAQ system.

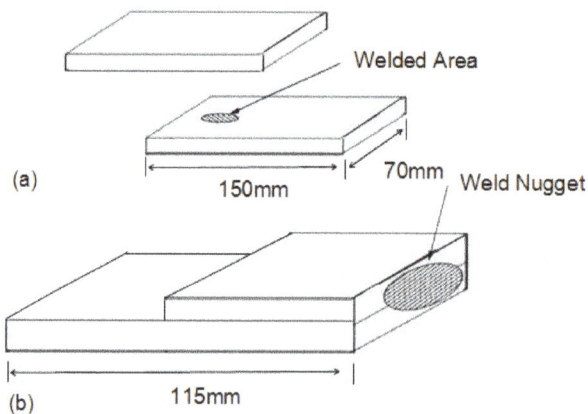

**Figure 4.** (a) SEM and (b) cross-sectional microscopic view of weld nugget.

mensions of metal sheet were obtained by using laser cutting to achieve good surface flatness and edge finishing. Furthermore, these metal sheets were rinsed by ethanol to remove contaminants such as oil, grease and etc. from the surface. The over-lapped area for the sheets is maintained at $70 \times 70$ mm.

RWMA Class "A" standard was followed and a pair of C-type, cone tip Chromium Copper electrodes was used in the experiments. The electrode force and water flow rate was set at 2250 N and 2.5 litres per minute. In order to investigate the weld nugget growth with respect to different welding time, the SEM graphs of faying interface at the center of welded area were taken at the welding time, 0.02 s, 0.04 s and 0.06 s. On the other hand, the cross-sectional macroscopic graphs of the weld nugget along its centerline were taken at 0.04 s and 0.08 s (see **Figure 4(b)**).

## 3. Measurement of $Z_{in}(t)$

As shown in **Figure 5** and **Figure 6**, the voltage and current signal based on 8.8 kA root mean square (RMS) welding current are presented. Basically, all the welding current presented in this paper is in RMS value. In order to obtain the analytic voltage and current signal, the Hilbert transform of both signals was calculated. As mentioned above, the product of Hilbert transform is to shift the original real-valued signal by +90° and then form the imaginary part of analytic signal. After taking the quotient of analytic voltage to current signal, the $Z_{in}(t)$ was calculated and presented in **Figure 7**. To illustrate the relationship between the $Z_{in}(t)$ with weld

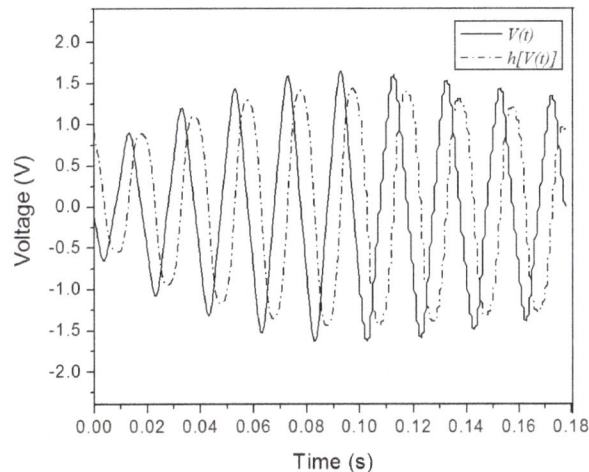

**Figure 5.** Welding voltage and its product of Hilbert transform.

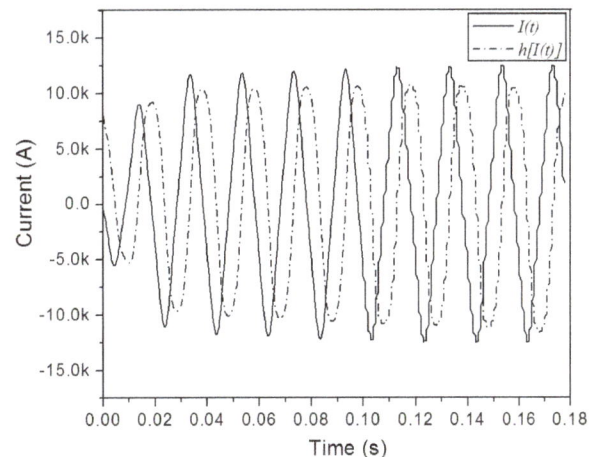

**Figure 6.** Welding current and its product of Hilbert transform.

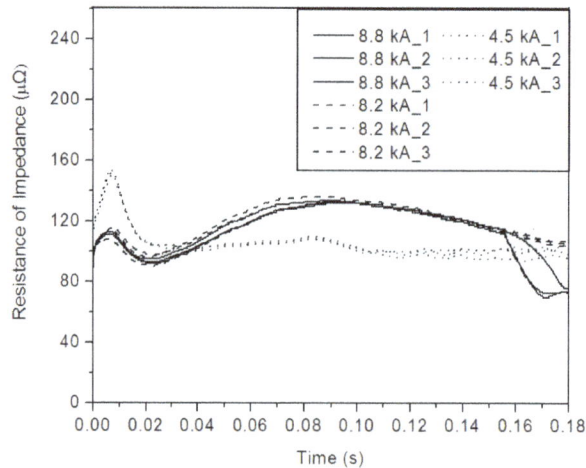

**Figure 7.** Time varying $Z_r(t)$ curves at different welding currents.

ing current, the $Z_r(t)$ curves based on 4.5 kA, 8.2 kA and 8.8 kA welding current are presented. We observe that the $Z_r(t)$ increases sharply and reaches the first peak before 0.01 s. After that, it decreases rapidly and the decrement stops at 0.02 s. For 8.2 kA and 8.8 kA welding current, the $Z_r(t)$ increases gently to the second peak at 0.10 s. However, the $Z_r(t)$ drops sharply in case of 8.8 kA welding current due to weld expulsion. On the other hand, no second peak appears for 4.5 kA welding current because "no weld" (no weld nugget) is made. In the followings, the time variation of $Z_r(t)$ for 8.2 kA welding current is evaluated by comparing with the conventional dynamic contact resistance.

## 4. Evaluation of $Z_r(t)$

At the beginning of weld, the $Z_r(t)$ increases rapidly from 90 μΩ to 112 μΩ because the contact resistance at faying interface is high. Due to the continuous heating, the $Z_r(t)$ reaches the first peak when the contact resistance is at its maximum value at 0.01 s. By comparing with the dynamic contact resistance (as shown in **Figure 2**) and other researchers' work [15] [16], this first peak is present right at the beginning of weld. However, based on our finding and the experimental result obtained by Tan *et al.*, the first peak of resistance should not be present at the beginning of weld. The incorrect location of the first peak could be caused by the rough measurement of dynamic contact resistance, which only considers the instant voltage and current at its peak value within every half cycle. Subsequently, the dynamic contact resistance has to lose the information at the beginning of weld since the quotient of voltage to current cannot be done when the voltage is zero. Furthermore, the dynamic contact resistance curve proposed by Dickinson *et al.* did not include the effect of contact electrical resistivity at the faying interface which has been proven to be important at the beginning of weld stage.

After the first peak, the $Z_r(t)$ drops to 93 μΩ at 0.02 s. This significant drop in resistance is because the contact resistance is reduced to zero where the metal sheets are in intimate contact. As shown in **Figure 8(a)**, some localized dents are found at the beginning of weld even though the metal sheets are assumed to be in intimate contact. However, these localized dents disappear and only dark color regions (recognized as burnt marks) are observed as result of softened metal sheets at high temperature after the welding process is started (see **Figure 8(b)**). Therefore, the metal sheets are truly in intimate contact and the contact resistance vanishes. Similarly, the same observation was reported by Tan *et al.*

As shown in **Figure 8(c)** and **Figure 8(d)**, the melting of metal starts at the faying interface from 0.04 s and 0.06 s due to the continuous heating of metal sheets. It reveals that the melting of metal initially happens at the faying interface where the contact resistance is high. As the metal sheets are continuously heated and reach its melting temperature, the melting of metal propagates vertically towards the upper and lower electrodes. This can be seen by the increment of $Z_r(t)$ from 93 μΩ (at 0.02 s) to 133 μΩ (at 0.10 s).

**Figure 9** illustrates the history of weld nugget growth in correlation with the time variation of $Z_r(t)$. Different welding time was set at 0.02 s, 0.04 s, 0.06 s, 0.08 s, 0.10 s, 0.12 s, 0.14 s, 0.16 s and 0.18 s in order to

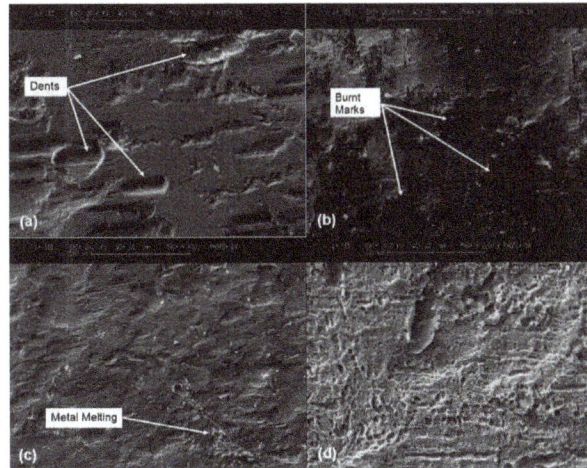

**Figure 8.** SEM graphs of faying interface at the center of welded area, welding time = (a) 0 s; (b) 0.02 s; (c) 0.04 s; (d) 0.06 s.

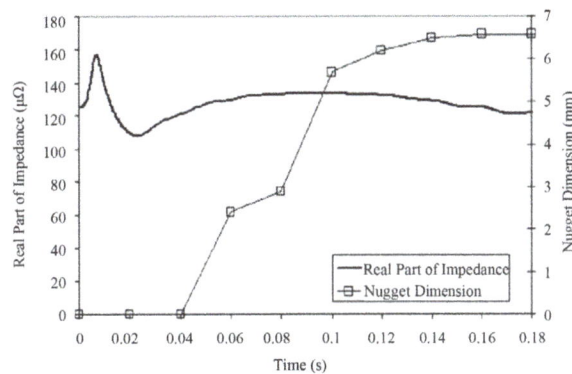

**Figure 9.** Relationship between the weld nugget growth and $Z_r(t)$ curve along welding time.

make the spot welds at the specified instant moment. After the spot welds were done based the aforementioned welding times, the diameter of weld nuggets was measured according to the BS standard [17]. Firstly, the top plate of spot weld was torn away before the measurement. Two diagonal diameters of weld nugget was measured and then averaged to obtain the final value of weld nugget diameter. It is found that the weld nugget starts to grow in between 0.04 s to 0.06 s. From 0.10 s onwards, the size of weld nugget does not change significantly until the end of welding process. In fact, good agreement is found when examining the time variations of $Z_r(t)$. Therefore, this observation suggests that the $Z_r(t)$ can be used for the characterization of weld nugget growth.

## 5. In-Process Characterizing of Weld Nugget Growth

In order to demonstrate the characterization of in-process weld nugget growth, the $Z_r(t)$ based on the spot weld at 8.8 kA welding current is divided into 5 stages (as shown in **Figure 10**): (I) initial heating; (II) contact resistance breakdown; (III) bulk heating; (IV) weld nugget growing and (IV) end of welding and expulsion. In the followings, each stage will be discussed and compared with the findings from Tan *et al.* and dynamic contact resistance to confirm the correlations between the features of $Z_r(t)$ with the weld nugget growth.

Stage I. At 0 s, the contact resistance is usually unstable because it is mainly depended on the contact conditions at the faying interface. As shown in **Figure 8(a)**, the metal sheets are not in full contact due to the presence of localized dents with air, which the resistivity of air is considered extremely high. Therefore, the $Z_r(t)$ increases rapidly to the alpha peak at 0.01 s.

As shown in **Figure 2**, this alpha peak is not seen from the dynamic contact resistance reported by Dickinson

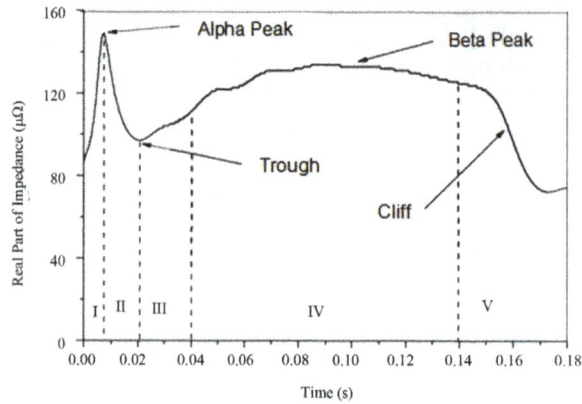

**Figure 10.** 5 stages of $Z_r(t)$ curve.

*et al.* In fact, there were no reports to reveal this phenomenon until the joining of thin Ni sheets done by DC RSW was carried out and reported by Tan *et al.* Since DC current was used, it is possible to obtain the alpha peak because the time-varying resistance curve is easily calculated based on Ohm's Law. Given the possible reasons for having no alpha peak, it could be due to extremely high increasing rate of current and the presence of thick surface film at the faying interface. However, both possibilities are not true based on our findings. Firstly, the increasing rate of current for large scale RSW is much higher as compared to small scale RSW. For an example, the increasing rate of current for 8.8 kA welding current is rated at 147,000 A/s based on an average ramp-up time of 60 ms. Despite of having such a high increasing rate, the alpha peak still appears in the $Z_r(t)$ curve. On the other hand, the surface of metal sheets used in the experiments was not mechanically treated. It means that a thin layer of oxide film is present during the welding process. In other words, the alpha peak can be identified easily by the proposed method.

Stage II. The $Z_r(t)$ drops sharply to a lower value where the "trough" is defined. As mentioned earlier, the metal sheets are softened and deformed so that there are no more gaps at the faying interface. Therefore, only bulk resistance is obtained at the trough and the contact resistance at the faying interface is zero.

Stage III. The bulk resistance increases as the temperature of metal sheets is increasing. Without the abruptly change of contact resistance, the $Z_r(t)$ is able to increase gently. Local melting at the faying interface is observed at 0.02 s and the heating of metal sheets is still continuous. Therefore, we should define this stage as "bulk heating" only which is not clearly shown in the dynamic contact resistance.

Stage IV. This is the most important stage because it shows the formation of weld nugget. As shown in **Figure 11(a)** and **Figure 11(b)**, the weld nugget starts to form from the faying interface and fully develop to its final shape. Basically, the cooling water in the electrode removes the excessive heat from the metal sheets. Therefore, it cools down the electrodes and the outer surface of metal sheets so that the molten metal will not penetrate out the enclosure of liquid weld nugget. During this stage, the $Z_r(t)$ reaches the beta peak (133 $\mu\Omega$) where the size of weld nugget is confirmed. After 0.10 s, the $Z_r(t)$ reduces gently from the beta peak to 127 $\mu\Omega$ because of the indentation of electrodes resulting in decreasing of the bulk resistance after the metal sheets are softened. In case of "no weld", it is expected to have no beta peak (as shown in **Figure 7** for welding current 4.5 kA) because no weld nugget is formed eventually.

Stage V. The $Z_r(t)$ reduces abruptly at 0.16 s which is defined as "Cliff". Due to the weld expulsion, the molten metal is ejected from the weld nugget at the faying interface. Therefore, it causes very serious indentation of electrode which results into a shorter electrical path and then lower resistance. According to the observations from the results, the higher amplitude of "cliff" is measured, the more severe weld expulsion can be.

## 6. Conclusions

In this paper, the time varying $Z_r(t)$ of input electrical impedance was discussed by comparing with the dynamic contact resistance curve as reported in the literatures. It showed that the proposed method is capable to obtain an accurate and time-varying resistance because the $Z_r(t)$ is inherently isolated from the influence of inductive noise which is only reflected by the $Z_r(t)$. Furthermore, the measurement of $Z_r(t)$ is done by tap-

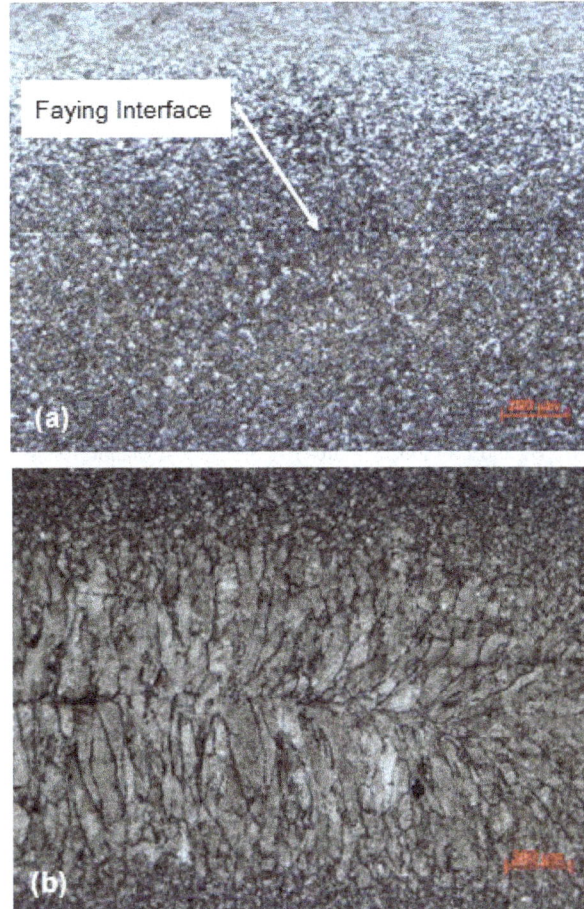

**Figure 11.** Microscopic view of weld nugget, welding time = (a) 0.04 s; (b) 0.08 s.

ing the current and voltage probe on the electrodes of welding machine without jeopardizing the normal welding operations. Therefore, it is an accurate, easy-to-use, robust and cost effective method to investigate the in-process changes of weld nugget growth.

After the investigation was done on a typical spot weld with weld expulsion, several important findings were concluded as followed:

A new characteristic curve based on the $Z_r(t)$ of input electrical impedance was established. Several feature points of $Z_r(t)$ were identified to characterize the in-process changes of weld nugget growth.

The "alpha peak" can be identified easily from the $Z_r(t)$ curve despite of reasons given by Tan *et al.*, 1) extremely high increasing rate of current; 2) the presence of thick surface film at the faying interface.

Based on the $Z_r(t)$ curve, a new stage "(III) bulk heating" has been identified clearly to better reflect the change of temperature during welding process.

Since the $Z_r(t)$ curve is truly time varying, it is possible to determine the instantaneous temperature of welding process based on the $Z_r(t)$ curve because the resistance is proportional to the resistivity of metal sheets and air gap which is sensitive to the change of temperature.

## Acknowledgements

The authors thank Dr. Tan Tong Tat for his kind supports and valuable comments of this manuscript.

## References

[1]    Scharff, R. and Caruso, D. (1990) Complete Automotive Welding Metals and Plastics. Albany, New York.

[2]   Harlin, N., Jones, T.B. and Parker, J.D. (2003) Weld Growth Mechanism of Resistance Spot Welds in Zinc Coated Steel. *Journal of Material Processing Technology*, **143-144**, 448-453. http://dx.doi.org/10.1016/S0924-0136(03)00447-3

[3]   Dickinson, D.W., Franklin, J.E. and Stanya, A. (1980) Characterization of Spot Welding Behavior by Dynamic Electrical Parameter Monitoring. *Welding Journal*, **59**, 170-176.

[4]   Gedeon, S.A., Sorensen, C.D., Ulrich, K.T. and Eagar, T.W. (1987) Measurement of Dynamic Electrical and Mechanical Properties of Resistance Spot Welds. *Welding Journal*, **66**, 378-385.

[5]   Cho, Y. and Rhee, S. (2002) Primary Circuit Dynamic Resistance Monitoring and Its Application to Quality Estimation during Resistance Spot Welding. *Welding Journal*, **81**, 104-111.

[6]   Wang, S.C. and Wei, P.S. (2001) Modeling Dynamic Electrical Resistance during Resistance Spot Welding. *Transactions of the ASME, Journal of Heat Transfer*, **123**, 576-585. http://dx.doi.org/10.1115/1.1370502

[7]   Garza, F. and Das, M. (2001) On Real Time Monitoring and Control of Resistance Spot Welds Using Dynamic Resistance Signatures. *Proceeding of the 44th IEEE-Midwest Symposium on Circuits and Systems*, Dayton, 14-17 August 2001, 41-44.

[8]   Tan, W., Zhou, Y., Kerr, W. and Lawson, S. (2004) A Study of Dynamic Resistance during Small Scale Resistance Spot Welding of Thin Ni Sheets. *Journal of Physics D: Applied Physics*, **37**, 1998-2008. http://dx.doi.org/10.1088/0022-3727/37/14/017

[9]   Ling, S.-F., Luan, J., Li, X.C. and Ang, W.L.Y. (2006) Input Electrical Impedance as Signature for Nondestructive Evaluation of Weld Quality during Ultrasonic Welding of Plastics. *NDT & E International*, **39**, 13-18. http://dx.doi.org/10.1016/j.ndteint.2005.05.003

[10]  Ling, S.-F., Zhang, D., Yi, S. and Foo, S.W. (2006) Real-Time Quality Evaluation of Wire Bonding Using Input Impedance. *IEEE Transactions on Electronics Packaging Manufacturing*, **29**, 280-284. http://dx.doi.org/10.1109/TEPM.2006.887400

[11]  Fu, L.Y., Ling, S.-F. and Tseng, C.-H. (2007) On-Line Breakage Monitoring of Small Drills with Input Impedance of Driving Motor. *Mechanical Systems and Signal Processing*, **21**, 457-465. http://dx.doi.org/10.1016/j.ymssp.2005.04.004

[12]  Ling, S.-F., Wan, L.-X., Wong, Y.-R. and Li, D.-N. (2010) Input Electrical Impedance as Quality Monitoring Signature for Resistance Spot Welding. *NDT & E International*, **43**, 200-205. http://dx.doi.org/10.1016/j.ndteint.2009.11.003

[13]  Julius, S.B. and Allan, P.G. (2000) Random Data: Analysis and Measurement Procedures. 3rd Edition, Wiley, New York, 518-543.

[14]  Boctor, S.A. and David, A.B. (1992) Electrical Circuit Principles. Prentice Hall Inc., Englewood Cliffs, 489-536.

[15]  Cho, Y. and Rhee, S. (2000) New Technology for Measuring Dynamic Resistance and Estimating Strength in Resistance Spot Welding. *Measurement Science & Technology*, **11**, 1173-1178. http://dx.doi.org/10.1088/0957-0233/11/8/311

[16]  Chen, J.Z. and Farson, D.F. (2006) Analytical Modeling of Heat Conduction for Small Scale Resistance Spot Welding Process. *Journal of Material Processing Technology*, **178**, 251-258. http://dx.doi.org/10.1016/j.jmatprotec.2006.03.175

[17]  BS 1140:1993 (1993) Specification for Resistance Spot Welding of Uncoated and Coated Low Carbon Steel. BSI Publications, London.

# 17

# Respiratory Sensitization & Sickness from Welding/Burning Isocyanate Containing Paints

Terrence Stobbe, Ryan Westra

College of Public Health, University of Arizona, Tucson, USA
Email: tjs9@email.arizona.edu

## Abstract

The purpose of this paper is to make the environmental and occupational health community aware of a serious health risk associated with the common practice of burning industrial paint off of metal surfaces during or prior to welding. On four occasions bystanders and welder/burner personnel have experienced illness as a result of being exposed to the combustion products of isocyanate paints that were being burned off metal surfaces. In each case, the burning and the exposed people were outside in an open environment where the health risk was thought to be minimal due to the open environment with nominal wind movement through the work area. In one case, the person (a burner) developed permanent sensitization to phthalic anhydride as a result of the exposure. Phthalic anhydride was determined to be decomposition product of burned isocyanate paint. In the other three cases (which involved very short exposures), between two and six people became ill but did not develop sensitization. Their symptoms included dizziness, nausea, headache, and breathing difficulty the severity of which varied from very uncomfortable to temporarily incapacitating. This paper discusses the circumstances associated with each event, the approach used to determine that phthalic anhydride was a decomposition product, and some practical things that can be done to avoid having employees become victims of exposure.

## Keywords

Welding on Epoxy Paint, Burning Epoxy Paint, Welding Health Hazards, Burning Health Hazards

## 1. Introduction

A common industrial problem both in manufacturing and in maintenance is the fact that metal surfaces (usually iron or steel) must be welded together. Often, particularly in maintenance situations, these surfaces have been painted or coated with some form of corrosion resistant material. Often the material is either epoxy based or polyurethane based. These materials are used for coating because they are "self-priming, single coat corrosion protecting coatings, that provide excellent chemical and abrasion resistance, unusual flexibility, tolerance of damp substrates, and they can be applied by brush, roller, or spray on hand-tool cleaned surfaces" (various coating manufacturer SDS and product sheet websites). Another words they are quick and effective to use and they do the coating job well. The recommended approach (from a health and safety perspective) is to use either a wire brush or a grinder to remove the coating in the area to be welded. On some occasions, removal is attempted with some form of paint stripper, but the stripping materials are generally quite toxic and require special handling and work conditions. A third, and commonly used way of removing the coating is to use a welding torch to burn the paint off the area to be welded.

From a "production" standpoint, this is the most efficient way to remove the coating. It is quick. Compared to wire brushing or grinding, it typically less than 10% of the work time. It is effective, since it removes all of the coating which, if left on the metal, would impair the integrity of the weld. It is easy—you just start the torch, apply the flame, and watch the coating literally "go up in smoke". It requires no special training or chemical handling skills (anyone who can do production or maintenance welding can do it). It is particularly useful in maintenance work because that is often done in field (away from the production line) situations where chemicals and grinders are not readily available, and where a lot of worker time would be used going to and from the shop to get tools, chemicals etc. In some cases, it is also done to remove paint which was applied to a surface incorrectly, and there is a large area that needs to have the paint removed quickly. In this situation, wire brushing, grinding, and/or chemicals would work, but the same time constraints would apply.

So, here we have the classic workplace "occupational health & safety" problem: what is more important, to do it quickly and efficiently or to do it the safe but slow way. In considering this question please remember that in most cases doing the unsafe way will not result in an injury or immediately observable health effect. This fact, combined with the fact that the company having the work done is in business to make money and not to provide the safest of workplaces, means that usually the quick cheap way is chosen. Similarly, for the worker, in most cases the choice will be to use the "quick and easy" way rather than the correct way because the correct way means more work for the worker with the added risk of being reprimanded for working too slowly.

Part of the reason the workers choose the "quick and easy way" is that when it comes to health hazards, they are not aware of the hazard, they have not been educated about it, typically they cannot see the effect immediately (much like cigarette smoking), and they do not think about how difficult life will be later in time when have developed lung disease or some form of cancer from their work exposures. In some case management is also unaware of a specific health hazard associated with a work activity (like burning a protective coating off a piece of metal). This is where the occupational health and environmental health professionals come into the picture. Their job is to "anticipate, recognize, evaluate, and control" workplace and environmental health hazards that may affect the workers or the general population. In this case we are talking primarily about workers, so it is the occupational health professionals (typically referred to professionally as either Occupational Hygienists or Industrial Hygienists depending on what country you are located in) that need to act to protect the workers. They can only do this to the extent that they are aware of the potential health hazards. This paper discusses some aspects of a poorly understood worker health problem that may result from the burning of coatings prior to, or during welding.

## 2. Situation

This health hazard initially came to our attention when the supervisor of a painting line in a railroad car manufacturing facility presented at an occupational health clinic complaining of severe respiratory distress. His symptoms were initially diagnosed as asthma, later to occupational asthma, and then to isocyanate exposure related occupational asthma. The third diagnosis was assumed because during further medical evaluation and industrial hygiene investigation, it was determined that some of the paints he was exposed to on the paint line were isocyanate based. At this point he was sent to a pulmonary medicine specialist who tested him for an allergic reaction to isocyanates, and it was found that he was NOT allergic to isocyanates. This led to further testing, which eventually determined that he was allergic to phthalic anhydride. Phthalic anhydride (sometimes referred to as phthalic acid anhydride) is known to be a pulmonary irritant and is capable of causing hypersensitivity pneumonitis (which has many symptoms in common with isocyanate related occupational asthma). This then led to a final diagnosis of hypersensitivity pneumonitis secondary to pthtallic anhydride exposure. The only problem with the diagnosis was the source of exposure. There was no phthalic anhydride in the paint being used in the paint shop. Here again is where the industrial hygienist enters the picture. When a worker is diagnosed with a disease thought to be work related, it is their job to determine the source of exposure and to implement control measures that will eliminate or at least significantly reduce exposures.

Industrial hygiene follow up consisted of a series of worksite visits to evaluate possible exposure sources, and a series of interviews with the injured person and his co-workers. during this process, it was determined that the day before becoming ill, the man had used an oxyacetylene welding torch to burn the paint of a large section of a railroad car. This was done because the paint had not cured properly, and a repaint was needed. The quickest way to get the old paint off was to burn it off. The work was done outside on a side rail away from the building. The man was working alone and he was not wearing a respirator. (No one saw a need to use a respirator because burning paint is a common industrial process, it was being done outside where there was presumed to more than adequate ventilation, and no one was aware of the possible health hazard.)

A literature search quickly revealed that little was published about the combustion products created by burning epoxy and/or polyurethane paints and coatings. The situation at hand was a one-off event, so there was no easy way to duplicate it at work. Thus, there was no immediate follow up. There was receiving medical treatment but he due to his respiratory problems, he was deemed medically unable to work and placed on long term disability. This might have been the end of the story, but then at another location the burning of these paints again made people sick.

In these later cases, the following scenario applied. On three different occasions men workers as maintenance contractors doing welding at industrial facilities were directed to replace the wear strips on front end loaders. The front end loader is a type of tractor or bull dozer used to move bulk materials. The bucket on the front end where the materials is picked up and carried is made of high strength steel, and is initially coated with a corrosion resistant paint. The most forward part of the bucket has an extra section of steel welded in place to act as a wear surface so that the bucket itself does not wear out. As the wear surface is destroyed by constant friction, it can be replaced by cutting off the old wear strip and welding a new one in its place. The new strips are also coated with the corrosion resistive paint. As previously indicated, in the area to be welded, the paint must be removed prior to welding. The "quick and easy" way to do this is to burn it off. The contractors are paid by the job (not by the hour) so they want to get the work done as fast as possible. The work is done outside in the "open air" where adequate ventilation is presumed. The work is of short duration, typically less than 30 minutes. Neither the contractor nor the company hiring the contractor was aware of the potential health hazard.

Unfortunately, in each of these cases, the welder and his helper got sick. The welder was sicker than the helper suggesting a dose-response relationship. Their symptoms included dizziness, nausea, headache, and breathing difficulty the severity of which varied from very uncomfortable to temporarily incapacitating. In each case it was severe enough to have the site's safety personnel called to the scene and in one case the affected men were taken to a nearby urgent care clinic for evaluation. All of the men recovered within a couple hours of exposure and were able to return to work. No effort was made to determine if there were any long term health effects.

When this information became available to the authors, a pilot study was undertaken to try to better understand the illness producing process. This involved both a literature search, and a epoxy paint burn test. The literature search is briefly discussed in the next section. The burn test was conducted as follows. A sample of the paint was purchased from the manufacturer. The paint was applied to a set of steel sheets. A box was built to house the sheets. The box was designed with places to insert industrial hygiene sampling devices (sorbent tubes, filters, evacuated canisters, and impingers) to collect air samples which could be analyzed by an accredited laboratory. The box formed a 95% enclosure around the steel sheets, and had a glove box type opening thru which a welder could use an oxyacetylene torch to burn off the paint. A sample of the combustion products released during the burning was then collected by the air sampling equipment. One of the authors did the actual burning while the other author and a few other people observed the process. Even though the burning was conducted outside in the open air, and it was done inside a box which retained most of the combustion products, all of the people involved except the welder experienced symptoms similar to, but not as severe as the men involved in paint burning around the wear plates. Only the other author was aware of the welding contractors response to the burning they had done.

## 3. Discussion

As indicated, the literature search revealed that little is known about the combustion products of these coatings. The most useful references identified phenol, Biphenyl A, and phthalic anhydride as common combustion products (Eckerman, 1990; Engstromm 1990; Cook, 2002). These combustion products are similar to those found in our study, with the exception of the BPa which we did not find. This may be due to a different but similar paint being used in their studies. It may also be due to a difference in combustion temperatures, since studies have shown that combustion products can vary as a function of combustion temperature (Herpol 1976; Lepchek, 2004). In all cases, the contaminant concentrations were less than 20% of the OSHA and ACGIH recommended workplace exposure levels (ACGIH, 2013; OSHA Regulation 1910.1000, 2014). Clearly, in the case of this type of exposure, the applicable exposure standards are not low enough to protect exposed workers.

It is interesting to note that the MSDS (or SDS) supplied by the manufacturer (it lists 22 compounds that are initially present as the paint is applied including two epoxy compounds and an "amine adduct" which is a trade secret) does an excellent job of complying with the applicable hazard communication standard (OSHA 1910.1200 in the USA) without mentioning anything about the combustion products associated with burning the paint. While it is true that the exact products would be difficult to predict given their variability due to combustion conditions, it does seem that it would be helpful to warn workers and safety persons about the most severe of the hazards and how they might be avoided.

## 4. Recommendation

OSHA has a construction standard that applies to welding work (1926.354 Welding, cutting, and heating in way of preservative coatings). It states: "*In enclosed spaces*, all surfaces covered with toxic preservatives shall be stripped of all toxic coatings for a distance of at least 4 inches from the area of heat application, or the employees shall be protected by air line respirators, meeting the requirements of Subpart E of this part."

This is a correct directive, but "in enclosed spaces" is misleading. All of the events previously described occurred in the open air, and all but the first one involved short duration exposures. People reading the directive could assume that outside an enclosed area precautions are un necessary.

The balance of the OSHA directive addresses the issue of flammability which while a relevant safety issue, is not related to the problem at hand. The OSHA General Industry Standard, 1910. 252 addresses a variety of safety and fire related issues, but does not address the toxic coating issue.

In modern times, people who are curious about health hazards, usually turn to the internet. A review of internet comments about welding/burning on epoxy paints/coatings is mostly unhelpful. Much of it is devoted to the hazards of welding on lead containing materials, which is not relevant to the toxic coatings issue. One site briefly addressed the industrial hygiene issues with the following comments: "Every effort should be made to try and remove all protective coatings. There are instances where metal is sandwiched together and it is impossible to access the backside of the metal or the beam extends into the building structure. In these instances as much paint as possible must be removed, proper respiratory protection worn, and proper ventilation must be used to capture fume at the point of operation."

Here again, the information is correct but misleading. Since it suggests proper exhaust ventilation be used to capture fumes, a typical user might assume that since the work is being done outside there is proper ventilation and the hazard is removed (proper exhaust ventilation is a meaningless concept to the average person, but outside is usually assumed to be adequate).

Our recommendation is that all workers who may find themselves welding/burning on painted surfaces of any type be told about the potential hazards of burning paint. Since production demands will require them to do this work, they either need local exhaust ventilation designed by an industrial hygienist or qualified engineer, or at least they to position themselves where they will receive the least exposure (upwind). In this situation they must wear a respirator. A supplied air respirator may not be available, but a 1/2 mask organic respirator with a HEPA pre filter will significantly reduce exposure.

## 5. Summary

This paper has briefly discussed the poorly understood problem of the health hazards associated with welding/burning painted metal surfaces. It briefly described the primary reasons these exposure will continue to occur. Phthalic anhydride in particular was identified as the cause of permanent disability in one worker, and the combustion products of paints were identified as making multiple persons sick on a number of occasions. The lack of useful or properly protective regulatory guidelines was identified. A specific recommendation for worker protection was made.

## References

ACGIH (2013). Threshold Limit Values for Chemical Substances & Physical Agents.

Henriks-Eckerman, M.-L., Engström, B., & ÅNÄs, E. (1990). Thermal Degradation Products of Steel Protective Paints. *American Industrial Hygiene Association Journal, 51,* 241-244. http://dx.doi.org/10.1080/15298669091369592

Engström, B., Henriks-Eckerman, M.-L., & Ånäs, E. (1990). Exposure to Paint Degradation Products When Welding, Flame Cutting, or Straightening Painted Steel. *American Industrial Hygiene Association Journal, 51,* 561-565. http://dx.doi.org/10.1080/15298669091370103

Isenstein, M., & Cook, L. (2002). Health Hazards from Burning a Chemical Agent Resistant Coating, a Polyurethane Paint. *Poster Session Presented at the Annual American Industrial Hygiene Association (AIHCE) Conference No. 328.*

Herpol, C. (1976). Biological Evaluation of the Toxicity of Products of Pyrolysis and Combustion of Materials Fire and Materials, 1, 29 to 35.

Levchik, S., & Weil, E. (2004). Review Thermal Decomposition, Combustion and Flame-Retardancy of Epoxy Resins—A Review of the Recent Literature. *Polymer International, 53,* 1901-1929. http://dx.doi.org/10.1002/pi.1473

OSHA 1910.1000 (2014) Permissible Exposure Limits for Air Contaminants, Subpart Z.

# 18

# An Electrothermal Model Based Adaptive Control of Resistance Spot Welding Process

**Ziyad Kas, Manohar Das**

Department of Electrical and Computer Engineering, Oakland University, Rochester, USA
Email: zrkas@oakland.edu, das@oakland.edu

## Abstract

**Resistance Spot Welding (RSW) is a process commonly used for joining a stack of two or three metal sheets at desired spots. The weld is accomplished by holding the metallic workpieces together by applying pressure through the tips of a pair of electrodes and then passing a strong electric current for a short duration. Inconsistent weld and insufficient nugget size are some of the common problems associated with RSW. To overcome these problems, a new adaptive control scheme is proposed in this paper. It is based on an electrothermal dynamical model of the RSW process, and utilizes the principle of adaptive one-step-ahead control. It is basically a tracking controller that adjusts the weld current continuously to make sure that the temperature of the workpieces or the weld nugget tracks a desired reference temperature profile. The proposed control scheme is expected to reduce energy consumption by 5% or more per weld, which can result in significant energy savings for any application requiring a high volume of spot welds. The design steps are discussed in details. Also, results of some simulation studies are presented.**

## Keywords

**Resistance Spot Welding, Adaptive Control, Nugget Formation, Energy Saving**

## 1. Introduction

In resistance spot welding, the welding process begins by applying pressure on a stack of metal sheets, held together between a pair of electrodes. A weld current is then passed through the electrodes, causing resistive heating of the metal workpieces and the formation of a welded joint or nugget, as shown in **Figure 1**. The formation of a weld nugget strongly depends on the electrical and thermal properties of the sheet and coating materials [1]. Since the contact resistance near the faying surface is much higher than the resistance of the sheets and electrodes, most of the heating is concentrated near the faying surface, causing melting and formation of a nugget

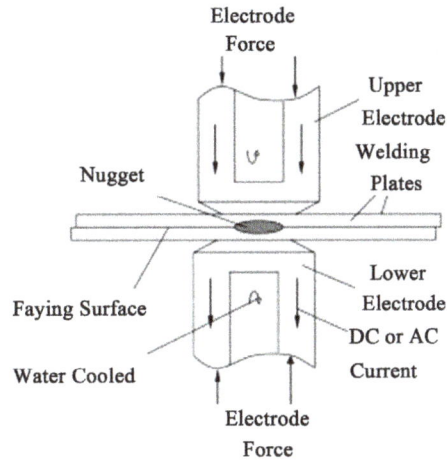

**Figure 1.** Resistance spot welding system.

there. Depending on the thickness and type of material, welding current ranges from 1,000 to 20,000 amperes or more, while the voltage typically is between 1 and 30 volts [2].

A Resistance Spot Welding cycle consists of three main stages as follows:

Stage 1: Squeeze time, which is the time when electrodes press the welded workpieces together.

Stage 2: Weld time, which is the time when welding current is applied producing heat at the faying surface of the workpieces and thus creating a weld nugget.

Stage 3: Hold time, which is the time when electrode force still presses the workpieces together and cools the weld down after the welding current is switched off.

One of the most common applications of resistance spot welding is in the automobile manufacturing industry, where it is used almost universally to weld the sheet metals to form the car body and parts. A typical automotive vehicle today requires about 4000 - 6000 spot welds per vehicle. Considering a worldwide annual production volume of 80 million automotive vehicles, an energy saving RSW controller can result in significant energy savings and reduce carbon footprint accordingly.

During the past two decades, a number of studies have been carried out to improve the RSW process, which focuses on monitoring and control of weld parameters to improve weld quality. The RSW control techniques proposed to date include Proportional-Integral (PI) [3], Proportional-Derivative (PD) [4], Proportional-Integral-Derivative (PID) [5], Fuzzy [6]-[8], Neural Networks (NN) [9] [10], or a combination of Fuzzy and NN [11]. The main drawback of these techniques is that they do not take into account the thermal dynamics of the RSW process, *i.e.* they do not utilize dynamical models that govern the heat transfer and nugget formation in the RSW process. Also, these systems don't take into account any welding process variations, such as variations in coating materials, electrode degradation, and weld force variations.

In this paper, a novel approach to RSW control is presented. This approach has not been explored by other researchers. We start with a simplified heat balance model of a RSW process proposed in [12] and [13], and then use it to design a controller. This thermal model of the heat balance is a function of nugget growth and it determines the temperature variation during welding time. This model is used later to design an adaptive-one-step-ahead (AOSA) controller and an adaptive-weighted one-step-ahead (AWOSA) controller that compensate for unknown process variations and track a desired reference temperature profile. Finally, some simulation results that show the performance of the proposed controllers are presented and compared to the performance of a PID controller. Simulation results show that AOSA and AWOSA controllers are capable of tracking a reference temperature profile when the weld parameters are unknown, as well as reduce the energy needed to make a weld by 6%.

The organization of this paper is as follows. Section 2 presents a simplified electrothermal dynamical model of a RSW nugget formation process. The design of adaptive OSA and WOSA controllers is discussed in Section 3. Section 4 presents the results of some simulation studies, and finally some concluding results are provided in Section 5.

## 2. Electrothermal Dynamical Model of a RSW Nugget Formation Process

To start with, we consider a simplified heat balance model of a RSW process, presented in [13]. The simplified dynamical model of a RSW process determines the heat balance in the system as a function of nugget temperature. For a simplified nugget model, shown in **Figure 2**, the heat balance can be described by the following equations:

The total heat generation rate, $\dot{Q}_g(t)$ is given by

$$\dot{Q}_g(t) = I^2(t)R(t) \tag{1a}$$

$$R(t) = R_w + R_c + R_e \tag{1b}$$

where $I(t)$ denotes the welding current, and $R(t)$ denotes the total resistance consisting of the resistance of work pieces, $R_w$, contact resistance, $R_c$, and electrode resistance, $R_e$. Since $R_w$ and $R_e$ are very small compared to the total contact resistance $R_c$, $R_w$ and $R_e$ can be neglected in (1b).

The total contact resistance can then be described as,

$$R_c = R(t)_{\text{electrode-sheet}} + R(t)_{\text{sheet-sheet (faying surface)}} \tag{1c}$$

A linear relationship between the resistance and temperature is assumed to model the heat generated as a function of temperature. Thus,

$$R(t)_{\text{electrode-sheet}} = \rho \frac{2l_1}{A} \tag{1d}$$

$$R(t)_{\text{sheet-sheet(faying surface)}} = \rho \frac{2p}{A} \tag{1e}$$

$$\rho = \rho(T) = \rho_\circ \left[1 + \alpha_r(\theta - \theta_\circ)\right] \tag{1f}$$

Electrodes and Sheets Setup

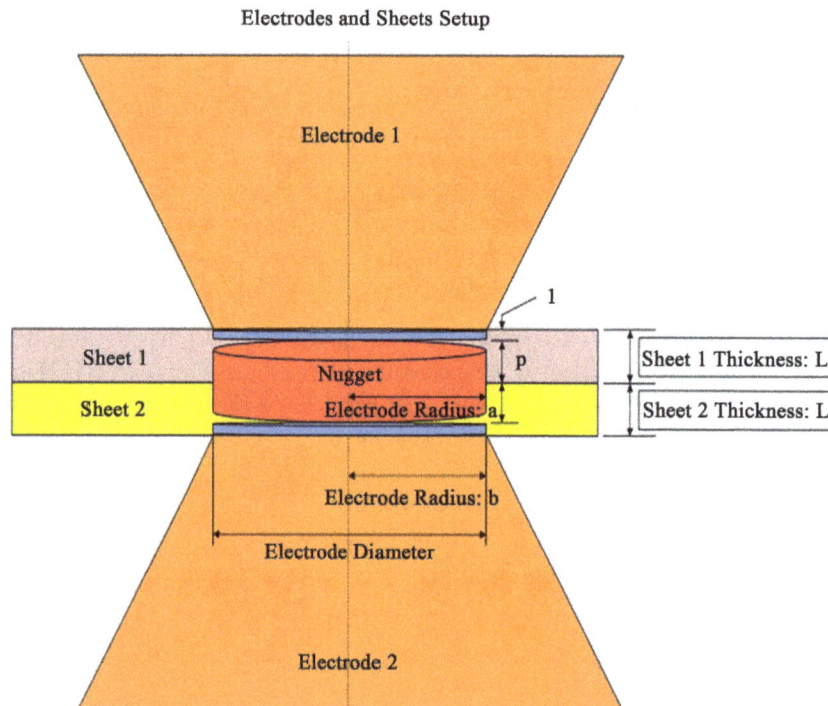

**Figure 2.** A simplified model of a weld nugget.

where $\rho$ denotes the resistivity of the material, $l_1$ denotes the distance from the melting interface to electrode contact surface, $p$ denotes the penetration, $A$ is the cross sectional area, $\rho_\circ$ denotes the resistivity at reference temperature $\theta_\circ$, $\theta$ and $\alpha_r$ are the temperature to be controlled and the temperature coefficient respectively.

Substituting (1f) in (1d) and (1e) we get

$$R(t)_{\text{electrode-sheet}} = c_1\theta(t) + c_2 \tag{1g}$$

$$R(t)_{\text{sheet-sheet(faying surface)}} = c_3\theta(t) + c_4 \tag{1h}$$

where

$$c_1 = \frac{2l_1\rho_\circ\alpha}{A} \tag{1i}$$

$$c_2 = \frac{2l_1\rho_\circ}{A}(1 - \alpha_r\theta_\circ) \tag{1j}$$

$$c_3 = \frac{2p\rho_\circ\alpha}{A} \tag{1k}$$

$$c_4 = \frac{2p\rho_\circ}{A}(1 - \alpha_r\theta_\circ) \tag{1l}$$

Substituting (1g) and (1h) in (1a) we get

$$\dot{Q}_g(t) = I^2(t)\left(c_1\theta(t) + c_2 + c_3\theta(t) + c_4\right) \tag{1m}$$

$$= c_5 I^2(t)\theta(t) + c_6 I^2(t) \tag{1n}$$

where

$$c_5 = c_1 + c_3 \tag{1o}$$

$$c_6 = c_2 + c_4 \tag{1p}$$

The heat of fusion required for nugget formation is given by:

$$H_f = H\Delta V_n \tag{2a}$$

$$\Delta V_n = \pi a^2 p \tag{2b}$$

where $H$ denotes the heat of fusion per unit volume, $\Delta V_n$ denotes the nugget volume, and $p$, $a$ denote the penetration and nugget radius respectively. Substituting (2b) in (2a) and normalizing over the weld duration, $\Delta t$, we get the heat of fusion per unit time:

$$\frac{H_f}{\Delta t} = H\pi a^2 p = c_7 \tag{2c}$$

Neglecting the heat loss in the surroundings and the electrodes, the heat required to raise temperature by $d\theta(t)$ is given by

$$dQ_T(t) = \rho C_p d\theta(t)\Delta V \tag{3a}$$

where $\rho$ denotes the density, $C_p$ denotes the specific heat, $V$ is the volume, and $d\theta(t)$ is the temperature rise. We rewrite (3a) as:

$$dQ_T(t) = c_8 d\theta(t) \tag{3b}$$

where

$$c_8 = \rho C_p \pi a^2 p \tag{3c}$$

The total heat loss rate is given by

$$\dot{Q}_L(t) = \dot{Q}_a(t) + \dot{Q}_r(t) \tag{4a}$$

$$= k_1 \pi a^2 \left[ \frac{\theta(t) - \theta_1}{l_1} + \frac{10\theta(t)\beta L}{b\sqrt{\alpha}} \right]$$

$$= \left( \frac{k_1 \pi a^2}{l_1} + \frac{10 k_1 \pi a^2 \beta L}{b\sqrt{\alpha}} \right) \theta(t) - \frac{k_1 \pi a^2 \theta_1}{l_1}$$

$$= c_9 \theta(t) - c_{10} \tag{4b}$$

where

$$c_9 = \left( \frac{k_1 \pi a^2}{l_1} + \frac{10 k_1 \pi a^2 \beta L}{b\sqrt{\alpha}} \right) \tag{4c}$$

$$c_{10} = \frac{k_1 \pi a^2 \theta_1}{l_1} \tag{4d}$$

In the above equations, $\dot{Q}_a(t)$ and $\dot{Q}_r(t)$ denote the axial and radial loss rates, respectively; $k_1$ represents thermal conductivity, $a$ is the nugget radius; $\theta(t)$, $\theta_1$, represent the melting temperature and the interface temperature at the work piece respectively; $l_1$ is the distance from the melting interface to the electrodes contact area; $\beta$ represents the final penetration to work piece thickness ratio; $L$ is the sheet thickness; $b, \alpha$ represent the electrode radius and thermal diffusivity of work piece respectively.

The heat balance equation over time $(t, t + \mathrm{d}t)$ is given by

$$\dot{Q}_g(t) = \frac{H_f}{\Delta t}\mathrm{d}t + \mathrm{d}Q_T(t) + \dot{Q}_L(t)\mathrm{d}t \tag{5}$$

Substituting (1n), (2c), (3b), and (4b) in (5) and rearranging it, we get

$$c_8 \frac{\mathrm{d}\theta(t)}{\mathrm{d}t} = c_5 I^2(t)\theta(t) + c_6 I^2(t) - c_9\theta(t) + c_{10} - c_7 \tag{6a}$$

or, equivalently,

$$\frac{\mathrm{d}\theta(t)}{\mathrm{d}t} = c_{11} I^2(t)\theta(t) + c_{12} I^2(t) - c_{13}\theta(t) + c_{14} \tag{6b}$$

where

$$c_{11} = c_5 / c_8 \tag{6c}$$

$$c_{12} = c_6 / c_8 \tag{6b}$$

$$c_{13} = c_9 / c_8 \tag{6c}$$

$$c_{14} = (c_{10} - c_7) / c_8 \tag{6d}$$

For the sake of notational convenience, let $y(t) = \theta(t)$ and $u(t) = I^2(t)$. Then (6b) can rewritten as

$$\frac{dy(t)}{dt} = c_{11}u(t)y(t) + c_{12}u(t) - c_{13}y(t) + c_{14} \tag{7}$$

Equation (7) represents a bilinear electrothermal dynamical model of a RSW process. Note that this simplified model neglects the heat required to raise the temperature of the electrodes and the nugget surroundings. Also, it assumes that most of the heating occurs near the faying surface due to its high contact resistance. The size of the workpieces is assumed to be infinite in the radial direction and the nugget shape is assumed to be a disk growing radially and axially in the same proportions. The nominal nugget diameter is assumed to be $4.5\sqrt{L}$, where $L$ is the sheet thickness.

Using a first order Euler approximation for $\frac{dy}{dt}$ with a sampling period $T_s$, the following discrete time equation is derived from the system Equation (7):

$$\frac{y(k+1) - y(k)}{T_s} = c_{11}u(k)y(k) + c_{12}u(k) - c_{13}y(k) + c_{14} \tag{8a}$$

or

$$y(k+1) = Ay(k) + Bu(k) + Cu(k)y(k) + D \tag{8b}$$

where

$$A = 1 - c_{13}T_s \tag{8c}$$

$$B = c_{12}T_s \tag{8d}$$

$$C = c_{11}T_s \tag{8e}$$

$$D = c_{14}T_s \tag{8f}$$

Also, $k$ denotes the discrete time index $(k = 0, 1, 2, \cdots)$ and $kT_s$ denote the sampling instances. The above electrothermal model is characterized by four unknown parameters, namely, $A$, $B$, $C$, and $D$.

## 3. Design of a RSW Controller

To develop a control scheme for controlling the nugget temperature of the RSW model presented by Equation (8a), we realize that it presents a bilinear system characterized by some unknown parameters. These parameters can vary from weld to weld, and in most cases we have no prior knowledge of the parameter values. In view of this, we propose to use an adaptive OSA and WOSA controllers.

The proposed adaptive control scheme involves measurement of the inputs and outputs of the system, estimation of unknown system parameters using a recursive least squares (RLS) parameter estimation algorithm, and computation of a control signal based on the estimated parameter values. Also, the temperature of the weld nugget is monitored indirectly by assuming it to be proportional to the contact resistance.

### 3.1. Adaptive OSA and WOSA Controllers

In an adaptive controller, the sampled measurements, $u(k)$ and $y(k)$, are used to estimate the model parameters, $A, B, C$ and $D$ in Equation (8b), using a recursive parameter estimation method, such as recursive least square (RLS). The estimated values of these parameters are then used to compute the OSA/WOSA control signals.

### 3.2. Parameter Estimation

First we write model Equation (7) in the following form:

$$y(k+1) = \varphi(k)^{\mathrm{T}} X^* \tag{9a}$$

where

$$\varphi(k)^{\mathrm{T}} = \begin{bmatrix} y(k-1) & u(k-1) & y(k-1)*u(k-1) & 1 \end{bmatrix}^{\mathrm{T}} \tag{9b}$$

$$X^* = \begin{bmatrix} A & B & C & D \end{bmatrix}^{\mathrm{T}} \tag{9c}$$

Next, the estimated value of $\theta_\circ$ is computed recursively using the following RLS algorithm:

$$\hat{\theta}(k) = \hat{\theta}(k-1) + \frac{P(k-2)\varphi(k-1)}{1+\varphi(k-1)^{\mathrm{T}} P(k-2)\kappa(k-1)} \left[ y(k) - \varphi(k-1)^{\mathrm{T}} \hat{\theta}(k-1) \right]; \quad k \geq 1 \tag{10a}$$

$$P(k-1) = P(k-2) - \frac{P(k-2)\varphi(k-1)\varphi(k-1)^{\mathrm{T}} P(k-2)}{1+\varphi(k-1)^{\mathrm{T}} P(k-2)\varphi(k-1)} \tag{10b}$$

$$\hat{X}(0) = \begin{bmatrix} \gamma & 0 & 0 & 0 \end{bmatrix}^{\mathrm{T}} \tag{10c}$$

$$P(-1) = \sigma I \tag{10d}$$

where $\gamma > 0$ is a small number and $\sigma > 0$ is chosen to be large. Also, $\hat{C}(k)$ is always constrained to be non-negative, *i.e.*,

$$\hat{C}(k) > \varepsilon > 0 \quad \text{for all } k \tag{10e}$$

Given an estimate $\hat{X}(k)$ of $X^*$, we define the predicted output at time $k+1$ as:

$$\hat{y}(k+1) = \varphi(k)^{\mathrm{T}} \hat{X}(k) \tag{11}$$

## 3.3. Adaptive-One-Step-Ahead Tracking Controller

One-step-ahead (OSA) control scheme for linear systems has been well investigated in [14]. An OSA controller attempts to bring the predicted output, $y(k+1)$ at time $k+1$, to the desired value, $y^*(k+1)$ in one step. Thus, it minimizes the following cost function:

$$J_1(k+1) = \frac{1}{2} \left[ y(k+1) - y^*(k+1) \right]^2 \tag{12}$$

The corresponding OSA control law is given by [14]:

$$\bar{u}(k) = \frac{y^*(k+1) - Ay(k) - D}{B + Cy(k)} \tag{13}$$

The above control signal needs to be constrained by the maximum current delivery capacity of the controller, $u_{\max}$, as follows:

$$u(k) = \begin{cases} \bar{u}(k), & \text{if } 0 < \bar{u}(k) < u_{\max} \\ 0, & \text{if } \bar{u}(k) \leq 0 \\ u_{\max}, & \text{if } \bar{u}(k)) \geq u_{\max} \end{cases} \tag{14}$$

The adaptive OSA controller uses the estimate, $\hat{X}(k)$ in Equation (11) to compute the control signal, $u(k)$, from the following adaptive version of Equation (13) above:

$$\bar{u}(k) = \frac{y^*(k+1) - \hat{A}(k)y(k) - \hat{D}(k)}{\hat{B}(k) + \hat{C}(k)y(k)} \tag{15}$$

where $\hat{A}(k), \hat{B}(k), \hat{C}(k)$, and $\hat{D}(k)$ denote the estimated values of $A, B, C$, and $D$, respectively, at time $k$.

One of the potential drawbacks of OSA controllers is excessive control efforts that often result from attempting to bring $y(k+1)$ to $y^*(k+1)$ in one step. To address this potential problem, an AWOSA controller is discussed below.

## 3.4. Adaptive Weighted One-Step-Ahead Controller

The excessive effort to bring the output $y(k+1)$ to the desired value $y^*(k+1)$ in one step using AOSA may result in an unfavorable saturation of the input. The adaptive weighted one-step-ahead controller attempts to seek a tradeoff between tracking accuracy and control effort by considering a slight generalization of the cost function (12) to the form (16) given below. Thus, it minimizes the following cost function:

$$J_2(k+1) = \frac{1}{2}\left[y(k+1) - y^*(k+1)\right]^2 + \frac{\lambda}{2}u(k)^2 \tag{16}$$

where, $0 < \lambda < 1$ is chosen to provide a desired tradeoff.

The minimization of the cost function in (16) leads to the weighted one-step-ahead control law [14]:

$$\bar{u}(k) = \frac{\left(B + Cy(k)\right)\left(y(k+1) - Ay(k) - D\right)}{\left(B + Cy(k)\right)^2 + \lambda} \tag{17}$$

The above control law is also constrained by the maximum current delivery capacity, $u_{\max}$, as shown in Equation (14) above. The choice of $\lambda$ provides a desired tradeoff between tracking accuracy and control effort. A small $\lambda$ results in good tracking but requires high level of control effort. A large $\lambda$, on the other hand, reduces control efforts at the cost of tracking accuracy.

The adaptive WOSA controller uses the estimate, $\hat{X}(k)$, in Equation (11) to compute the control signal, $u(k)$ from the following adaptive version of Equation (17) above:

$$\bar{u}(k) = \frac{\left(\hat{B}(k) + \hat{C}(k)y(k)\right)\left(y(k+1) - \hat{A}(k)y(k) - \hat{D}(k)\right)}{\left(\hat{B}(k) + \hat{C}(k)y(k)\right)^2 + \lambda} \tag{18}$$

where $\hat{A}(k), \hat{B}(k), \hat{C}(k),$ and $\hat{D}(k)$ denote the estimated values of $A, B, C,$ and $D$, respectively, at time $k$.

## 4. Simulation Results and Discussion

This section presents the results of a simulation study showing the performance of the system with the proposed AOSA and AWOSA controllers and also compare them with a PID controller. Each controller is designed for tracking a reference temperature profile.

The reference temperature profile is a good indicator of the weld quality. Therefore, it is desirable to keep the temperature variation close to a desired variation curve, which may be experimentally predetermined for the good welds. A typical reference temperature profile for good weld is shown in **Figure 3** below [1]. Basically, such a curve is characterized by a fast rise of temperature to melting point, melting of the workpieces at the faying surface area which causes a slight drop in temperature, followed by a cooling zone that results from removal of weld current. The actual nugget temperature is measured during the weld cycle using the relationship described by Equation (1f). Depending on the tracking error signal, the welding current is adjusted so as to reduce the temperature error.

For these simulations, we have selected two sheets of mild steel with the same thickness as the materials to be welded. The force variation and electrode wear are considered as unknown process variables that impact the nugget size (diameter and penetration). The Figures below show the performance of the AOSA, AWOSA, and PID controllers due to 20% increase in nugget diameter and 50% increase in indentation from their desired values.

**Figure 4** shows the performance of the AOSA controller using $I_{max} = 12\,\text{KA}$, where $I_{max}$ denotes the maximum current delivery capacity of the weld controller. We can see that the AOSA controller adapts to the parameter change and force the output temperature profile to follow the desired temperature profile. Also, we can see that the energy required for the weld is lower than that of the PID controller.

**Figure 5** and **Figure 6** show the performance of AWOSA controller using $I_{max} = 12\,\text{KA}$ with $\lambda = 0.1$ and 1, respectively. Here we notice that when $\lambda$ is high, the output temperature profile does not follow the desired output temperature profile well. However, increasing $\lambda$ results in decreasing the total energy required for the weld.

**Figure 7** shows the performance of the PID controller prior to any parameter change using $I_{max} = 12\,\text{KA}$. After multiple trial and error attempts to get satisfactory results, the parameters of the PID controllers are: Proportional (P) = 0.5, Integral (I) = 26.56, Derivative (D) = 0.

In **Figure 8** we see that the PID controller looses track of the reference temperature profile due to weld parameters change. Also, we can see that PID controller requires more energy for the weld comparing to AOSA and AWOSA.

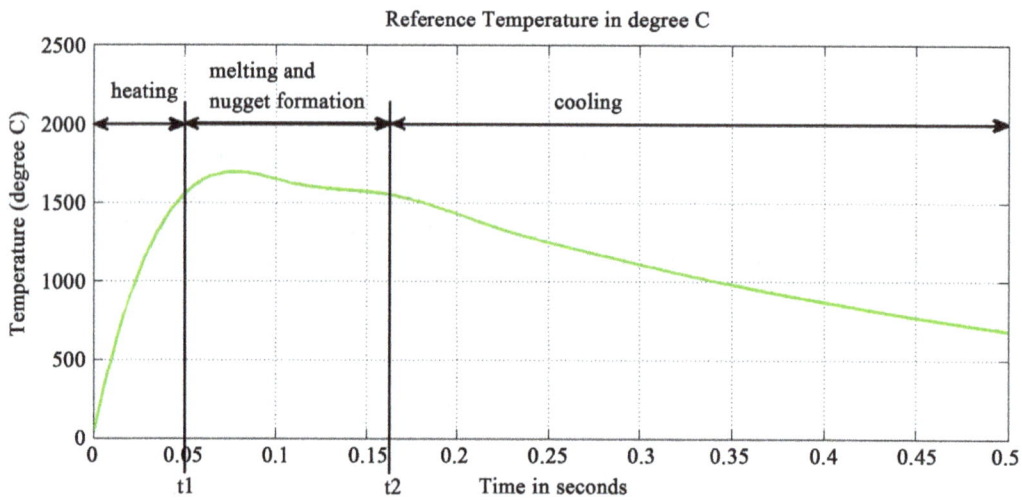

**Figure 3.** Desired reference temperature profile.

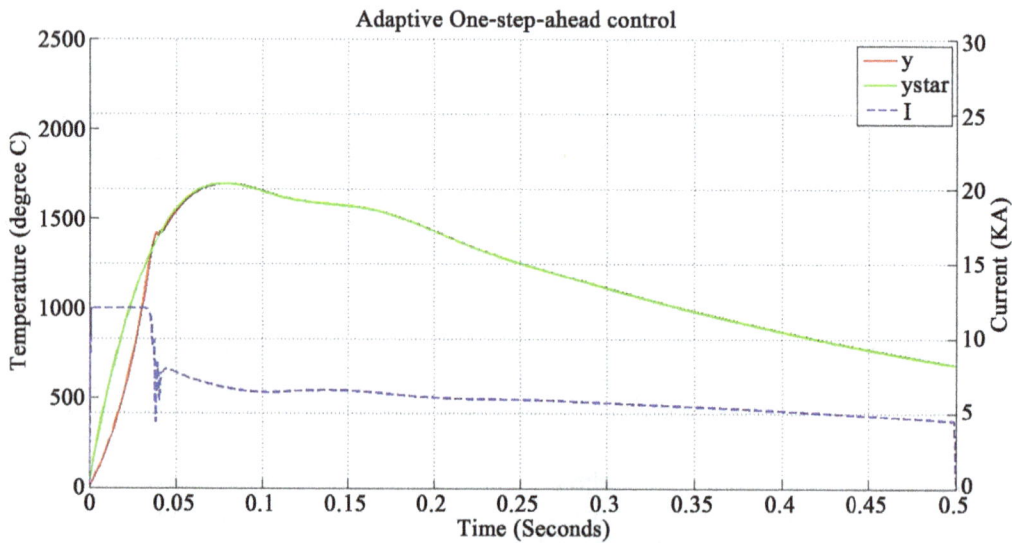

**Figure 4.** Performance of AOSA Controller with 20% increase in nugget diameter and 50% increase in indentation; $I_{max} = 12\,\text{KA}$, Energy = 2583 W.

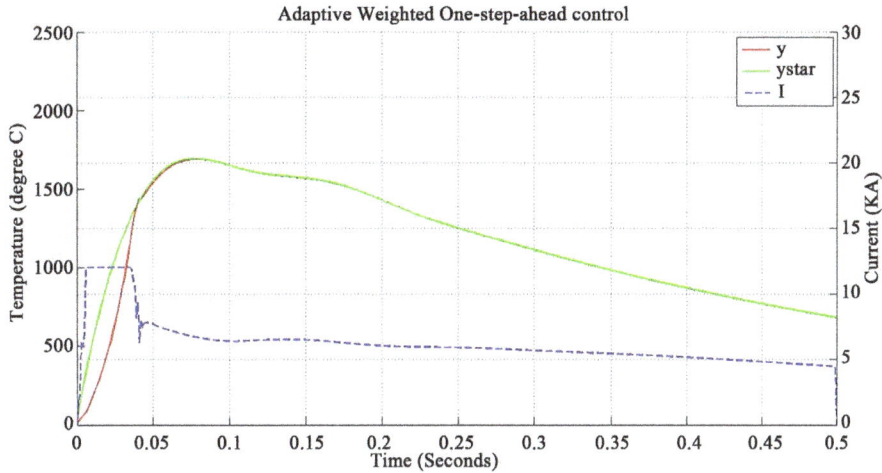

**Figure 5.** Performance of AWOSA Controller with 20% increase in nugget diameter and 50% increase in indentation; $\lambda = 0.1, I_{max} = 12$ KA, Energy = 2558 W .

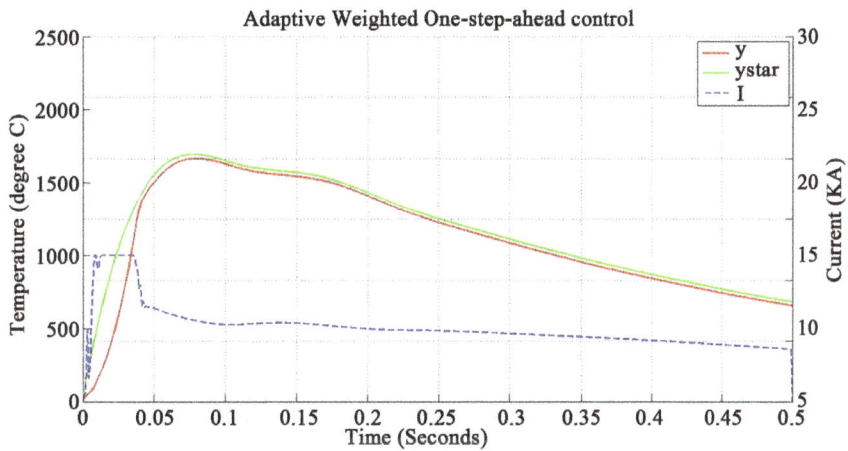

**Figure 6.** Performance of AWOSA Controller with 20% increase in nugget diameter and 50% increase in indentation; $\lambda = 1, I_{max} = 12$ KA, Energy = 2470 W .

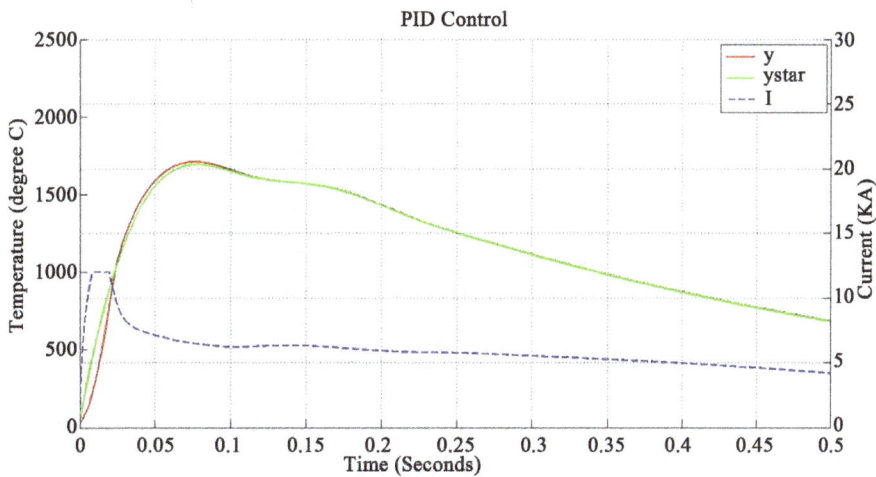

**Figure 7.** Performance of PID Controller prior to unknown parameter variations; $I_{max} = 12$ KA, Energy = 2393 W .

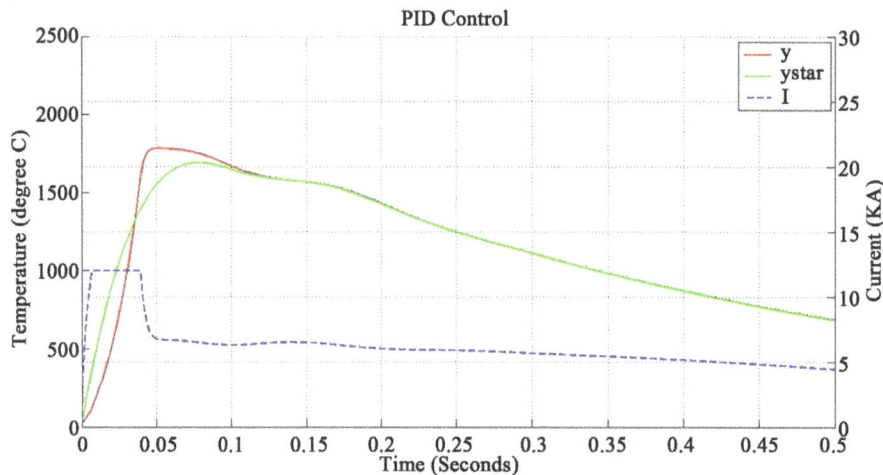

**Figure 8.** Performance of PID Controller with 20% increase in nugget diameter and 50% increase in indentation; $I_{max} = 12$ KA, Energy $= 2632$ W .

Comparing the simulation results for the three controllers, we can see that AOSA and AWOSA controllers compensate for the parameter variations and track the reference temperature profile quite well. Simulation results in **Figure 5** for the AWOSA controller show satisfactory performance and a good tradeoff between tracking error and total energy required for the weld regardless of change in weld parameters. The output temperature profile follows the desired temperature profile reasonably well during the heating stage prior to the melting point. Also, we can see that the total energy required to make a weld using AWOSA is reduced by 6% comparing to the PID controller when $I_{max} = 12$ KA . This can result in significant energy savings for applications requiring a high volume of spot welds, such as manufacturing of automotive vehicles.

## 5. Conclusion

This paper presents a new approach for designing adaptive OSA and WOSA controllers for resistance spot welding processes by utilizing a simplified electrothermal dynamical model of the process. Simulation results of AOSA and AWOSA performance are compared with those of a PID controller. These results indicate that using the proposed AOSA and AWOSA controllers, the nugget temperature profile is forced to track a desired reference temperature profile in presence of unknown parameter variations. Also, these controllers reduce the energy consumed to perform a spot weld, which can result in significant energy savings for applications requiring a high volume of spot welds, such as manufacturing of automotive vehicles.

## References

[1]   Zhang, H. and Senkara, J. (2012) Resistance Welding Fundamentals and Applications. Taylor & Francis Group, Boca Raton.

[2]   Govik, A. (2009) Modeling of the Resistance Spot Welding Process. M.S. Thesis, Institute of Technology, Linkopings University, Linkoping.

[3]   Won, Y.J., Cho, H.S. and Lee, C.W. (1983) A Microprocessor-Based Control System for Resistance Spot Welding Process. *Proceedings of ACC*, San Francisco, 22-24 June 1983, 734-738.

[4]   Zhou, K. and Cai, L. (2014) A Nonlinear Current Control Method for Resistance Spot Welding. *Proceedings of ASME Transactions on Mechatronics*, **19**, 559-569. http://dx.doi.org/10.1109/TMECH.2013.2251351

[5]   Salem, M. and Brown, L.J. (2011) Improved Consistency of Resistance Spot Welding with Tip Voltage Control. *Proceedings of CCECE*, Niagara Falls, 8-11 May 2011, 548-551. http://dx.doi.org/10.1109/ccece.2011.6030511

[6]   Chen, X., Araki, K. and Mizuno, T. (1997) Modeling and Fuzzy Control of the Resistance Spot Welding Process. *Proceedings of SICE*, Tokushima, 29-31 July 1997, 898-994.

[7]   El-Banna, M., Filev, D. and Chinnam, R.B. (2006) Intelligent Constant Current Control for Resistance Spot Welding. *Proceedings of IEEE Conference on Fuzzy Systems*, Vancouver, 16-21 July 2006, 1570-1577. http://dx.doi.org/10.1109/fuzzy.2006.1681917

[8]  Chen, X. and Araki, K. (1997) Fuzzy Adaptive Process Control of Resistance Spot Welding with a Current Reference Model. *Proceedings of IEEE Conference on Intelligent Processing Systems*, Beijing, 28-31 October 1997, 190-194.

[9]  Shriver, J., Peng, H. and Hu, S.J. (1999) Control of Resistance Spot Welding. *Proceedings of ACC*, San Diego, 2-4 June 1999, 187-191. http://dx.doi.org/10.1109/acc.1999.782766

[10]  Ivezic, N., Allen Jr, J.D. and Zacharia, T. (1999) Neural Network-Based Resistance Spot Welding Control and Quality Prediction. *Proceedings of IPMM*, Honolulu, 10-15 July 1999, 989-994. http://dx.doi.org/10.1109/ipmm.1999.791516

[11]  Messler Jr, R.W., Jou, M. and Li, C.J. (1995) An Intelligent Control System for Resistance Spot Welding Using a Neural Network and Fuzzy Logic. *Proceeding of IAC*, Orlando, October 1995, 1757-1763. http://dx.doi.org/10.1109/ias.1995.530518

[12]  Kim, E.W. and Eagar, T.W. (1988) Parametric Analysis of Resistance Spot Welding Lobe Curve. SAE Technical Paper Series, Warrendale.

[13]  Kas, Z. and Das, M. (2014) A Thermal Dynamical Model Based Control of Resistance Spot Welding. *Proceedings of IEEE EIT* 2014, Milwaukee, 5-7 June 2014, 264-269. http://dx.doi.org/10.1109/eit.2014.6871774

[14]  Goodwin, G.C. and Sin, K.S. (1983) Adaptive Filtering Prediction and Control. Prentice-Hall, Englewood Cliffs.

# Appendix

## Boundedness of Nugget Temperature

Since a RSW is a time limited process ( $\Delta t < 0.5$ sec usually), establishing a proof of asymptotic tracking would be meaningless. However, it is important to make sure that the nugget temperature remains bounded during time $(0, \Delta t)$. A theoretical upper bound of the nugget temperature rise, $\Delta \theta(t)$, during time, $(0, \Delta t)$, can be established as follows.

Notice the amount of heat absorbed = the amount of heat supplied – the amount of heat loss

Suppose

$$\Delta \theta = \text{rise in temperature during time, } (0, \Delta t)$$

Thus,

$$\text{Amount of heat obsorbed} = C_A \Delta \theta(t) \tag{19}$$

where $C_A$ is a constant.

$$\text{Amount of heat supplied} = \int_0^{\Delta t} I^2(t) R(t) \, dt \le \Delta t I_{max}^2 R_{max} \tag{20a}$$

where

$$R_{max} = R_\circ + \alpha_r \Delta \theta \tag{20b}$$

and $I_{max}$ denotes the maximum weld current.

$$\text{Amount of heat lost} = C_L \Delta \theta \tag{21}$$

where, $C_L$ is a constant.

Thus,

$$C_A \Delta \theta \le \Delta t I_{max}^2 \left( R_\circ + \alpha_r \Delta \theta \right) - C_L \Delta \theta \tag{22a}$$

or,

$$\Delta \theta \left( C_A - \Delta t I_{max}^2 \alpha_r + C_L \right) \le \Delta t I_{max}^2 R_\circ \tag{22b}$$

or,

$$\Delta \theta \le \frac{\Delta t I_{max}^2 R_\circ}{C_A - \Delta t I_{max}^2 \alpha_r + C_L} \tag{22c}$$

which proves the boundedness of the nugget temperature rise during weld time, $(0, \Delta t)$.

# 19

# Microstructure and Corrosion Properties of the Plasma-MIG Welded AA5754 Automotive Alloy

**A. Abouarkoub[1*], G. E. Thompson[2], X. Zhou[2], G. Scamans[3]**

[1]The Libyan Petroleum Institute, Tripoli, Libya
[2]School of Materials, The University of Manchester, Manchester, UK
[3]Innoval Technology Limited, Banbury, UK
Email: *abdalhadi_aboargoub@yahoo.com

## Abstract

The influence of heating cycles during plasma metal inert gas (MIG) welding on the microstructure and corrosion properties of the AA5754 automotive alloy has been investigated. The high heat input during plasma-MIG welding results in a significant modification in the microstructure of the AA5754 alloy adjacent to the fusion boundaries. As a consequence of partial melting of the Al-Fe-Mn-(Si) intermetallics at the partially melted zone (PMZ) and segregation of the high melting point elements (particularly Fe and Mn) toward the fusion zone, severe galvanic corrosion attacks can be enhanced along the PMZ of the AA5754 weld during exposure to aqueous corrosion environments.

## Keywords

Plasma-MIG Welding, AA5754, Microstructure, Corrosion, Partially Melted Zone

## 1. Introduction

In the automotive industry, welding is one of the most critical elements of the body assembly process, which determines the structural integrity and quality of the vehicles being manufactured. The microstructure of the automotive aluminium wrought alloys can be significantly modified by the high heat input employed during fusion welding techniques, such as the plasma MIG welding [1]-[3]. However, different from the heat treatable aluminium-alloys of the 2xxx, 6xxx and 7xxx series, the change in the mechanical properties of aluminium-magnesium based 5xxx alloys is not significant except for the cold worked alloys, where the mechanical

---

*Corresponding author.

strength can be slightly reduced. There are, however, a number of corrosion issues associated with the formation of various weld zones through fusion welding of the automotive aluminium-magnesium alloys that may deteriorate the performance of the car body connections during service. One of the most critical defects induced by fusion welding, which may contribute to the loss of mechanical and corrosion properties, is the inhomogeneity near the fusion boundary. This phenomenon is often encountered in dissimilar welding of aluminium alloys, where filler metal of different composition from the base metal is used, or when two base metals of different composition are fusion welded together. In such a case, the chemical composition, microstructure and mechanical properties of the partially melted zone near the fusion boundary may differ significantly from the weld and the parent metals [4] [5]. The modification in the chemical composition of the partially melted zone relative to the base metal can be related to solute segregation during the initial transient of solidification, which has been frequently observed in various welds, including steel, stainless steels and aluminium alloys [6] [7]. The composition gradients through the base metal, partially melted zones and the fusion zone of the commercial 5xxx alloys, such as AA5754, are often caused by the rejection of high melting point impurities, for example iron and manganese, from the newly forming solid weld metal into the melt during the initial stage of solidification. This phenomenon, in turn, is expected to modify the corrosion behaviours and, thus, the overall performance of these alloys during service.

## 2. Experimental

The weld samples in the present work were provided by Jaguar Land Rover in the form of a plasma butt weld made of 2 mm AA5754-O panels using AA4043A filler wire "as shown in **Figure 1(a)**".

For metallographic examination, the investigated welds were first cross-sectioned, polished using conventional grinding and polishing methods followed by etching for 15 - 20 s at ambient temperature in Keller's reagent solution (85 ml $H_2O$, 10 ml $H_2SO_4$ and, 5 ml HF) or 15 - 20 s of electro-etching at 20 V in Barker's reagent (4 ml hydrofluoric acid and 200 ml $H_2O$).

The influence of the welding cycle on the mechanical properties of the plasma welded alloys was investigated by conducting Vickers microhardness testing across the polished weld sections (parallel to the weld interface) using a Buehler microhardness tester unit. The microhardness measurements were performed using a 0.25 kg load with a 15 s dwell time.

For corrosion investigation, several 10 × 2 cm samples of the AA5754 plasma weld were cut. Further, in order to eliminate any possible interference of sample geometry with the immersion test results, and to simulate the final surface preparation (prior to painting) in the automotive industry, parts of cut AA5754 plasma welds were mechanically polished until the upper surface was completely flattened (the weld cup was removed), as schematically "shown in **Figure 1(b)**".

Prior to the immersion test, the transverse sections of the cut samples were mechanically polished to 5 μm finish, degreased, washed with water, rinsed with acetone, and then masked with a suitable chemically resistant lacquer (lacomite solution), which was allowed to dry at room temperature for 48 h.

The localized corrosion behaviour of the AA5754 plasma welds was then assessed using a standard accelerated corrosion test in accordance to the G66 ASTM standard by a continuous immersion in in a solution containing 1.0 M $NH_4Cl$, 0.25 M $NH_4NO_3$, 0.01 M $(NH_4)_2C_4H_4O_6$, and 0.09 M $H_2O_2$. The immersion test was conducted at 65°C ± 1°C in a one litre glass vessel and the solution temperature was precisely controlled during experiments by using a thermostatically controlled water bath.

After the immersion test, samples were directly cleaned with a mixture of methanol and deionized water in an ultrasonic bath for 2 min. Samples for study by scanning electron microscopy, to reveal the morphology and severity of corrosion attack, were cross-sectioned and mechanically polished to a 1 μm surface finish.

## 3. Results and Discussion

### 3.1. Microstructure Characterization of the AA5754 Plasma-MIG Weld

The optical micrographs of **Figure 2(a)** and **Figure 3(b)** show the cross-sectional features of the AA5754 plasma-MIG butt weld after, respectively, electro-etching at 20 V in Barker's reagent solution (4 ml hydrofluoric acid and 200 ml $H_2O$) and chemical etching in Keller's reagent (85 ml $H_2O$, 10 ml $H_2SO_4$ and, 5 ml HF). In general, the following three different weld zones can be readily distinguished in the etched weld section: 1) the fusion zone "**Figure 2 (b)**"; 2) the partially melted zone "**Figure 2(c)**"; 3) the parent metal "**Figure 2(d)**".

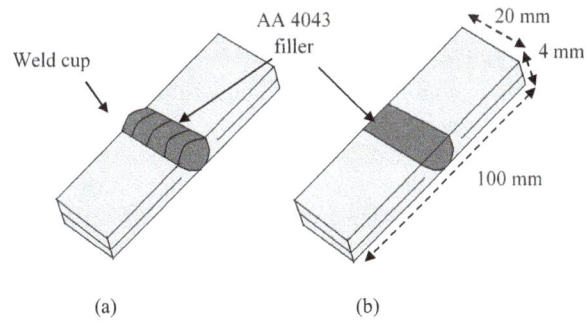

**Figure 1.** Schematic diagram of immersion test samples of the AA5754 plasma butt weld: (a) as received sample; (b) mechanically polished sample.

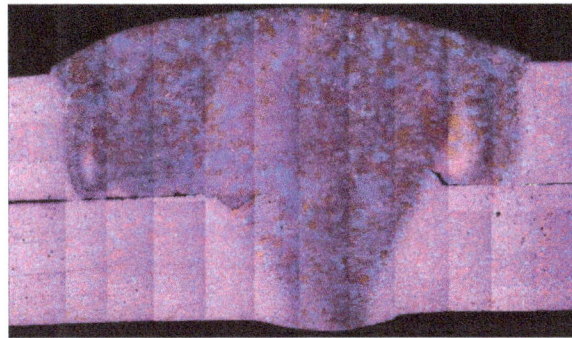

(a)

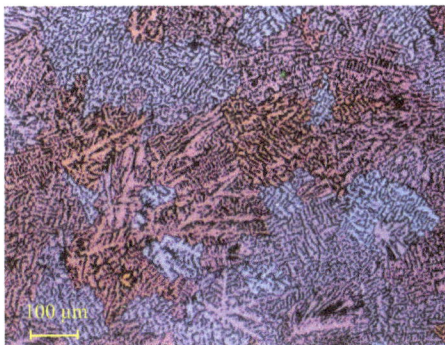

(b)

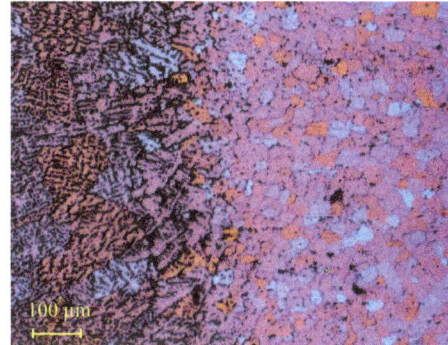

(c)

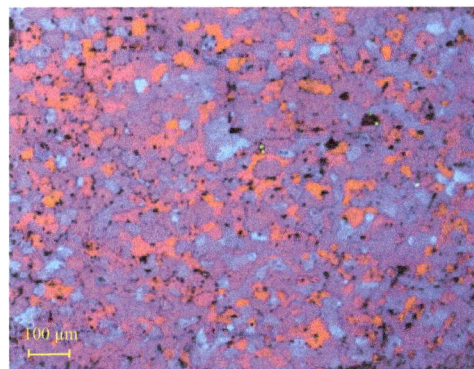

(d)

**Figure 2.** Optical cross-sectional micrographs of the plasma-MIG AA5754 weld after electro-etchingat 20 V in Barker's reagent (4 ml hydrofluoric acid and 200 ml $H_2O$): (a) an over-view; (b) fusion zone; (c) the partially melted zone; (d) the parent metal.

### 3.1.1. The Fusion Zone

The fusion zone of the plasma-MIG weld is the region where the parent metal is completely melted and mixed with the filler alloy material. Upon cooling, solidification of the fusion zone takes place first at the partially melted or solid grains along the fusion boundary (*i.e.* at the solid-liquid interface). The growth of the solidified material then proceeds toward the fusion centre (*i.e.* epitaxial growth opposite of the thermal gradient across the weld regions [2]), resulting in the formation of a cellular dendritic Al-Mg-Si structure "**Figure 3(b)** and **Figure 3(c)**" of around 150 μm width along the fusion zone boundaries. As solidification continues toward the fusion zone, heterogeneous nucleation of new grains [4] leads to the formation of a dendritic structure of equiaxed grains at the weld centre "**Figure 3(d)**".

### 3.1.2. The Partially Melted Zone

Next to the fusion boundary of the AA5754 plasma-MIG weld, there is a partially melted structure of about 300 μm width "**Figures 3(a)-(c)**". The partial melting within this zone resulted in the formation of eutectic-phases at the grain boundaries of the parent material. The peak temperature in this region during welding is expected to exceed the equilibrium solidus or the eutectic temperature for the Al-Mg-Si eutectic constituents. However, the temperature across this region was not sufficient to fully dissolve phases containing high melting point elements (particularly those containing Fe and Mn). As a consequence, partial dissolution of Fe-rich constituents in this zone was detected "**Figure 3(e)**". Such phenomena may lead to the migration of undissolved solutes (solid state atomic diffusion) toward the fusion zones. In turn, this may cause severe variation in the chemical composition between the PMZ and the adjacent areas.

### 3.1.3. The Parent Metal

The parent metal adjacent to the partially melted zone of the AA5754 plasma-MIG weld showed no evidence of over-heating (e.g. excessive grain growth, dissolution or coarsening of second phase particles "**Figure 2(d)**"and/ or softening "**Figure 4**".

## 3.2. Corrosion Assessment of the AA5754 Plasma-MIG Weld

The optical micrographs in **Figures 5(a)-(d)** show the appearance of the as-received "**Figure 5(a)** and **Figure 5(c)**" and mechanically polished "**Figure 5(b)** and **Figure 5(d)**" AA5754 plasma weld after room temperature immersion in an acidified/chloride solution for 24 h. In general two types of localized attack were detected in the plasma welded samples, namely severe localized corrosion following the upper edges, and large deep pits randomly distributed over the parent metal.

Based on its severity and shape, the first type of localized corrosion attack was identified as a "knife-like attack". Detailed investigation using scanning electron microscopy revealed that the knife-like attack was preferentially initiated and propagated more than 300 μm within specific areas just outside and parallel to the fusion zone "**Figure 6(a)** and **Figure 6(b)**". The preferential propagation of severe localized attack parallel to the AA5754 alloy weld can be related to the presence of a narrow partially melted zone within which high melting point elements, particularly iron and manganese, were ejected into the melt during the initial stage of solidification. This was evident from the microstructural analysis of the AA5754 plasma weld prior to the immersion test. Partial dissolution of Fe-Mn rich constituents in this region followed by segregation of iron and manganese towards the fusion zone cause severe composition gradients and, thus, potential gradients through the weld zones "**Figures 7(a)-(d)**". These introduce a galvanic corrosion susceptibility of the plasma welded 5754 alloy in corrosive environments. Form the corrosion view point, the partially melted zone, which is more depleted in iron and manganese compared with the fusion and heat affected zones, is more prone to galvanic attack because its open circuit potential must be shifted to more negative values. Further, the saturation of the fusion zone matrix with elemental silicon (from the 4xxx filler) normally shifts the open circuit potential to more anodic values. As a consequence, this area can be sacrificially corroded in the presence of a conducting, corrosive electrolyte, where the galvanic circuit elements are all established. Furthermore, in the immersion test, the exposed surface area of the PMZ is very small compared with the more noble adjacent areas; this means that the cathodic galvanic currents may be extremely high, which certainly enhance the severity of galvanic corrosion.

The second type of attack was detected only under areas covered with loose, white scale, which suggests that the scale build up over certain areas during the immersion test somehow initiates such localized corrosion. Based

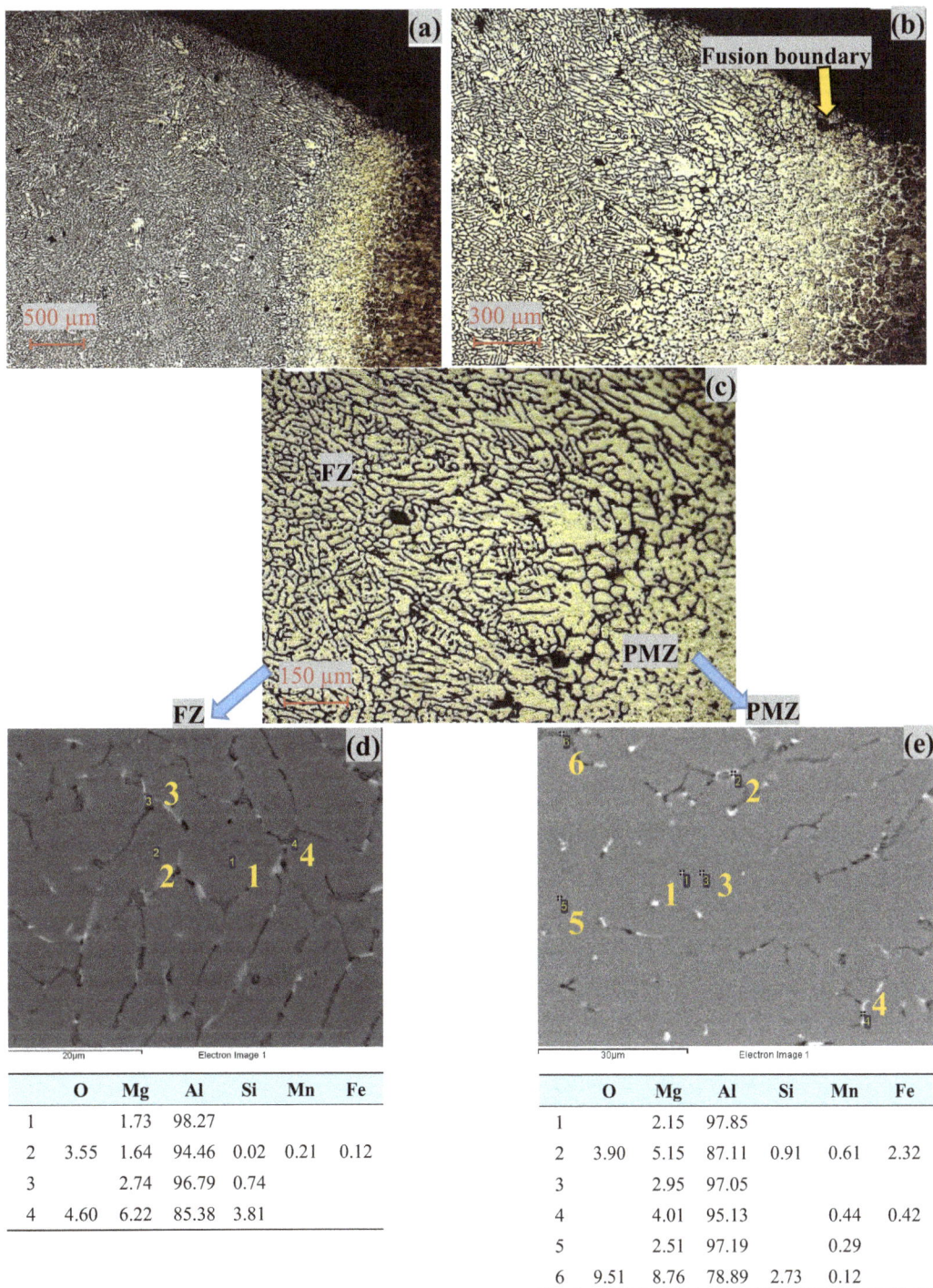

**Figure 3.** ((a)-(c)) Optical micrographs of increased magnifications of the partially melted zone in the plasma-MIG AA5754 weld after etching in Keller's reagent (85 ml $H_2O$, 10 ml $H_2SO_4$ and, 5 ml HF)); and ((d), (e)) back scattered electron micrographs and corresponding EDX spot analysis of the eutectic phases at the fusion and partially melted zones, respectively.

on these findings, the second type of corrosion was identified as "under deposit pitting attack". The severity of under deposit attack is better represented in **Figure 6(c)** and **Figure 6(d)**, where a pit extended horizontally and vertically, and reached a final depth of approximately 1 mm. In this investigation, however, no evidence of any microstructural or chemical variations across the tested samples can be currently related to the initiation or

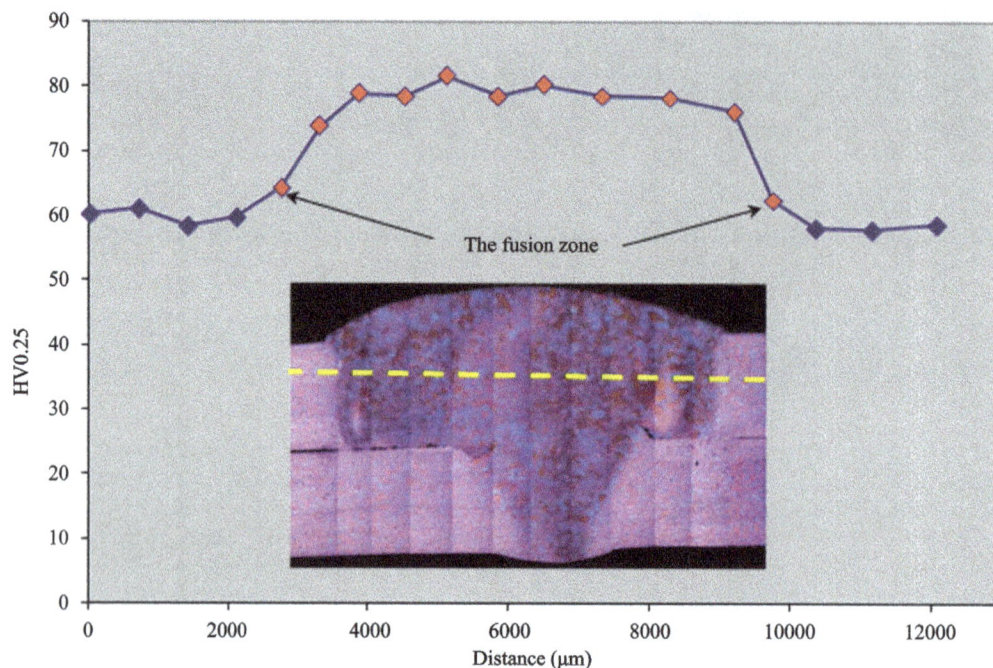

**Figure 4.** Vickers microhardness profile along the cross-section of the AA5754 plasma-MIG weld.

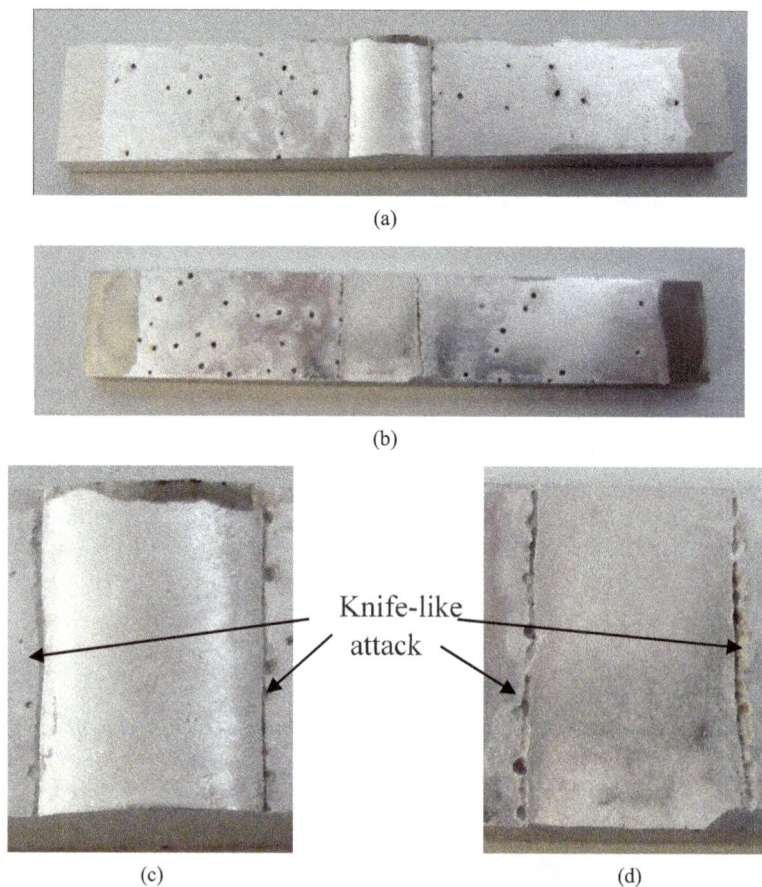

**Figure 5.** General view of the AA5754 plasma-MIG butt weld after 24 hours of room temperature immersion in an acidified/chloride solution: ((a), (c)) as-received; ((b), (d)) mechanically polished.

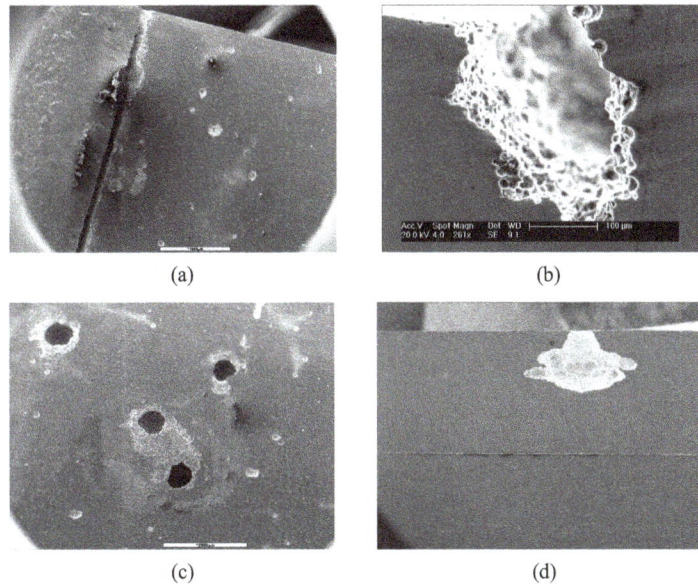

**Figure 6.** Top and cross-sectional views of : ((a), (b)) the knife-like attack at the weld edges; and ((c), (d)) severe pitting attack at the parent metal of the AA5754 at plasma butt weld after 24 hours of room temperature immersion in an acidified/chloride solution.

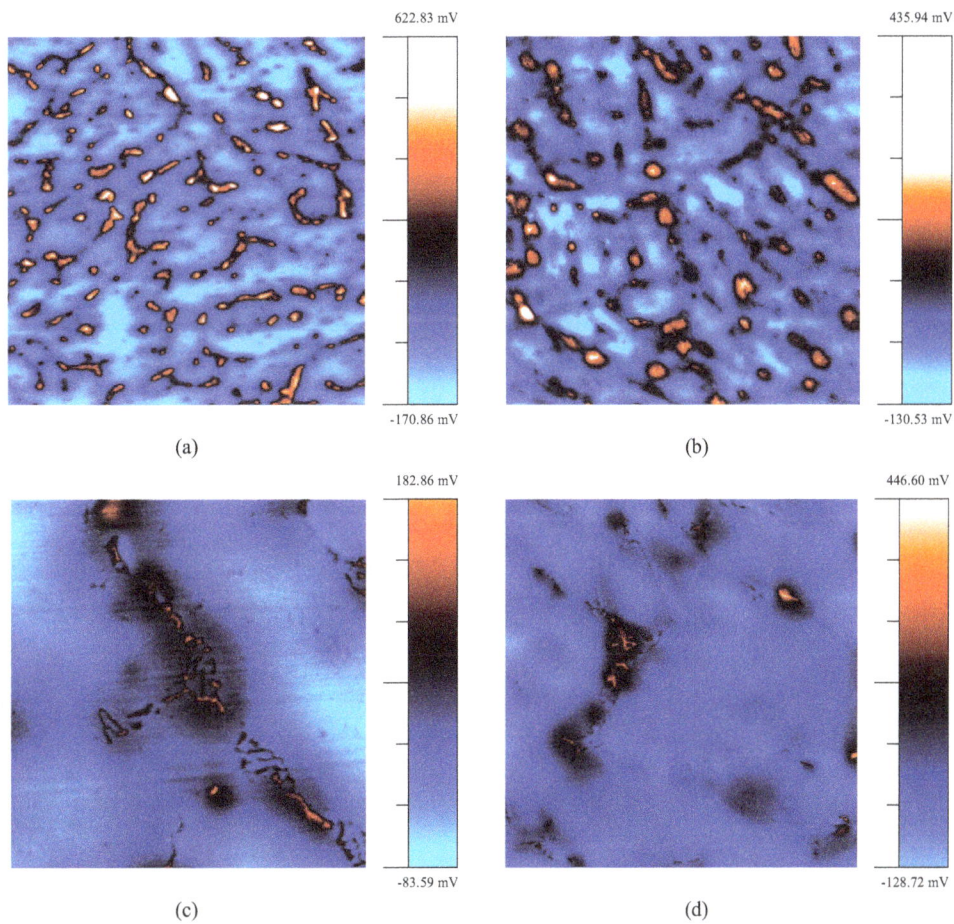

**Figure 7.** AFM mapping across the various weld zones of a polished AA5754 plasma-MIG weld: (a) the fusion zone; (b) the fusion/partially melted zone boundary; (c) the partially melted zone; and (d) partially melted zone/parent metal boundary.

propagation of the under deposit attack at the parent metal. Most probably, the scale deposition and, thus, the initiation of this attack are related to the local alkalinity around cathodic intermetallics at the samples surface.

## 4. Conclusion

Plasma-MIG welding of the AA5754 alloy using silicon-rich filler wire (AA4043A) results in the formation of a partially melted zone adjacent to the fusion boundaries. Partial melting of the Al-Fe-Mn-(Si) intermetallics at the partially melted zone, and segregation of the high melting point elements (particularly Fe and Mn) towards the fusion zone increase the open circuit potential difference between the weld and its adjacent partially melted zones. Consequently, severe galvanic corrosion attacks take place preferentially along the PMZ at each side of AA5754 weld during exposure to aqueous corrosion environments.

## Acknowledgements

The authors thank EPSRC for support of the LATEST 2 Programme Grant.

## References

[1]  Totten, G.E. and MacKenzie, D.S., Eds. (2003) Handbook of Aluminium: Vol. I: Physical Metallurgy and Processes. CRC Press, New York, 481-532.

[2]  Kou, S. (2003) Welding Metallurgy. 2nd Edition, Wiley Interscience, New Jersey.

[3]  Mathers, G. (2002) The Welding of Aluminium and Its Alloys. Woodhead Publishing Ltd., Cambridge.

[4]  Yang, Y. and Kou, S. (2007) Mechanisms of Microsegregation Formation near Fusion Boundary in Welds Made with Dissimilar Filler Metals. *Materials Science & Technology* 2007 *Conference and Exhibition*, Detroit, September 16-20 2007, 329-340.

[5]  Kou, S. and Yang, Y. (2007) Fusion-Boundary Macrosegregation in Dissimilar-Filler Welds. *Welding Journal*, **86**, 303-s-312-s.

[6]  Baeslack III, W., Lippold, J. and Savage, W. (1979) Unmixed Zone Formation in Austenitic Stainless Steel Weldments. *Welding Journal*, **58**, 168s-176s.

[7]  Savage, W., Nippes, E. and Szekeres, E. (1976) A Study of Fusion Boundary Phenomena in a Low Alloy Steel. *Welding Journal*, **55**, 260s-268s.

# 20

# Fatigue Strength and Modal Analysis of Bogie Frame for DMUs Exported to Tunisia

Wei Tang*, Wenjing Wang, Yao Wang, Qiang Li

School of Mechanical Electronic and Control Engineering, Beijing Jiaotong University, Beijing 100044, China
Email: *wtang@bjtu.edu.cn

## Abstract

The equivalent stress at key positions of Bogie Frame for DMUs Exported to Tunisia is obtained by using simulation analysis. The evaluation of static strength and fatigue strength is checked referring to UIC specification and Goodman sketch for welding materials. In addition, the modal analysis of the frame is made, and the vibrational modal of frame in given frequency domain is predetermined to evaluate the dynamical behavior of the frame in order to meet the dynamical design requirements. The results show that the key points of the calculated frame of the equivalent stress are less than allowable stress, and thus it could provide a theoretical foundation for the optimized design of frame structure and safety of industrial production.

## Keywords

Fatigue Strength, Diesel Multiple Units, Frame, Finite Element, Modal Analysis

## 1. Introduction

The bogie frame is the main load bearing components and power transmission components of the vehicle, when the vehicle is in motion the process, not only to the bogie frame to withstand loads, but also need to pass a variety of forces between the body and the wheel. Due to the fatigue test costs are expensive, the fatigue strength assessment of key components in the bogie frame using finite element model can find out the fatigue strength of the weak parts, can reduce the risk of fatigue testing prototypes, shorten development cycles, reduce trial costs. In addition, the current domestic commonly uses Electric Multiple Units [1], lacks of bogie products of Diesel Multiple Units; Diesel Multiple Units still have a large market in many countries such as Tunisia for its poor line conditions and economic factors. Therefore, strength analysis and dynamic assessment for the bogie frame of Diesel Multiple Units is of great significance.

This paper is to understand the export Tunisia DMUs bogie basic components, infrastructure characteristics, determined the type of bogie frame load sources and calculated in accordance with the relevant specifications to determine the load; then to use the Hyper mesh software architecture network entities meshing, to re-use the ANSYS finite element analysis software for finite element analysis of the bogie frame. The evaluation of static strength and fatigue strength is checked referring to UIC specification and Goodman sketch for welding materials. In addition, the modal analysis of the frame is made [2].

*Corresponding author.

## 2. Bogie Frame Structure and Finite Element Modeling

### 2.1. Bogie Frame Structure

The bogie Frame for DMUs Exported to Tunisia is adopted by welded structure, **Figure 1** demonstrates that the main framework architecture is H-shaped in the horizontal plane, which is composed of two box-shaped side sills, the overall composition of the box beam welding, by the central concave belly of the fish box structure composed of a spring seat side beam welding, basic brake mounts, anti-roll torsion bar seat, etc., the cavity has a thickness of 10 mm stiffener plate [3]. Box beam structure for the central opening, the transverse beam welding has ended with stopper seat, traction rod seat, motor bracket, gearbox bracket and secondary lateral damper seat and so on.

### 2.2. FEM Model of Bogie Frame

Considering calculating workload, precision and the actual situation in structure of the entire bogie frame, this research selects 10-node solid element of solid 92. Based on the model, the entire bogie frame is discrete with the software Hyper mesh and analyzed with the large generally used finite element software ANSYS [4]. In order to simulate the real boundary conditions of the bogie frame, axle box spring in the bogie frame mount simulated by a series of axle box spring unit Combine14 spring means, consistent with the axle box spring stiffness of the spring element stiffness. In the end, the finite element discrete nodes of 110,368, the number of units to 341,334, finite element discrete model shown in **Figure 2**.

### 2.3. Evaluating Standard of Bogie Frame Strength

In the fatigue strength of welded bogie frame has now formed the international standard UIC 615-4 [5] as the representative of the design, evaluation system. Bogie frame structure strength assessment generally includes three aspects, namely, the role of analysis to determine the load, static strength analysis and assessment, analysis and evaluation of the fatigue strength.

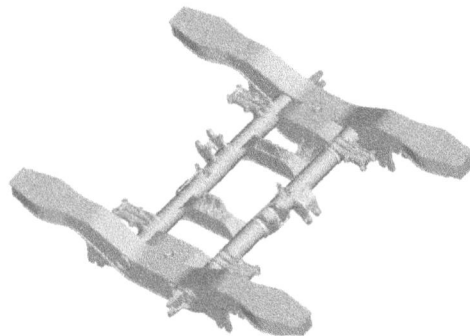

**Figure 1.** Welded bogie frame.

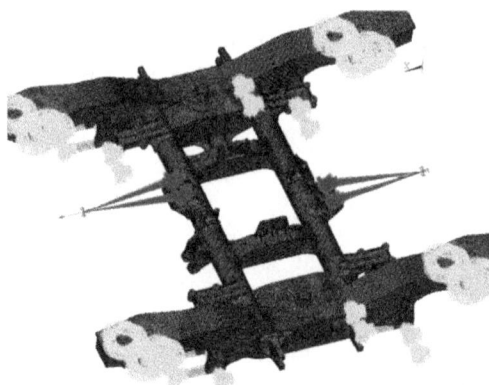

**Figure 2.** The FEM model of the bogie frame.

According to the UIC 615-4 regulations, we can calculate the appropriate supernormal load, simulated operational load and special operational load. Supernormal load when the maximum load operations may occur; simulate actual operating load refers to the load operations occur frequently; special operational load refers to the load frame by a special device caused. In the practical constraints are consistent with the principles of the frame, and the constraints of the axial knuckle arm spring constraints loads. Then, referring to the UIC 615-4 regulations on load conditions are combined to get the final five groups exceptional load cases, four groups of special load cases and 13 groups of operational load cases. **Tables 1-3** lists the typical cases of supernormal loads and operating loads.

## 3. Results and Analysis

### 3.1. Calculation and Analysis of Static Strength

The conditions of supernormal loads are used to verify that there is no permanent deformation when the bogie frame experiences supernormal loads, which can be used to evaluate static strength of the bogie frame. In the

**Table 1.** Main extraordinary load case combinations table. KN

| Loading point | Conditions | | | | |
| --- | --- | --- | --- | --- | --- |
| | 1 (K = 1.4) | 2 (K = 2.0) | 3 | 4 | 5 |
| Left air spring vertical loads | −168.3 | −240.4 | −168.3 | −168.3 | −168.3 |
| Right air spring vertical loads | −168.3 | −240.4 | −168.3 | −168.3 | −168.3 |
| Lateral load stopper | | | 108.2 | | 108.2 |
| Air springs lateral load | | | 16.4 | | 16.4 |
| The left side of the anti-roll load | | | | | 79.6 |
| The right side of the anti-roll load | | | | | −79.6 |
| Buckling load/mm | | | | +24.0 | +24.0 |
| A series of vertical damper load | | | | | −9.0 |
| Secondary lateral damper load | | | | | 8.0 |
| Anti-snake damper load | | | | | 24.0 |

Note : K is a safety factor.

**Table 2.** Extraordinary special load case combinations table. KN

| Loading point | Emergency braking condition | Shunting impact conditions | Equipment inertia conditions | Derailment conditions |
| --- | --- | --- | --- | --- |
| Left air spring vertical loads | −168.3 | −168.3 | −168.3 | −120.2 |
| Right air spring vertical loads | −168.3 | −168.3 | −168.3 | −120.2 |
| Traction rod seat longitudinal load | 168.8 | 367.5 | | |
| 1st gearbox reaction rod load Flank | | | | |
| 2nd gearbox reaction rod load Flank | | | | |
| 1st unit brake load | 24.0 | | | Three-point support analog derailment |
| 2nd unit brake load | 24.0 | | | |
| Gearbox vertical vibration | | | 47.6 | |
| Gearbox lateral vibration | | | 9.2 | |
| Gearbox longitudinal vibration | | | 9.2 | |

**Table 3.** Typical operating conditions load combination table.                                        KN

| Conditions | Vertical load | | Lateral load | Longitudinal load | Distorting load | Braking load z/y | Motor load z/y/x | Gearbox load | Two series shock absorbers x/z |
|---|---|---|---|---|---|---|---|---|---|
| | The left side of the beam | The right side of the beam | | | | | | | |
| 1 | $F_z$ | $F_z$ | | | | | $-mg$ | | |
| 3 | $F_z(1+\alpha-\beta)$ | $F_z(1-\alpha-\beta)$ | $+F_y$ | $-15.5$ | | 27.4/61.2 | 31.5/21/21 | $-27.7$ | 8.7/5.63 |
| 7 | $F_z(1-\alpha-\beta)$ | $F_z(1+\alpha-\beta)$ | $-F_y$ | $-15.5$ | | 27.4/61.2 | 31.5/21/21 | $-27.7$ | 8.7/5.63 |
| 8 | $F_z(1-\alpha+\beta)$ | $F_z(1+\alpha+\beta)$ | | 15.5 | | 27.4/61.2 | 31.5/21/21 | 27.7 | 8.7/5.63 |
| 10 | $F_z(1+\alpha-\beta)$ | $F_z(1-\alpha-\beta)$ | $+F_y$ | 15.5 | 5.75 | 27.4/61.2 | 31.5/21/21 | 27.7 | 8.7/5.63 |
| 11 | $F_z(1+\alpha+\beta)$ | $F_z(1-\alpha+\beta)$ | $+F_y$ | $-15.5$ | $-5.75$ | 27.4/61.2 | 31.5/21/21 | $-27.7$ | 8.7/5.63 |
| 12 | $F_z(1-\alpha-\beta)$ | $F_z(1+\alpha-\beta)$ | $-F_y$ | 15.5 | 5.75 | 27.4/61.2 | 31.5/21/21 | 27.7 | 8.7/5.63 |
| 13 | $F_z(1+\alpha-\beta)$ | $F_z(1+\alpha+\beta)$ | $-F_y$ | $-15.5$ | $-5.75$ | 27.4/61.2 | 31.5/21/21 | $-27.7$ | 8.7/5.63 |

Note : $\alpha$ is roll coefficient is taken as 0.1; $\beta$ coefficient for the ups and downs , taken as 0.2; m motor quality ; g is the gravitational acceleration.

supernormal main loads conditions, the maximum stress occurs at the welded joint of Cross-side beam connections under lateral beam support beams in the cover plate and cover plate in condition 5, and the maximum value is 295.2 MPa; In the supernormal special loads conditions, the maximum stress occurs at the welded joint of Cross-side beam connections under lateral beam support beams in the cover plate and cover plate when the bogie derails, and the value is 256.5 MPa. All these stress analyzed above is less than the yield stress of P355NL1 steel (355 MPa), which satisfies the UIC standards static strength requirements [6].

### 3.2. Calculation and Analysis of Fatigue Strength

According to the framework structure and analysis of static strength, fatigue crack tends to happen on 13 major parts that endure larger stress. Finite element analysis is carried out on these 13 major parts in different conditions, as shown in **Table 3**. Corresponding maximum stress $\sigma_{max}$ and minimum stress $\sigma_{min}$ is found. The mean stress $\sigma_m$ can be found with the standards of UIC:

$$\sigma_m = \frac{\sigma_{min} + \sigma_{max}}{2}, \sigma_a = \frac{\sigma_{min} - \sigma_{max}}{2} \tag{1}$$

**Table 4** shows the calculation results of mean stress and dynamic stress amplitude in strong stress areas. Selective analysis is carried out on critical points, which are selected according to the framework structure. **Figure 3**/**Figure 4** show the overall architecture , and the high stress amplitude of dynamic stress nephogram when the fatigue strength is of the greatest effect conditions, the maximum stress is found at the welded joint of Longitudinal beams and beams, and the value is 79.5 MPa.

## 4. Evaluation of Fatigue Strength

Fatigue strength is evaluated with Goodman line. Import the mean stress and dynamic stress amplitude in strong stress areas into the fatigue limit diagram of frame materials (**Figure 5**), we can find that all these representative and dangerous points are located inside of the Goodman line, which means that all these mean stress and dynamic stress amplitude are less than the fatigue limit of P355NL1 steel. Therefore, the bogie frame meets the design requirements of the fatigue strength.

## 5. Modal Analysis

In consideration of the influence of practical operation constrains on the modal, we apply horizontal constraint and vertical constraint on locating seat of axle box rotary arm, and we also apply vertical elastic constraint on the bottom of the axle box. In order to determine whether there is resonance or other vibration mode that against the operating of vehicles, we used the subspace iteration method provided by the ANSYS software to carry out the modal analysis on the frame. In general, there is no high-frequency vibration during the operation of trains,

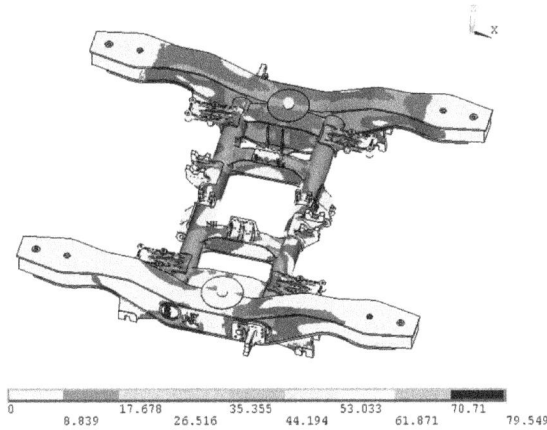

**Figure 3.** Overall dynamic stress amplitude cloud.

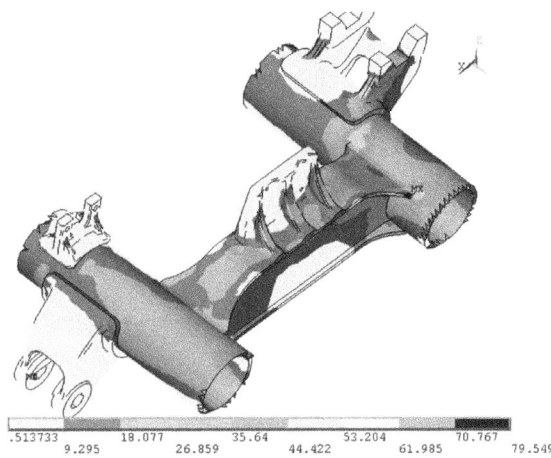

**Figure 4.** Partial cloud dynamic amplitude stress in large stress parts.

**Table 4.** Synthesis of the results mean stress/dynamic stress amplitude on the frame big stress area.                    MPa

| Part name | No. | Location | Average stress | Dynamic stress amplitude | Materials area |
|---|---|---|---|---|---|
| Beam and side beams connecting area | 1 | Within a support beam and side sill beam weld connection | 47.4 | 64.3 | Weld |
| | 2 | beams and side beams connecting welds | 66.4 | 51.6 | Weld |
| | 3 | Cover plate with the support of the beam connecting the beams and side beams under three side beams connecting welds Department | 116.1 | 58.5 | Weld |
| Side sill area | 4 | positioning seat upright plate portion of the opening arc bends | 28.4 | 60.1 | Base metal |
| | 5 | positioning seat cover is connected with the lower side beam welds | 67.6 | 47.6 | Weld |
| | 6 | Under positioning seat cover parts connected with the vertical plate welds | 76.1 | 53.5 | Weld |
| | 7 | Anti-nake-seat legislature damper plate | 0 | 51.0 | Base metal |
| | 8 | Anti-snake damper seat and side sill outer webs connecting portion | 33.8 | 28.4 | Weld |
| Beam area | 9 | Brake bracket vertical plate | 0 | 55.8 | Base metal |
| | 10 | Brake bracket and beam connection area | 25.4 | 37.4 | Weld |
| | 11 | Anti-roll torsion bar seat ribs | 0 | 78.1 | Base metal |
| | 12 | Longitudinal beams and beam weld connection | 41.8 | 79.5 | Weld |
| | 13 | Gearbox boom stand upright plate | 0 | 60.3 | Base metal |

so when to analyze framework of free mode, only to take the first six modal characteristics. **Table 5** shows the inherent frequency and vibration shape for each modal.

From **Table 5** we can find that the first-order characteristics is two side beams nod reversing, which means that the torsional stiffness of the bogie frame is small; this helps trains to overcome the vertical irregularity of lines. The six-order characteristic is that beams in the vertical plane of the first bending with a larger frequency, which means that the stiffness of the beam is pretty big; this helps the beam to bear load and keep connection to other parts. As a conclusion, the vertical stiffness and transverse stiffness of the bogie frame is ideal. Both of them meet the design requirements and the smooth running of vehicles.

## 6. Conclusions

According to the UIC615-4 specification, this research analyzes the static strength and fatigue strength of bogie frame for DMUs exported to Tunisia. The result shows that all the stress amplitudes are less than fatigue limit, which means that the bogie frame meets the requirements of fatigue strength.

ANSYS software is used to calculate the inherent frequency and vibration shape of bogie frame, and the results reveal that the torsional stiffness of the bogie frame is small. Trains benefits from the low torsional stiffness to come over lines with vertical irregularity, and bogie frame can avoid other excitation frequency.

With the help of CAD/CAE, people can do the simulation and analysis on bogie frame of high speed train effectively, which contributes a lot to shorten the development cycle, reduce cost and raise efficiency.

## Acknowledgements

This research was financially supported by the National Natural Science Foundation of China (NO.51205017)

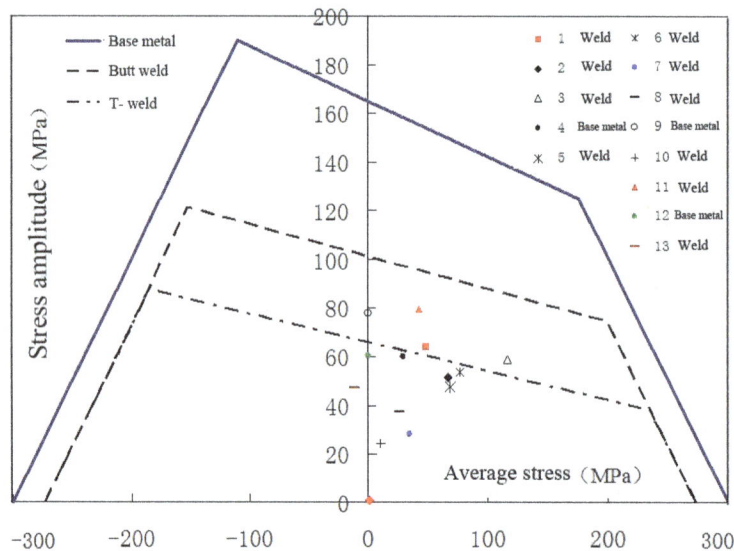

**Figure 5.** Fatigue limit diagram for base metal/welded joint of steel P355NL1.

**Table 5.** Frame modal analysis results.

| Order | Frequency/Hz | Modal characteristics |
| --- | --- | --- |
| 1 | 43.3 | Two side beams reverse nod |
| 2 | 76.2 | First bending of the frame beams |
| 3 | 84.7 | Two side beams reverse bend in the horizontal plane |
| 4 | 87.0 | Two side beams in the same direction in the horizontal plane first bending |
| 5 | 102.3 | Two side beams in the same direction in the horizontal plane of second-bending |
| 6 | 103.2 | Beams in the vertical plane of the first bending |

and the National High Technology Research and Development Program (863) Fund Project (Key technology research lineage system of high-speed trains.NO.2012AA112001-01).

# References

[1] Wang, W.J., Liu, Z.M., Li, Q., *et al.* (2009) CRH2 EMU Fatigue Strength of the Bogie Frame Analysis [J]. Beijing Jiaotong University, 2.

[2] Yan, J.M. (1993) Vehicle Engineering [M]. China Railway Publishing House, Beijing.

[3] Yu, W. (2008) Finite Element Analysis of Locomotive Bogie Frame Strength [J]. Machinery Manufacturing, 8.

[4] Hu Y.M. (2009) The Basic Theory of the Structural Strength of the Vehicle and CAE Analysis Technology [M]. Chongqing University Press, Chongqing.

[5] UIC615-4. Motive Power Units-Bogies and Running Gear-Bogie Frame Structure Strength Tests. International Union of Railways (UIC), 2003.

[6] Miao, L.X. (2005) Fatigue Strength of the Base Vehicle Structure [M]. Beijing Jiaotong University Press, Beijing.

[7] EN13749. Railway Applications-Wheel Sets and Bogies-Methods of Specifying Structural Requirements of Bogie Frames. European Standard, 2005.

# 21

# Corrosion of Steel Pipelines Transporting Hydrocarbon Condensed Products, Obtained from a High Pressure Separator System: A Failure Analysis Study

Gerardo Zavala Olivares[*], Mónica Jazmín Hernández Gayosso

Instituto Mexicano Del Petróleo. Eje Central Lázaro Cárdenas Norte,Col. San Bartolo Atepehuacan, C.P., México D.F., México
Email: [*]gzavala@imp.mx

## Abstract

In this paper, the corrosion of steel pipelines transporting hydrocarbon condensed products was studied. Different activities of sampling and analysis were carried out to diagnose the failure causes and to establish a control system for the corrosion problem. The combination of three types of corrosion, including erosion corrosion, galvanic corrosion and microbiologically induced corrosion, was synthetically considered. A serial of experiments were designed to research those types of corrosion. This type of failure was observed in characteristics sites of the pipeline, mainly in direction changes and welding joints. Additionally, localized corrosion was observed in the inner steel wall and distributed along the pipeline, although a tendency was not detected.

## Keywords

Erosion Corrosion, Galvanic Corrosion, Microbiologically Induced Corrosion

## 1. Introduction

Nowadays, hydrocarbon transportation in the oil industry is accomplished through pipelines. Huge volume of gas and liquid can be transported in an efficient and safe way. During the gas-oil separation processes, the operational parameters may come out of control and some operational problems may occur, including corrosion failures.

---

[*]Corresponding author.

A corrosion problem was observed in a 2 inches diameter steel pipeline, transporting hydrocarbon condensed products obtained from a high pressure separator system located in an offshore platform in the Gulf of Mexico. This situation generated critical conditions that favored the pipeline corrosion development and resulted in some leaks. Initially, it was presumed that the failure could be caused by effect of microbiologically induced corrosion. However, as the failures did not exhibit a regular pattern, it was considered that others types of corrosion were involved in the problematic. According to this, several activities, mainly directed to diagnose the failure conditions, were carried out [1]-[3].

Therefore, it was intended to diagnose the causes for the corrosion failures in the pipeline at the output of the high pressure separator system. With this, recommendations for corrosion control can be made.

## 2. Activities

The activities carried out were divided in two stages: Stage 1 focused on the characterization of the hydrocarbon streams, at the sites where the leaks occurred. This was in order to determine the main corrosive agents in the system. Stage 2 directed to analyze the corrosion products at the metal surface, as well as the type and morphology of the corrosion process. With these activities, the causes for the corrosion failures could be established.

### 2.1. Stage 1

Two monitoring points were selected, considering streams that were incorporated to the system that exhibited the leak. These points were named P-I and P-II.

Field activities: Initially, an inspection of the system was carried out, to identify the monitoring points and to collect both, water and condensed products. Four samples were taken from the monitoring points, leaving a gap of 24 hours between each sampling [4].

For water samples, different parameters were measured *in situ*: Temperature, Pressure, pH, $O_2$ content, $CO_2$ content, $H_2S$ content, conductivity and presence of sulfate reducing bacteria (SRB) [5].

Laboratory activities: Physical and chemical analysis were carried out to water samples. The analysis for the condensed products samples were: hydrocarbon characterization, $O_2$ content, $CO_2$ content, $H_2S$ content, humidity, Sulfur compounds [6].

### 2.2. Stage 2

A leak occurred in a 2 inches steel pipeline transporting hydrocarbon condensed products. A section of the pipeline, where the failure occurred, was cut and replaced. The sample was prepared and sent to the laboratory, for its respective analysis. Once the sample was at the laboratory, the failure was located and corrosion products were obtained from the adjacent area. Different analysis were carried out, including X-ray diffraction and fluorescence, Mössbauer spectroscopy, surface analysis, hardness, chemical analysis and identification of sulfate reducing bacteria, among others.

## 3. Results and Discussion

### 3.1. Stage 1

#### 3.1.1. Water Analysis

During the sampling procedure, it was observed that the hydrocarbon obtained exhibited high water content. Water is necessary when a corrosion process is taking place and the extent of the damage depends on its corrosive characteristics.

In all cases, the samples were identified as condensation water, with low conductivity. However, the Langelier index indicated a corrosive tendency for all samples [7]. Values between −3 and −5 were observed, as shown in **Figure 1**. This situation implies that water may become very aggressive to the metallic structures.

Additionally, it must be indicated that the water corrosiveness for these systems was also related to the content of dissolved gases, including $H_2S$, $CO_2$ and $O_2$. For this case, the $H_2S$ concentration was between 5 and 45 ppm, $O_2$ between 0 and 2.6 ppm and $CO_2$ between 28 and 75 ppm (**Figures 2-4**). The presence of these gases in water and hydrocarbon represents a corrosion risk for the metallic structures, and a prevention system must be considered [8]-[10].

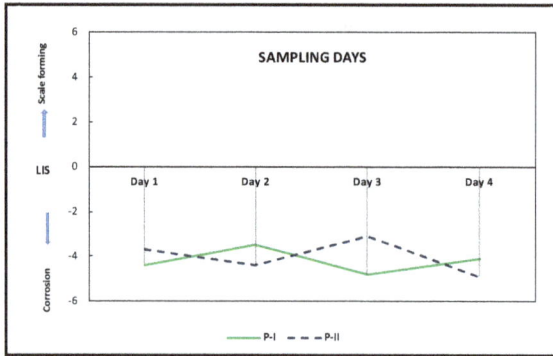

**Figure 1.** Langelier Index Stability (LIS) for water samples.

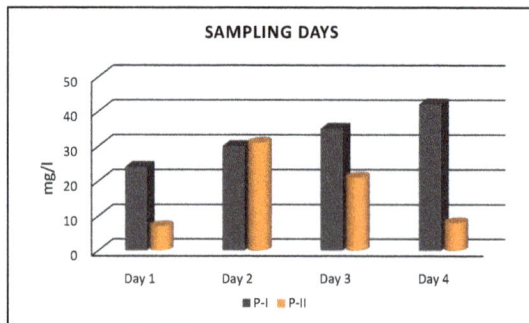

**Figure 2.** $H_2S$ concentration in water samples.

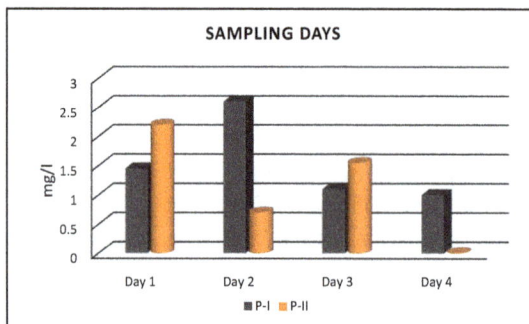

**Figure 3.** $O_2$ concentration in water samples.

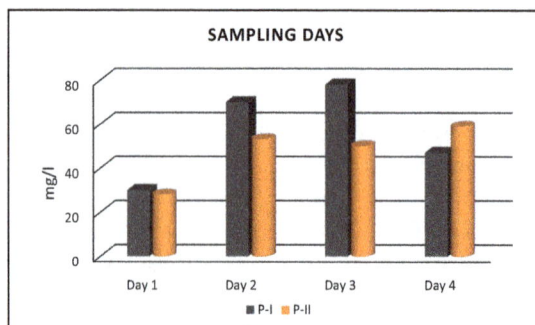

**Figure 4.** $CO_2$ concentration in water samples.

It is important to point out that the variations observed between the analyzed samples for P-I and P-II are slight; therefore, their corrosive tendency is similar. At the same time, the $O_2$ and $CO_2$ contents exhibit some differences that are considered as normal, due to the fluid characteristic variations.

One of the parameters that present bigger differences is the $H_2S$ content. Here, an increase with time for samples taken from P-I was observed, reporting values between 20 and 40 mg/l. For samples taken from P-II, lower values were reported, in the range of 5 and 30 mg/l. It must be noted that the concentration of corrosive gases in water depends upon the general characteristics of the hydrocarbon, specially pressure and temperature. Therefore, the results allow establishing the presence of corrosive gases in the analyzed water samples.

In the same way, it may be indicated that iron content in the water samples is a reference to establish whether the corrosion process is taking place or not, and to determine the necessity of a control system [11]. For this case, both fluids P-I and P-II exhibited corrosive characteristics, although a difference in the iron content could be observed (**Figure 5**).

The samples taken in P-I had values around 0.5 ppm, while the samples from P-II reported values between 3.3 and 4.2 ppm.

These differences can be related to some specific system conditions and parameters:
• Chemical composition and/or material resistance;
• Temperature;
• Flow;
• System life time;
• Addition of chemicals, such as corrosion inhibitors, scale inhibitors, among others;
• Process efficiency.

On the other hand, the presence of microorganisms was determined in all water samples. Sulfate reducing bacteria populations around 100 bacteria/$cm^3$ were observed. This situation is considered as a potential problem of localized corrosion.

### 3.1.2. Hydrocarbon Analysis

The fluid is mainly composed of light hydrocarbons, such as methane, ethane and propane. This is considered as a normal situation. However, there were some corrosive gases in the hydrocarbon composition. Contents of $H_2S$, $CO_2$ and $O_2$ were determined, as shown in **Figure 6** and **Figure 7**. Here, both points P-I and P-II, show similar gases proportion.

Considering their effect on the corrosion processes, these gases represent a continuous source of corrosive compounds for the aqueous phase. Once the products are consumed in the reaction, the gases dissolve in water to keep the corrosion process going on.

According to the fluid characteristics, the parameters identified as the responsible for the corrosion process in the systems are:
1) Water content;
2) Presence of dissolved corrosive gases;
3) Presence of microorganisms, mainly Sulfate reducing bacteria (SRB);

## 3.2. Stage 2

The analyzed sample was taken at the exit of a high pressure separator, as shown in **Figure 8**. The pipeline had

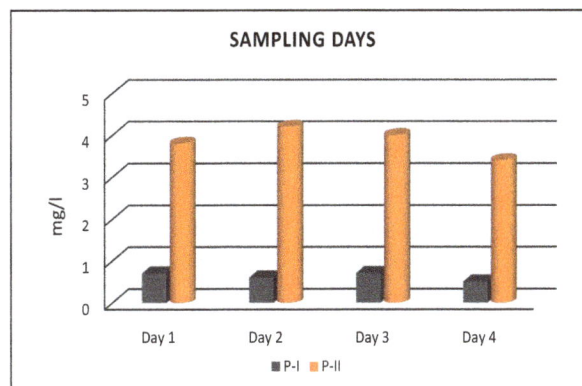

**Figure 5.** $Fe^{2+}$ Concentration in water samples.

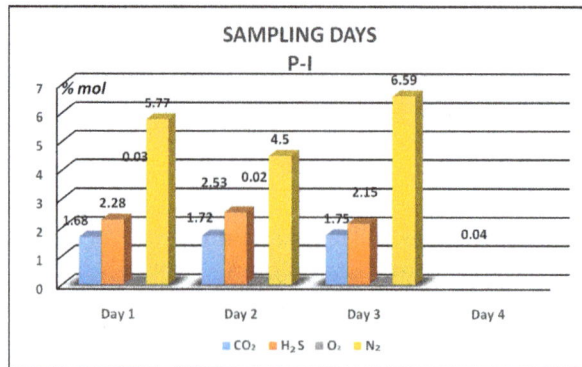

**Figure 6.** Concentration of corrosive gases in hydrocarbon composition (P-I).

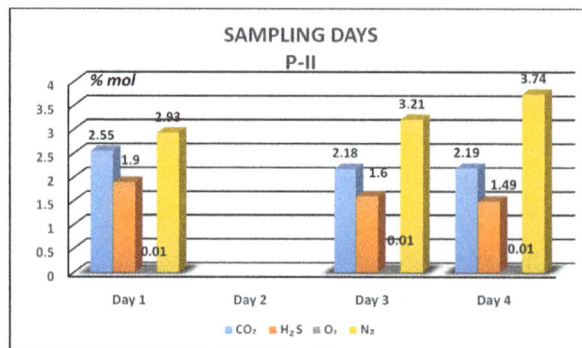

**Figure 7.** Concentration of corrosive gases in hydrocarbon composition (P-II).

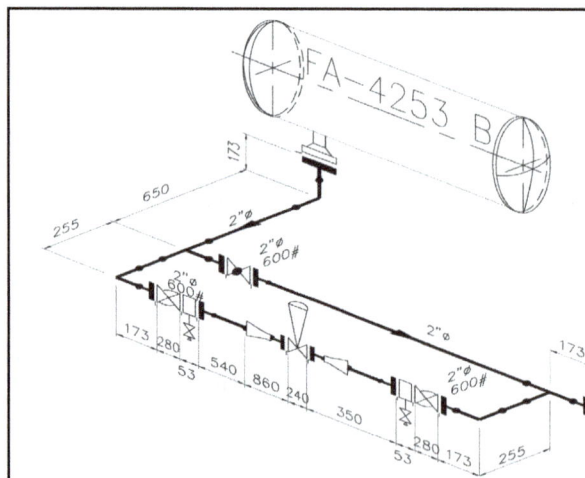

**Figure 8.** Diagram of the high pressure separator system.

closed valves at the ends and nitrogen gas was bubbled inside, to assure anoxic conditions.

The failure was located at the welding joint with the first elbow, according to the fluid flow and at the 6 technical hours position, as shown in **Figure 9**.

The sample analyses were carried out as follows.

### 3.2.1. Microbiological Analysis

The presence of sulfate reducing bacteria (SRB) was identified inside the pipeline, next to the failure and in dif-

**Figure 9.** Location of the site where failure occurs.

ferent areas along the inner wall. SRB have been considered as the main agent for the microbiological induced corrosion occurring in hydrocarbon transport and distribution systems [12]-[14].

SRB presence was determined using culture media, according to API Recommended Practice No. 38 [15]. This is a specific culture for this kind of microorganisms.

When SRB are present in the samples, they reduce the sulfate of the media to sulfide, which reacts with iron to produce a black precipitate of iron sulfide. This is an indicative of SRB presence. When there is no presence of SRB, the culture media remains transparent with no change (**Figure 10**).

The formation of FeS deposits at the metal surface is one of the main characteristics of this type of corrosion. These results are in good agreement with those obtained in Stage 1, where the presence of SRB in water was indicated.

It is important to point out that the presence of SRB constitutes a risk for the integrity of metallic structures, as these microorganisms induce localized corrosion. Therefore, its identification and control becomes necessary.

### 3.2.2. X-Ray Diffraction, X-Ray Fluorescence and Mössbauer Spectroscopy Analyses

The corrosion products obtained from the metal surface were analyzed by different techniques: X-ray diffraction, X-ray fluorescence, and Mössbauer spectroscopy. The following was observed:

Mössbauer Spectroscopy: This analysis is specific to determine iron compounds. The spectrum obtained is shown in **Figure 11**. The compounds found mainly correspond to iron sulfides: Troilite (FeS), Mackinawite (FeS$_{0.9}$) and FeS$_2$.

X-Ray Fluorescence: This technique is directed to establish the presence of chemical elements in the sample. The results indicate different elements, mainly S, Fe and O.

X-Ray Diffraction: Using this technique, the presence of diverse compounds with crystalline structure can be identified. The obtained diagram exhibited the following compounds: Mackinawite (FeS$_{0.9}$), Marcasite (FeS$_2$), Pyrite (FeS$_2$), Troilite (FeS) and Pyrrhotite (FeS).

According to theses analyses, it may be possible to establish that the main corrosion products formed at the inner metal surface are iron sulfides, which may result from:

• The presence of H$_2$S in the hydrocarbon and the associated water;

• The activity of SRB, which was identified in the system. Iron sulfide is a sub-product of the microorganism metabolism.

It is important to mention that although the X-ray fluorescence analysis indicated oxygen content, no oxides were identified by X-ray diffraction nor Mösbauer spectroscopy. In this way, it must be said that during the first inspection of the corrosion products at the metal surface, some "reddish" products characteristics for oxide compounds were observed. These red products were entrusted in the interstices of the inner metal wall, underneath the metal-corrosion products interface. For this reason, the oxides were not detected, even though its presence was visually corroborated.

### 3.2.3. Surface Analysis

Several surface analyses, using the Scanning Electron Microscope, were carried out at the site where the failure

(a)                                   (b)                                   (c)

**Figure 10.** Presence of sulfate reducing bacteria, indicated by the formation of FeS deposits. a) Presence of SRB; b) Presence of SRB; c) Absence of SRB (Blank).

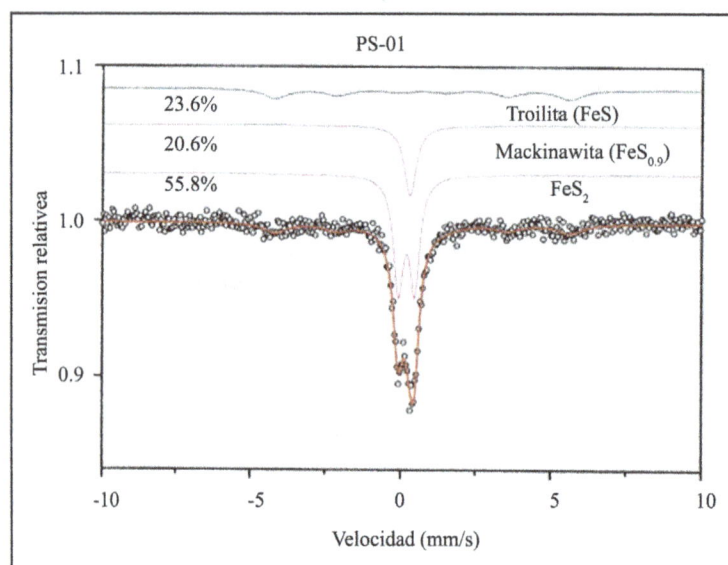

**Figure 11.** Spectrum obtained from Mössbauer spectroscopy.

occurred, before and after the corrosion products were removed from the metal surface [16]. The results corresponding to the analysis with corrosion products at the surface, exhibited a rectangular failure, with side dimensions of 253 μm and 629 μm, located at the welding joint (**Figure 12**). An elemental analysis indicated a typical mild steel composition, including Fe, C, Ni, Cr, Si, and Mn.

Moreover, S and O were observed. These elements are associated to the steel composition, but also could be related to the corrosion products formed at the metal surface. Once the corrosion products were removed, a surface analysis was carried out. A uniform corrosion process was observed in the entire metal surface, with a type of localized corrosion in specific sites (**Figure 13**).

In this case, the characteristics elements for carbon steel are still observed. However, there is lower oxygen content and the presence of sulfur was detected. These results corroborate the fact that the corrosion products are mainly formed by sulfides and oxides. Additionally, as the failure occurred in a welding joint, a galvanic corrosion effect was considered and a metallographic analysis was suggested.

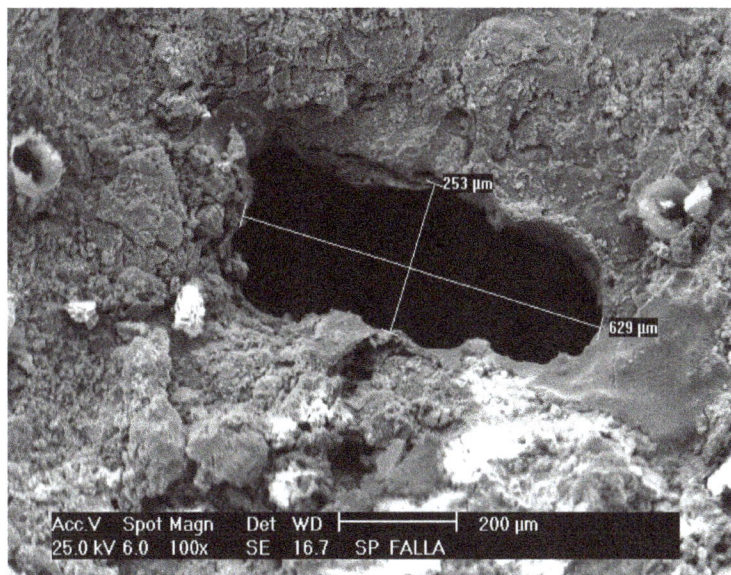

**Figure 12.** Micrograph of failure. Corrosion products at the metal surface.

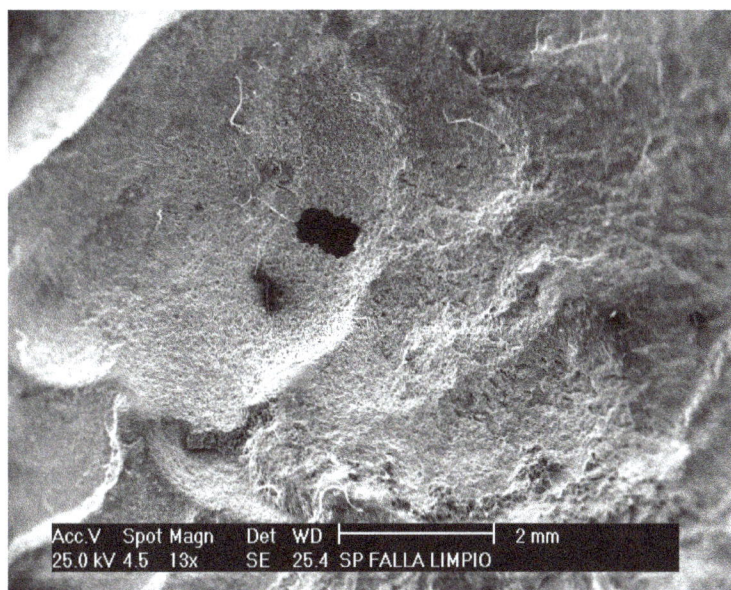

**Figure 13.** Micrograph of failure. Metal surface free of corrosion products.

### 3.2.4. Metallographic Analysis

A metallographic analysis and a hardness profile were carried out at the site where the failure occurred, considering the tube, the welding joint and the elbow. During the sample preparation, inclusions at the metal surface were evaluated. Sulfides and oxides were observed in the tube and oxides were noticed at the elbow surface. These results corroborate previous analysis, where the presence of these two compounds was reported. Regarding the hardness profile, the results indicate similar behavior for the three regions analyzed: tube, welding joint and elbow. Values around 74 Rockwell units (HRBW) were measured. However, the heat affected regions (HAR) exhibited higher hardness values, around 86 HRB, as shown in **Figure 14**.

On the other hand, according to the metallographic analysis, the tube—elbow microstructure arrangement indicated the presence of Pearlite and Ferrite, as shown in **Table 1**.

It is important to point out that the steel physical properties and its behavior depend mainly upon the carbon content and its distribution into the iron matrix. Most of the steels are a combination of three phases: pearlite,

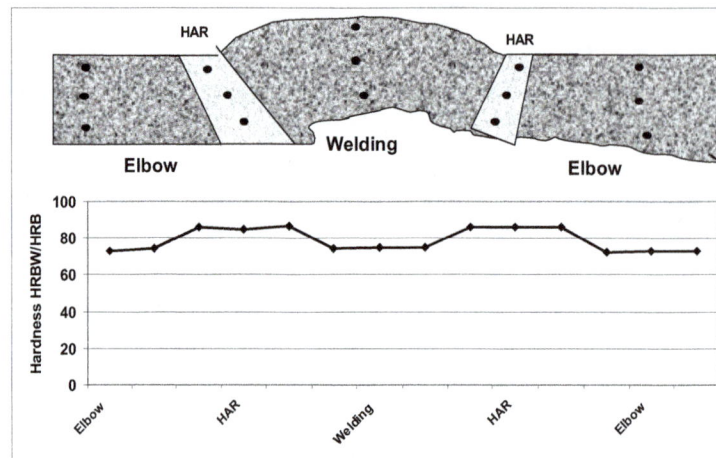

**Figure 14.** Hardness profile for the metal sample.

**Table 1.** Microstructures present in the materials.

| Analysis Region | Microstructure |
|---|---|
| Elbow | Pearlite + Ferrite[a] |
| HAR | Proeutectoid pearlite + fine pearlite |
| Welding joint | Proeutectoid pearlite + dendrites |
| HAR | Proeutectoid pearlite + fine pearlite |
| Tube | Pearlite + Ferrite[b] |
| Tube in corroded region | Pearlite + Ferrite |
| Elbow in corroded region | Pearlite + Ferrite |

(a) Grain size 9; (b) Grain size 7 - 8

cementite and ferrite. The strength and hardness of steel with no heat treatment depend on the proportion of these three phases.

For this specific case, a typical carbon steel microstructure formed by ferrite and pearlite was observed for the tube and elbow. The welding joint material had a microstructure formed mainly by proeutectoid pearlite that can be considered as typical for low carbon steel. The presence of dendrites in the welding joint may be related to a material heat effect. The regions identified as HAR presented microstructures constituted mainly by proeutectoid pearlite and fine pearlite. The different microstructures shown by each region could indicate a galvanic effect that contributes to the corrosion process at the failure site, which also corresponds to the welding joint region.

### 3.2.5. Chemical Analysis

To complete the metallographic analysis and to identify any difference between the chemical composition of the tube, elbow and welding joint, a chemical analysis was carried out, using the Atomic Absorption Technique. This was done to determine the elements present in the metals and the results are summarized in **Table 2**. According to this table, the tube and elbow had similar chemical composition, indicating the same type of steel. However, the welding joint material presented some significant differences, mainly related to the content of Fe, C, Mn, Si and S.

A galvanic effect could be explained by these differences and the different microstructures observed in the metals. It seems that the welding joint material acts as an anode and the adjacent regions as a cathode.

### 3.2.6. Corrosion Process Morphology

The results obtained at this moment indicate that there are right conditions in the system for the development of a corrosion process in the inner wall of the pipeline. There is high water content, in addition to the presence of corrosive agents and appropriate material conditions. For the failure considered in this work, different corrosion processes were observed:

1) The first visual inspection of the pipeline inner wall, after the first metal cut and before any metal cleaning, indicated a severe uniform corrosion, identified by a reduction of the pipeline thickness in specific sites, mainly located at the 6 hours position (**Figure 15**).

This type of corrosion was generally observed at the elbows and "T" joints, where the fluid changes its flow direction. It is expected that during the hydrocarbon transport, heavier fluids are located at the bottom of the pipeline. For this specific case, water phase is in this position, generating aggressive conditions for the metal.

Due to the morphology, location and distribution of this type of uniform corrosion, an erosion corrosion effect is considered. In this type of corrosion, the hydrocarbon flow is enough to remove corrosion products from the metal surface, decreasing their protective effect and increasing the corrosion rate. Corrosion processes due to fluids flow usually induced a localized impact pattern. The failure was located at the welding joint, in a site where an erosion corrosion effect was also observed, as shown in **Figure 16**.

In this way, it is very important to verify the fluid velocity, as it should not exceed the recommended material limits. From a corrosion point of view, a smooth flow is always preferable to a turbulent flow. At the same time, gases and solid particles must be eliminated from the fluid as possible.

**Figure 15.** Pipeline transverse cutting. First elbow after the high pressure separator.

**Table 2.** Chemical composition of metal samples.

| Metal | Chemical Composition | | |
|---|---|---|---|
| | Elbow (%) | Welding joint (%) | Tube (%) |
| Cr | 0.03 | 0.05 | 0.03 |
| Mn | 0.79 | 1.19 | 0.92 |
| Mo | 0.02 | <0.01 | <0.01 |
| Ni | 0.08 | 0.02 | 0.03 |
| Si | 0.31 | 0.62 | 0.19 |
| C | 0.195 | 0.043 | 0.163 |
| S | 0.003 | 0.024 | 0.002 |
| P | 0.018 | 0.023 | 0.022 |
| Fe | 98.60 | 97.96 | 98.66 |

**Figure 16.** Site where failure occurs. Erosion corrosion goes in flow direction.

2) A galvanic corrosion process may be considered at the welding joints regions. This type of corrosion occurs because of potential differences between metallic materials, when in contact and in presence of an electrolyte. The material with more negative potential acts as an anode and exhibited a corrosion process. For this specific case, apparently the welding joint acted as anode in the corrosion reaction. However, the galvanic corrosion was less evident in the site where the failure occurred, because of the effect of the erosion corrosion. At the 12 technical hours position, the erosion corrosion effect is less evident and the galvanic corrosion is clearly observed (**Figure 17**).

3) Microbiological induced corrosion, which can be associated to localized corrosion processes, was observed along the metal sample. It was more evident at longitudinal regions, between 5 and 7 technical hours positions, although this type of corrosion was also detected in most of the pipeline inner metal surface, as shown in **Figure 18**. Corrosion products were removed from some pits and SRB populations were detected inside the cavities.

Additionally, microorganisms were also detected at the region where the failure occurred. This situation indicated that microbiological induced corrosion had also an effect on the metal failure, although the evidences were not clear, because of the presence of the other types of corrosion.

## 4. Conclusions

• The hydrocarbon condensed products transported by the 2 inches diameter steel pipeline, at the exit of the high pressure separator system, exhibited aggressive conditions for the inner metal wall. The main corrosive agents identified in the fluid are:
1) Water content;
2) The presence of dissolved gases ($H_2S$, $CO_2$ and $O_2$);
3) Microorganisms population, mainly sulfate reducing bacteria.
   • The problem was considered as a combined effect of three types of corrosion:
1) Erosion corrosion, caused by the fluid flow and changes in the fluid direction;
2) Galvanic corrosion, mainly caused by differences in the chemical composition and microstructures of the metallic materials;
3) Microbiologically induced corrosion, caused by the presence of sulfate reducing bacteria.
   • The corrosive agents in the system, such as $CO_2$, $H_2S$ and $O_2$, participate in the cathodic reaction during the corrosion process.
   • This type of failure occurs in characteristic sites of the pipelines path, mainly in direction changes and in welding joints.
   • However, localized corrosion processes must also be considered. This type of corrosion does not follow a specific pattern, and becomes more important in sites where flow does not have influence.

**Figure 17.** Corrosion process at the interface tube-elbow.

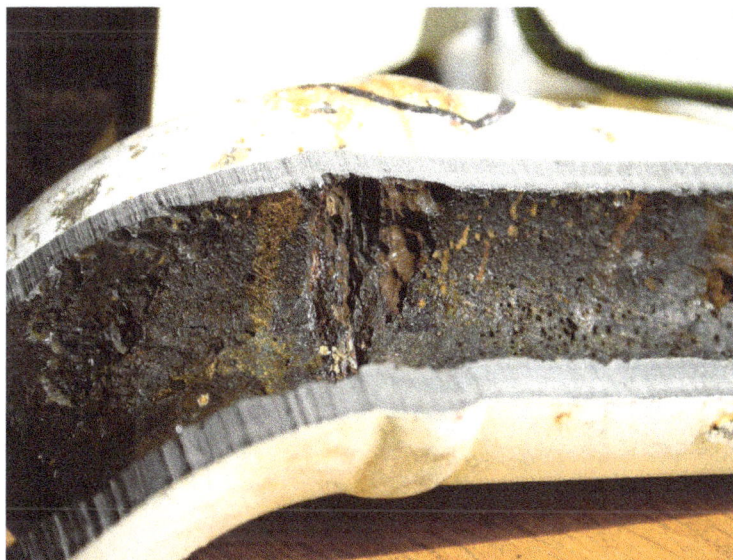

**Figure 18.** Localized corrosion at the elbow inner wall.

## 5. Recommendations

The cause for the failure observed in the system is a combined effect of different corrosion types and therefore several actions must be considered:

1) To install an inhibitor injection system. The inhibitor considered must remove the oxygen in the system and form a film in the metal surface;

2) To consider a biocide injection program;

3) To verify the specifications of the system, relating to the hydrocarbon transport, in order to control the fluid flow, according to the separator system design;

4) To eliminate the water content, as it represents one of the main corrosive agents. A modification on the separation equipment should be considered;

5) To carry out a fluid quality control program;

6) To review the welding procedures, in order to eliminate any discontinuity in the inner pipeline wall;

7) To consider a heat treatment for stress relieve, after heat treatment;

8) To maintain a constant flow in the pipelines.

## References

[1]    (2006) NACE Standard SP0106-2006: Control of Internal Corrosion in Steel Pipelines and Piping Systems. NACE International, Houston.

[2]    (1991) API RP 14E: Recommended Practice for Design and Installation of Offshore Production Platform Piping Systems. API, Washington DC.

[3]    (2002) NACE Standard RP0102: In Line Inspection of Pipelines. NACE International, Houston.

[4]    (2010) ASTM D 3370. Sampling Water from Closed Conduits.

[5]    (2014) NACE Standard TM0194-2014: Field Monitoring of Bacterial Growth in Oil and Gas Systems. NACE International, Houston.

[6]    (2012) ASTM D5504-12: Standard Test Method for Determination of Sulfur Compounds in Natural Gas and Gaseous Fuels by Gas Chromatography and Chemiluminescence.

[7]    Pisigan Jr., R.A. and Singley, J.E. (1985) Evaluation of Water Corrosivity Using the Langelier Index and Relative Corrosion Rate Models. *Materials Performance*, **24**, 26-36.

[8]    Sastri, V.S. (1998) Corrosion Inhibitors, Principles and Applications. John Wiley and Sons, New York.

[9]    Callister, W.D. (2000) Materials Science and Engineering: An Introduction. John Wiley and Sons, New York.

[10]   Byars, H. (1999) Corrosion Control in Petroleum Production. 2nd Edition, NACE International, Houston.

[11]   Atkinson, J.T.N. (1985) Corrosion and Its Control: An Introduction to the Subject. NACE International, Houston.

[12]   (2002) ASTM D 4412-84: Standard Test Method for Sulfate Reducing Bacteria in Water Formed Deposits.

[13]   Ghazy, E.A., Mahmoud, M.G., Asker, M.S., Mahmoud, M.N., Abo Elsoud, M.M. and Abdel Sami, M.E. (2011) Cultivation and Detection of Sulfate Reducing Bacteria (SRB) in Sea Water. *Journal of American Science*, **7**, 604-608.

[14]   (1995) ASTM F488-95: Standard Test Method for On-Site Screening of Heterotrophic Bacteria in Water.

[15]   (1982) API Recommended Practice No. 38: Biological Analysis for Subsurface Injection Waters. API, Philadelphia.

[16]   (2011) ASTM G1-03: Standard Practice for Preparing, Cleaning, and Evaluating Corrosion Test Specimens.

# Permissions

All chapters in this book were first published by Scientific Research Publishing; hereby published with permission under the Creative Commons Attribution License or equivalent. Every chapter published in this book has been scrutinized by our experts. Their significance has been extensively debated. The topics covered herein carry significant findings which will fuel the growth of the discipline. They may even be implemented as practical applications or may be referred to as a beginning point for another development.

The contributors of this book come from diverse backgrounds, making this book a truly international effort. This book will bring forth new frontiers with its revolutionizing research information and detailed analysis of the nascent developments around the world.

We would like to thank all the contributing authors for lending their expertise to make the book truly unique. They have played a crucial role in the development of this book. Without their invaluable contributions this book wouldn't have been possible. They have made vital efforts to compile up to date information on the varied aspects of this subject to make this book a valuable addition to the collection of many professionals and students.

This book was conceptualized with the vision of imparting up-to-date information and advanced data in this field. To ensure the same, a matchless editorial board was set up. Every individual on the board went through rigorous rounds of assessment to prove their worth. After which they invested a large part of their time researching and compiling the most relevant data for our readers.

The editorial board has been involved in producing this book since its inception. They have spent rigorous hours researching and exploring the diverse topics which have resulted in the successful publishing of this book. They have passed on their knowledge of decades through this book. To expedite this challenging task, the publisher supported the team at every step. A small team of assistant editors was also appointed to further simplify the editing procedure and attain best results for the readers.

Apart from the editorial board, the designing team has also invested a significant amount of their time in understanding the subject and creating the most relevant covers. They scrutinized every image to scout for the most suitable representation of the subject and create an appropriate cover for the book.

The publishing team has been an ardent support to the editorial, designing and production team. Their endless efforts to recruit the best for this project, has resulted in the accomplishment of this book. They are a veteran in the field of academics and their pool of knowledge is as vast as their experience in printing. Their expertise and guidance has proved useful at every step. Their uncompromising quality standards have made this book an exceptional effort. Their encouragement from time to time has been an inspiration for everyone.

The publisher and the editorial board hope that this book will prove to be a valuable piece of knowledge for researchers, students, practitioners and scholars across the globe.

# List of Contributors

**Mikihito Hirohata and Yoshito Itoh**
Graduate School of Engineering, Nagoya University, Nagoya, Japan

**Guoqing Gou and Hui Chen**
School of Materials Science and Engineering, Southwest Jiaotong University, Chengdu, China

**Yuping Yang**
EWI, Columbus, USA

**M. A. Bodude**
Department of Metallurgical and Materials Engineering, University of Lagos, Lagos, Nigeria

**I. Momohjimoh**
Department of Metallurgical Engineering, Yaba College of Technology, Lagos, Nigeria

**Mikhail Sokolov**
Laboratory of Laser Materials Processing, Lappeenranta University of Technology, Lappeenranta, Finland

**Antti Salminen**
Laboratory of Laser Materials Processing, Lappeenranta University of Technology, Lappeenranta, Finland
Machine Technology Centre Turku Ltd, Turku, Finland

**E. M. Anawa, M. F. Bograrah and S. B. Salem**
Industrial and Manufacturing Department, Faculty of Engineering, University of Benghazi, Benghazi, Libya

**Milan Dvořák and Emil Schwarzer**
BUT, Faculty of Mechanical Engineering, IMT, Brno, Czech Republic

**Miloš Klíma**
MU, Faculty of Science, PF MU, Brno, Czech Republic

**Joseph Achebo**
Department of Production Engineering, University of Benin, Benin City, Nigeria

**William Ejenavi Odinikuku**
Department of Mechanical Engineering, Petroleum Training Institute, Effurun, Nigeria

**Joseph Achebo**
Department of Production Engineering, University of Benin, Benin, Nigeria

**Sule Salisu**
Department of Mechanical Engineering, Petroleum Training Institute, Effurun, Nigeria

**Jie Cui, Qiuya Zhai and Jinfeng Xu**
School of Materials Science and Engineering, Xi'an University of Technology, Xi'an, China

**Yahui Wang and Jianlin Ye**
Xi'an Unit Container Manufacturing Co., Ltd., Xi'an, China

**Oscar Andersson**
Department of Production Engineering, KTH Royal Institute of Technology, Stockholm, Sweden

**Arne Melander**
Department of Production Engineering, KTH Royal Institute of Technology, Stockholm, Sweden
Department of Production Engineering, KTH Royal Institute of Technology, Stockholm, Sweden
Swerea KIMAB, Kista, Sweden

**Balamurugan Sivaramakrishnan and Murugan Nadarajan**
Department of Mechanical Engineering, Coimbatore Institute of Technology, Coimbatore, India

**Yehia Abdel-Nasser**
Naval Architecture and Marine Engineering, Faculty of Engineering, Alexandria University, Alexandria, Egypt

**Ninshu Ma, Sherif Rashed and Hidekazu Murakawa**
Joining and Welding Research Institute, Osaka University, Osaka, Japan

**Qiuya Zhai, Jinfeng Xu, Tianyu Lu and Yan Xu**
School of Materials Science and Engineering, Xi'an University of Technology, Xi'an, China

**Georgios K. Triantafyllidis, Dimitrios I. Zagliveris, Dionysios L. Kolioulis, Christos S. Tsiompanis, Titos N. Pasparakis, Athanasios P. Gredis, Melina L. Sfantou and Ioannis E. Giouvanakis**
Department of Chemical Engineering, Aristotle University, Thessaloniki, Greece

**Y.-C. Kim**
Osaka University, Osaka, Japan

**M. Hirohata**
Graduate School of Engineering, Nagoya University, Nagoya, Japan

**K. Inose**
IHI Corporation, Yokohama, Japan

**Yoke-Rung Wong and Xin Pang**
School of Mechanical & Aerospace Engineering, Nanyang Technological University, Singapore City, Singapore

**Terrence Stobbe and Ryan Westra**
College of Public Health, University of Arizona, Tucson, USA

**Ziyad Kas and Manohar Das**
Department of Electrical and Computer Engineering, Oakland University, Rochester, USA

**A. Abouarkoub**
The Libyan Petroleum Institute, Tripoli, Libya

**G. E. Thompson and X. Zhou**
School of Materials, The University of Manchester, Manchester, UK

**G. Scamans**
Innoval Technology Limited, Banbury, UK

**Wei Tang, Wenjing Wang, Yao Wang and Qiang Li**
School of Mechanical Electronic and Control Engineering, Beijing Jiaotong University, Beijing 100044, China

**Gerardo Zavala Olivares and Mónica Jazmín Hernández Gayosso**
Instituto Mexicano Del Petróleo. Eje Central Lázaro Cárdenas Norte,Col. San Bartolo Atepehuacan, C.P., México D.F., México

www.ingramcontent.com/pod-product-compliance
Lightning Source LLC
Chambersburg PA
CBHW080651200326
41458CB00013B/4807